Rheology and Processing of Polymers

Rheology and Processing of Polymers

Editors

Khalid Lamnawar
Abderrahim Maazouz

MDPI • Basel • Beijing • Wuhan • Barcelona • Belgrade • Manchester • Tokyo • Cluj • Tianjin

Editors
Khalid Lamnawar
Ingénierie des Matériaux Polymères
Université de Lyon
INSA Lyon
France

Abderrahim Maazouz
Ingénierie des Matériaux Polymères
Université de Lyon
INSA Lyon
France

Editorial Office
MDPI
St. Alban-Anlage 66
4052 Basel, Switzerland

This is a reprint of articles from the Special Issue published online in the open access journal *Polymers* (ISSN 2073-4360) (available at: https://www.mdpi.com/journal/polymers/special_issues/Rheology_Processing_Polymers).

For citation purposes, cite each article independently as indicated on the article page online and as indicated below:

LastName, A.A.; LastName, B.B.; LastName, C.C. Article Title. *Journal Name* **Year**, *Volume Number*, Page Range.

ISBN 978-3-0365-5263-7 (Hbk)
ISBN 978-3-0365-5264-4 (PDF)

© 2022 by the authors. Articles in this book are Open Access and distributed under the Creative Commons Attribution (CC BY) license, which allows users to download, copy and build upon published articles, as long as the author and publisher are properly credited, which ensures maximum dissemination and a wider impact of our publications.

The book as a whole is distributed by MDPI under the terms and conditions of the Creative Commons license CC BY-NC-ND.

Contents

About the Editors vii

Preface to "Rheology and Processing of Polymers" ix

Khalid Lamnawar and Abderrahim Maazouz
Rheology and Processing of Polymers
Reprinted from: *Polymers* **2022**, *14*, 2327, doi:10.3390/polym14122327 1

Bo Lu, Huagui Zhang, Abderrahim Maazouz and Khalid Lamnawar
Interfacial Phenomena in Multi-Micro-/Nanolayered Polymer Coextrusion: A Review of Fundamental and Engineering Aspects
Reprinted from: *Polymers* **2021**, *13*, 417, doi:10.3390/polym13030417 9

Nick Weingart, Daniel Raps, Mingfu Lu, Lukas Endner and Volker Altstädt
Comparison of the Foamability of Linear and Long-Chain Branched Polypropylene—The Legend of Strain-Hardening as a Requirement for Good Foamability
Reprinted from: *Polymers* **2020**, *12*, 725, doi:10.3390/polym12030725 31

Xiaoxiao Guan, Bo Cao, Jianan Cai, Zhenxing Ye, Xiang Lu, Haohao Huang, Shumei Liu and Jianqing Zhao
Design and Synthesis of Polysiloxane Based Side Chain Liquid Crystal Polymer for Improving the Processability and Toughness of Magnesium Hydrate/Linear Low-Density Polyethylene Composites
Reprinted from: *Polymers* **2020**, *12*, 911, doi:10.3390/polym12040911 53

Cindy Le Hel, Véronique Bounor-Legaré, Mathilde Catherin, Antoine Lucas, Anthony Thèvenon and Philippe Cassagnau
TPV: A New Insight on the Rubber Morphology and Mechanic/Elastic Properties
Reprinted from: *Polymers* **2020**, *12*, 2315, doi:10.3390/polym12102315 69

Mei Fang, Na Zhang, Ming Huang, Bo Lu, Khalid Lamnawar, Chuntai Liu and Changyu Shen
Effects of Hydrothermal Aging of Carbon Fiber Reinforced Polycarbonate Composites on Mechanical Performance and Sand Erosion Resistance
Reprinted from: *Polymers* **2020**, *12*, 2453, doi:10.3390/polym12112453 85

Jie Wang, Daniel Adami, Bo Lu, Chuntai Liu, Abderrahim Maazouz and Khalid Lamnawar
Multiscale Structural Evolution and Its Relationship to Dielectric Properties of Micro-/Nano-Layer Coextruded PVDF-HFP/PC Films
Reprinted from: *Polymers* **2020**, *12*, 2596, doi:10.3390/polym12112596 97

Fouad Erchiqui, Khaled Zaafrane, Abdessamad Baatti, Hamid Kaddami and Abdellatif Imad
Reliability of Free Inflation and Dynamic Mechanics Tests on the Prediction of the Behavior of the Polymethylsilsesquioxane–High-Density Polyethylene Nanocomposite for Thermoforming Applications
Reprinted from: *Polymers* **2020**, *12*, 2753, doi:10.3390/polym12112753 109

Francisco Parres, Miguel Angel Peydro, David Juarez, Marina P. Arrieta and Miguel Aldas
Study of the Properties of a Biodegradable Polymer Filled with Different Wood Flour Particles
Reprinted from: *Polymers* **2020**, *12*, 2974, doi:10.3390/polym12122974 125

Geunyeop Park, Jangho Yun, Changhoon Lee and Hyun Wook Jung
Effect of Material Parameter of Viscoelastic Giesekus Fluids on Extensional Properties in Spinline and Draw Resonance Instability in Isothermal Melt Spinning Process
Reprinted from: *Polymers* **2021**, *13*, 139, doi:10.3390/polym13010139 . . . . **149**

Pang-Yun Chou, Ying-Chao Chou, Yu-Hsuan Lai, Yu-Ting Lin, Chia-Jung Lu and Shih-Jung Liu
Fabrication of Drug-Eluting Nano-Hydroxylapatite Filled Polycaprolactone Nanocomposites Using Solution-Extrusion 3D Printing Technique
Reprinted from: *Polymers* **2021**, *13*, 318, doi:10.3390/polym13030318 . . . . **159**

Angelica Avella, Rosica Mincheva, Jean-Marie Raquez and Giada Lo Re
Substantial Effect of Water on Radical Melt Crosslinking and Rheological Properties of Poly(ε-Caprolactone)
Reprinted from: *Polymers* **2021**, *13*, 491, doi:10.3390/polym13040491 . . . . **173**

Violette Bourg, Rudy Valette, Nicolas Le Moigne, Patrick Ienny, Valérie Guillard and Anne Bergeret
Shear and Extensional Rheology of Linear and Branched Polybutylene Succinate Blends
Reprinted from: *Polymers* **2021**, *13*, 652, doi:10.3390/polym13040652 . . . . **189**

Raffael Rathner, Wolfgang Roland, Hanny Albrecht, Franz Ruemer and Jürgen Miethlinger
Applicability of the Cox-Merz Rule to High-Density Polyethylene Materials with Various Molecular Masses
Reprinted from: *Polymers* **2021**, *13*, 1218, doi:10.3390/polym13081218 . . . . **207**

Najoua Barhoumi, Kaouther Khlifi, Abderrahim Maazouz and Khalid Lamnawar
Fluorinated Ethylene Propylene Coatings Deposited by a Spray Process: Mechanical Properties, Scratch and Wear Behavior
Reprinted from: *Polymers* **2022**, *14*, 347, doi:10.3390/polym14020347 . . . . **221**

Geraldine Cabrera, Jixiang Li, Abderrahim Maazouz and Khalid Lamnawar
A Journey from Processing to Recycling of Multilayer Waste Films: A Review of Main Challenges and Prospects
Reprinted from: *Polymers* **2022**, *14*, 2319, doi:10.3390/polym14122319 . . . . **235**

Geraldine Cabrera, Ibtissam Touil, Emna Masghouni, Abderrahim Maazouz and Khalid Lamnawar
Multi-Micro/Nanolayer Films Based on Polyolefins: New Approaches from Eco-Design to Recycling
Reprinted from: *Polymers* **2021**, *13*, 413, doi:10.3390/polym13030413 . . . . **269**

About the Editors

Khalid Lamnawar

Khalid Lamnawar is a full professor at the University of Lyon, INSA Lyon, France. Interests: rheology and flow; linear and nonlinear rheology; polymers and composites; polymer processing; recycling; multilayer; coextrusion; modeling; multiphase polymeric systems; biopolymers; biocomposites; composites.

Abderrahim Maazouz

Abderrahim Maazouz is a full professor at the University of Lyon, INSA Lyon, France. Interests: polymers and composites; polymer processing; rheology; reactive processing; multiphase polymeric systems; biopolymers; biocomposites; composites.

Preface to "Rheology and Processing of Polymers"

This book focuses on recent fundamental and applied research on rheology and processing of polymers, covering the latest developments in the field. It demonstrates that the field of rheology and polymer processing is still gaining attention. Included within is an editorial and a variety of papers and reviews that serve communities of both academia and industry alike. In summary, the chapters collectively cover a wide range of topics, from polymer science to recycling, recyclability, and the reuse of polymers and their multiphase systems. They highlight cutting-edge research dealing with the rheology, innovative processing, recycling, and characterization of biopolymers and polymer-based products, polymer physics, composites, and modeling and simulations.

Khalid Lamnawar and Abderrahim Maazouz
Editors

Editorial

Rheology and Processing of Polymers

Khalid Lamnawar [1,2,*] and Abderrahim Maazouz [3]

1 CNRS, UMR 5223, Ingénierie des Matériaux Polymères, INSA Lyon, Université de Lyon, F-69621 Villeurbanne, France
2 University Jean Monnet, F-42100 Saint-Étienne, France
3 Hassan II Academy of Science and Technology, Rabat 10100, Morocco; abderrahim.maazouz@insa-lyon.fr
* Correspondence: khalid.lamnawar@insa-lyon.fr

I am so glad to share with you our Special Issue entitled 'Rheology and Processing of Polymers', which covers the latest developments in the field of rheology and polymer processing, highlighting cutting-edge research focusing on the processing of advanced polymers and their composites. It demonstrates that the field of "Rheology and Polymer Processing" is still gaining attention.

This Special Issue promises to include a variety of papers to serve both academia and industry communities alike. After a meticulous review and editing process, we are delighted to have finally accepted 15 papers and 1 review. We thank the authors for their contributions and the peer reviewers for their thorough and thoughtful reviewing.

In sum, the published review and papers cover a wide range of topics, from polymer science to recycling, recyclability and the reuse of polymers and their multiphase systems.

The following keywords cover all related topics:

- polymer processing;
- rheology;
- polymers;
- natural polymers and biopolymers;
- biopolymers;
- polymer nanocomposites;
- advanced polymers;
- composites and biocomposites;
- biocomposites;
- modeling;
- numerical simulation;
- polymer physics;
- innovative processing;
- polymer melts;
- polymer engineering;
- recycling

Bo Lu et al. [1] reviewed interfacial phenomena in multi-micro/nanolayered polymer coextrusion from fundamental and engineering aspects. Different from traditional approaches, such as layer-by-layer assembly and spin coating which suffer from low productivity, multilayer coextrusion processing is a top-down approach able to industrially manufacture multilayered films. These phenomena, during processing, including interlayer diffusion, interlayer reaction, interfacial instabilities and interfacial geometrical confinement, frequently occur at the interfaces of multilayered polymers in multilayer coextrusion processing. For example, the spatial confinement of layers greatly alters the microstructure and dynamics of multilayer polymers, dramatically influencing macroscopic properties, including mechanical, electric and gas/liquid barrier properties. This review clearly illustrates the origin of these interfacial phenomena and how they can affect the microstructural

Citation: Lamnawar, K.; Maazouz, A. Rheology and Processing of Polymers. *Polymers* **2022**, *14*, 2327. https://doi.org/10.3390/polym14122327

Received: 26 May 2022
Accepted: 6 June 2022
Published: 8 June 2022

Publisher's Note: MDPI stays neutral with regard to jurisdictional claims in published maps and institutional affiliations.

Copyright: © 2022 by the authors. Licensee MDPI, Basel, Switzerland. This article is an open access article distributed under the terms and conditions of the Creative Commons Attribution (CC BY) license (https://creativecommons.org/licenses/by/4.0/).

development and resulting macroscopic properties. In the future, more comprehensive work should be carried out in both academia and the industry to further clarify these interfacial phenomena toward the scaling up of multilayer coextrusion.

Najoua Barhoumi et al. [2] designed a sort of fluorinated ethylene propylene (FEP) polymer coating with the aim of protecting the stainless-steel (SS304) mold substrate surface from corrosion and wear. Through an air spray process and curing, a compact and uniform film was obtained, and the mechanical, adhesion and corrosion, etc., properties were tested. The FEP coating exhibited a dense, poreless and homogeneous structure with a pseudo-hexagonal lattice crystalline structure. It showed good mechanical properties, hardness, Young's modulus and good scratch resistance to adhesive and cohesive failure, with a high adhesion to SS304. The multipass scratch test provided an easy and quick approach to study the FEP coating's wear resistance. The friction coefficient of the FEP coating did not exceed 0.13, and the wear coefficient was approximately 3.12×10^{-4} mm^3 N m^{-1}. The FEP coating enhanced the corrosion resistance of SS304 and provided significant protection during a 60-day immersion in a NaCl solution.

Raffael Rathner et al. [3] investigated the suitability of the Cox–Merz rule to linear high-density polyethylene (HDPE) materials with various molecular masses. The Cox–Merz rule is an empirical relationship that is commonly used in science and the industry to determine shear viscosity on the basis of an oscillatory rheometry test. However, it does not apply to all polymer melts. Three rheological measurement techniques were used to determine the shear-rate-dependent viscosity of three different HDPE materials: an oscillatory parallel plate, high-pressure capillary and extrusion slit rheometry. While in parallel plate rheometry the Cox–Merz relation was used to estimate the shear-dependent viscosity, in high-pressure capillary and extrusion slit rheometry, the Weissenberg–Rabinowitsch relation was employed. The results showed that for the HDPE grades tested, the viscosity data based on the capillary pressure flow of the high-molecular-weight HDPE described the pressure drop inside the pipe head significantly better than data based on parallel plate rheometry applying the Cox–Merz rule. For the low-molecular-weight HDPE, both measurement techniques were in good accordance.

Linear and branched polybutylene succinate (PBS) blends were created by Violette Bourg et al. [4] to investigate their molecular architecture and rheological behavior. Long-chain branched (LCB) structures have already demonstrated their efficiency in improving the melt strength compared to the linear structures. LCB polymers in general, and in particular newly developed (bio)polymers, are more expensive than linear polymers due to the more complex synthesis or additional reactive extrusion step to obtain them. Blending linear and branched PBS is, therefore, an interesting concept in regards to improving the processability of linear polymers, while decreasing their price. According to their results, when the branched PBS (BPBS) weight fraction increased, a clear increase in the weighted relaxation strengths was observed. Meanwhile, under elongational flow, the rheological response of blends changed from shear thinning to strain hardening with the increase in strain rate. This study provided a deep structural and rheological characterization of the long-chain branching and linear PBS, helping to understand the influence of their blending on linear and nonlinear viscoelastic properties.

Angelica Avella et al. [5] compared crosslinking efficiency between a water-assisted reactive melt processing (REx) and a slower PCL macroradicals diffusion in the melt state. The results showed that the presence of water increased the PCL molecular weight and gel content compared to the dry process, from 1% to 34% with 1 wt.% peroxide, confirming a more efficient radical propagation in water-assisted REx. After Rex, a differential scanning calorimetry showed increased crystallization temperatures and an easier crystallization process of reacted PCL, compared to neat PCL. From the dynamic mechanical analysis, higher branching/crosslinking slightly increased the transition temperatures and led to a reinforcement effect in the temperature range below the glass transition. After the glass transition, the mechanical reinforcement was limited. Instead, in the melt state, the effect of branching/crosslinking was more evident as shown by melt rheology. The rheological behavior of crosslinked PCL showed a transition from the typical viscous-like to a solid-like state. Shear storage moduli were increased by the reactive melt processing, confirming the desired improvement of PCL rheological properties. This work represents a relevant reference for the future controlled water-assisted reactive melt processing of polymers and composites, in which water could have the role of feeding media, e.g., for polysaccharides suspension, but also be useful for boosting peroxide-initiated reactions.

Geraldine Cabrera et al. [6] developed a novel approach for commonly found complex polyethylene/polypropylene (PE/PP) and polyethylene/polystyrene (PE/PS) systems in industrial waste streams from design to recycling. The method involves moving from eco-design to the mechanical recycling of multilayer films via forced assembly coextrusion. PE/PP and PE/PS multilayer films were prepared by coextrusion with various numbers of layers, and then recycled using a process including steps such as (i) grinding, (ii) monolayer cast film extrusion or (iii) injection molding, with or without an intermediate blending step through twin-screw extrusion. For the PE/PP system, the mechanical strength of the multilayer films was apparently enhanced when the most suitable configuration was chosen. For the PE/PS system, the multilayered films presented a brittle behavior with a nominal thickness at the microscale. However, as the nominal layer thickness was decreased down to the nanoscale, the multilayer films became more ductile as the elongation at the break increased. Regardless of the polymer system studied, it was demonstrated that the design of multi-micro/nanolayer films is a very promising solution for industrial issues that accompany the valorization of recycled materials, without the use of compatibilizers.

Because of the lack of commercial filaments for conventional fused deposition modeling (FDM) method, the three-dimensional (3D) printing of polycaprolactone/ nanohydroxylapatite (PCL/nHA) nanocomposites seems to be very tough work, whereas Pang-Yun Chou et al. [7] fabricated PCL/nHA nanocomposites using a lab-developed solution-extrusion printer. The effects of distinct printing variables on the mechanical properties of nanocomposites were investigated. The tensile property of printed nanocomposites increased with the fill density, yet diminished with a decrease in the ratio of PCL/nHA to DCM and print speed. Nanocomposite parts printed with a 90° orientation demonstrated the most superior mechanical properties. In addition, the drug-loaded PCL/nHA screws were also prepared via the same technique, which could provide an extended elution of high levels of vancomycin/ceftazidime over a 14-day period. As a result, this kind of solution-extrusion 3D printing technology may be used to print drug-loaded implants for various medical applications.

Geunyeop Park et al. [8] explained the nature of stability curves in the spinning process of Giesekus fluids—the initial stabilization and subsequent destabilization pattern with respect to D_e in the intermediate range of the material parameter. Various theoretical approaches were considered in this study to examine the changes in the stability curves with respect to the material parameter of Giesekus fluids in the isothermal spinning process without cooling, including steady velocity profiles and extensional deformation properties in the spinline, and kinematic waves traveling along the entire spinline. The material parameter α_G in this fluid model suitably depicted the extensional thickening (stabilizing effect of D_e) and extensional thinning (destabilizing effect of D_e) properties of viscoelastic fluids in extensional deformation processes. In the intermediate range of values of α_G (approximately $0.01 < \alpha_G < 0.4$), the effect of D_e on the stability was unusual—the system was stabilized in the low and medium D_e regions and then destabilized in the high D_e region. This tendency may be qualitatively interpreted by extensional flow characteristics in the spinline. When D_e increased at a fixed drawdown ratio (e.g., r = 25 condition applied in this study) in the intermediate α_G region, the system, starting from the unstable state became stable after the first onset for low or medium D_e, resulting in a higher level of spinline velocity and strain-hardening viscosity. It became unstable again beyond the second onset for high D_e, yielding a lowered spinline velocity and an insignificant strain-hardening feature. A combination of transit times of the kinematic waves penetrating the entire spinline and period of oscillation, i.e., the simple indicator from the linear stability analysis, predicted well the draw resonance onsets under different D_e and α_G conditions. It was confirmed that these transit times of kinematic waves for varying D_e adequately reflected the dependence of change in stability on the values of α_G.

Francisco Parres et al. [9] studied the influence of wood flour particle size, at the microscale level, on the properties of biodegradable composite materials based on Solanyl®-type bioplastic. Wood flour with different granulometries, ranging from 70 to 1100 μm, was assessed as a filler. The rheological study revealed an increase in viscosity as the filler percentage increased. Despite the low viscosity variation as the percentage of filler increased, a more detailed analysis showed that pressures rose more quickly, which could be attributed to the obstruction of the nozzle due to the presence of wood flour particles. This fact could increase the number of defective pieces due to the obstruction of the gate inside the mold. All these results indicated that the material could be processed by extrusion. The tensile test results showed that, in general, Young's modulus provided increasing values for all types of particles. Similarly, the tensile strength increased with an increasing amount of wood flour small-sized (CB 120) and medium-sized particles (BK 40–90) in the formulation. The elongation at break showed little variation with the addition of different percentages of medium-sized particles (BK 40–90) and the largest particles (Grade 9) to the Solanyl®-type bioplastic. Nevertheless, the incorporation of the smallest lignocellulosic microparticles (CB 120) increased the elongation at the break. The impact strength was only affected by the incorporation of small- and medium-sized particles (CB 120 and BK 40–90), showing a maximum value for biocomposites with a 20 wt.% wood flour filler. Of all the samples analyzed, those with a 10 or 20 wt.% filler of the smallest particles (CB120) and medium-sized particles (BK 40–90) showed the best combination for processing by injection molding, as well as leading to biocomposites with improved properties and, thus, ones of interest for sustainable industrial applications.

Fouad Erchiqui et al. [10] assessed the reliability of the experimental method for the viscoelastic identification of a nanocomposite reinforced with polymethylsilsesquioxane nanoparticles (PMSQ–HDPE). Two tests of different nature were used, one based on the free inflation of the membrane and the other on a dynamic mechanical test (DMA). The experiments were carried out at a temperature of 130 °C. The material constants for Christensen's model were determined by the least square optimization. The comparative study of the viscoelastic behavior of PMSQ–HDPE showed that the biaxial test was more appropriate for the construction of a behavior law for applications in thermoforming. Concerning the viscoelastic identification obtained from the rheological data of the DMA, it did not seem to be capable of representing the thermoforming of a part, which requires large deformations. Following this study, comparative studies between the DMA and free blowing should be carried out at temperatures above 130 °C for viscoelastic identification, making it possible to characterize the effect of temperature on the reliability of the tests in thermoforming.

Jie Wang et al. [11] investigated the multiscale structural development involving both the layer architecture and microstructure within layers of micro/nanolayer coextruded polymer films, as well as their relationship to dielectric properties based on the poly (vinylidene fluoride-co-hexafluoropropylene) (PVDF-HFP)/polycarbonate (PC) system. The layers were stable and continuous for microlayer coextruded films with nominal layer thicknesses, whereas for nanolayer coextruded films at nominal layer thicknesses below 160 nm, layer integrity was reduced by interfacial instabilities triggered by viscoelastic differences between component melts. Layers even broke into microdroplets and microsheets due to the coalescence of thin layers. Moreover, with the reduction in the nominal layer thickness, the films displayed an enhanced crystallization, with the formation of a 2D oriented spherulite structure in microlayer coextruded systems and highly oriented in-plane lamellae in nanolayer coextruded counterparts. Layer breakup in thinner layers further resulted in a crystallization and structural orientation similar to that found in microlayer films, which was attributed to the relaxation and phase coalescence of thin layers during processing. Furthermore, dielectric properties of films were strongly dependent on these multiscale structures. The gradually increasing storage permittivity upon reducing the layer thickness was ascribed to the enhanced molecular orientations that could facilitate dipole switching. The lower storage permittivity in the thin layers with breakup was caused by the reduced orientations. In addition, the dielectric loss of microlayer coextruded films was lower than that of nanolayer coextruded analogues. This was due to the increased hopping of charge carriers and the higher energy dissipation from dipole switching when layers were broken. The results of this study could allow for a better understanding of the multiscale structure evolution in micro/nanolayer coextrusion to optimize the target macroscopic properties.

Mei Fang et al. [12] studied the aging process of carbon fiber (CF)/polycarbonate (PC) composites under a humid and hot environment. The moisture absorption rate was measured and calculated during the aging process. Changes in mechanical properties before and after hydrothermal aging were measured through tensile and flexural tests. Underlying mechanisms for hydrothermal aging-induced changes in mechanical and sand erosion properties were surveyed. The moisture absorption rate of CF/PC composites grew linearly with the square root of aging time during the first five days, and stayed flat upon reaching saturation. On the premise of a sampling period and frequency every seven days, the tensile properties reached their maximum at the seventh day, and the peak value of bending flexural performance was reached on the fourteenth day. The effects of aging time on the storage modulus and solid particle erosion resistance were consistent with the stretching results. Hydrothermal aging caused holes to form on the surface of CF/PC composites and reduced the specimen's sand erosion resistance. Meanwhile, the physical property of CF/PC composites was changed.

Cindy Le Hel et al. [13] studied the evolution of morphology at a high-level crosslinked elastomer phase in a thermoplastic matrix, and explained how such a level of concentration could be reached. Meanwhile, the influence of the concentration of the elastomer (EPDM) phase on the recovery elasticity of these thermoplastic vulcanizates (TPVs) as well as the influence of reinforcing fillers such as carbon black were also investigated. The phase inversion between both phases took place when the EPDM phase reached a certain level of crosslinking (swelling ratio ~20 corresponding to a crosslinking density of 10 mol/m^{-3}), which allowed the stress applied by the thermoplastic phase (PP) matrix to disperse this phase by fragmentation. Maximal stress was observed at the melting temperature of the PP pellets. The crosslinking chemistry nature (radical initiator or phenolic resin crosslinker) had less influence on the elasticity recovery in the case of the TPV compared with a pure EPDM system. Morphology was the first-order parameter, whereas the crosslinking density of the EPDM phase was a second-order parameter. The morphologies of the peroxide-based TPV showed heterogeneous systems with large EPDM domains. However, this large distribution of EPDM particles is the only way to increase the maximal packing fraction of the crosslinked EPDM phase, which behave as solid particles. The addition of carbon black fillers allowed the mechanical properties at the break to significantly improve at the lowest PP concentration (20%). This improvement of mechanical properties is still not well understood. Finally, to reach the best compression set while keeping good mechanical properties, a peroxide-based TPV containing carbon black and as little PP as possible should be formulated. Therefore, the low quantity of PP allows for a good compression set; the peroxide permits the carbon black to be located at the interface between the two phases and, thus, the filler can reinforce the mechanical properties of the material.

To effectively improve the processability and toughness of magnesium hydroxide (MH)/linear low-density polyethylene (LLDPE) composites, a polysiloxane grafted with a thermotropic liquid crystal polymer (PSCTLCP) was designed and synthesized by Xiaoxiao Guan et al. [14]. The effect of PSCTLCP loading on the processability and tensile properties of MH/LLDPE/PSCTLCP composites was studied. Moreover, the modified mechanism of PSCTLCP on MH/LLDPE/PSCTLCP composites was investigated at different processing temperatures, rotation speeds and shear rates to guide the actual processing in industry. The balance melt torque of MH/LLDPE/PSCTLCP composites obviously decreased with the loading of PSCTLCP, and that of the composite (5.0 wt.% loading) decreased from 16.59 N·m of the baseline MH/LLDPE to 9.67 N·m. Moreover, PSCTLCP weakened the non-Newtonian property and decreased the flowing activation energy of MH/LLDPE/PSCTLCP composites, and, thus, broadened the processing window and improved the processability. When considering the flexible polysiloxane as the main chains of PSCTLCP, the elongation at the break of MH/LLDPE composites significantly increased, and the tensile strength and modulus slightly decreased with PSCTLCP loading.

The long-chain branching (LCB) of polypropylene (PP) is regarded as a game changer in foaming due to the introduction of strain hardening, which stabilizes the foam morphology. A thorough characterization with respect to the rheology and crystallization characteristics of a linear PP Sinopec HMS20Z, a PP/PE-block copolymer Sinopec E02ES CoPo and a long-chain branched PP Borealis WB140 HMS was conducted by Nick Weingart et al. [15]. Although only LCB-PP exhibited strain hardening and had five times the melt strength of the other grades, it did not provide the broadest foaming window or the best foam quality in terms of density (140 g/L) and cellular morphology. On the contrary, the linear Sinopec HMS20Z delivered low densities (<40 g/L) and superior foam morphology. The Sinopec E02ES CoPo performed rather negatively in terms of density and mediocre in terms of cellular morphology. The beneficial foaming behavior of the Sinopec HMS20Z-PP was attributed to a slower crystallization and a low crystallization temperature compared to the other two materials. Multiwave experiments were conducted to study the gelation due to crystallization. In isothermal multiwave experiments, the Sinopec HMS20Z exhibited a gel-like behavior over a broad time frame, whereas the other two PPs froze quickly. Nonisothermal multiwave tests also underlined the finding of a broad processing window for the linear PP (Sinopec HMS20Z) as the temperature (and time) difference between the onset of crystallization and the gel point was significantly broader compared to the other PP grades. These findings were also confirmed in DSC experiments, where crystal perfection occurred notably slower for Sinopec HMS20Z, which in turn led to a longer gel-like state before solidification. Again, it was noted that this PP grade exhibited no strain hardening. Thus, it was concluded that, besides sufficient rheological properties, the crystallization behavior is of utmost importance for foaming in terms of a broad processing window, especially when aiming to obtain low densities and good foam morphology. In detail, a broad temperature window between the onset of crystallization and gelation is preferred for cooling, and for isothermal processes, an extended time in the gel-like state appeared beneficial.

In conclusion, as editors of this Special Issue, we are very proud of the high-quality papers published and the rigorous review process conducted, providing cutting-edge research results and the latest developments in the fields of polymer science and engineering, including the innovative processing, biopolymers, and characterization of polymer-based products, polymer physics, composites, modeling, simulations and rheology. Therefore, contributions focus on fundamental and experimental results in a thematic range comprising conventional processing technologies as well as innovative processing and material-based macromolecular research. The Special Issue aims to compile the current state-of-the-art and highlight the wide range of applications.

Funding: This research received no external funding.

Acknowledgments: A special thank you to all the authors for submitting their studies to the present Special Issue and for its successful completion. I would like to acknowledge the efforts of the *Polymers* reviewers and editorial staff in enhancing the quality and impact of all submitted papers.

Conflicts of Interest: The author declares no conflict of interest.

References

1. Lu, B.; Zhang, H.; Maazouz, A.; Lamnawar, K. Interfacial Phenomena in Multi-Micro-/Nanolayered Polymer Coextrusion: A Review of Fundamental and Engineering Aspects. *Polymers* **2021**, *13*, 417. [CrossRef] [PubMed]
2. Barhoumi, N.; Khlifi, K.; Maazouz, A.; Lamnawar, K. Fluorinated Ethylene Propylene Coatings Deposited by a Spray Process: Mechanical Properties, Scratch and Wear Behavior. *Polymers* **2022**, *14*, 347. [CrossRef] [PubMed]
3. Rathner, R.; Roland, W.; Albrecht, H.; Ruemer, F.; Miethlinger, J. Applicability of the Cox-Merz Rule to High-Density Polyethylene Materials with Various Molecular Masses. *Polymers* **2021**, *13*, 1218. [CrossRef] [PubMed]
4. Bourg, V.; Valette, R.; Le Moigne, N.; Ienny, P.; Guillard, V.; Bergeret, A. Shear and Extensional Rheology of Linear and Branched Polybutylene Succinate Blends. *Polymers* **2021**, *13*, 652. [CrossRef]
5. Avella, A.; Mincheva, R.; Raquez, J.-M.; Lo Re, G. Substantial Effect of Water on Radical Melt Crosslinking and Rheological Properties of Poly(ε-Caprolactone). *Polymers* **2021**, *13*, 491. [CrossRef] [PubMed]
6. Cabrera, G.; Touil, I.; Masghouni, E.; Maazouz, A.; Lamnawar, K. Multi-Micro/Nanolayer Films Based on Polyolefins: New Approaches from Eco-Design to Recycling. *Polymers* **2021**, *13*, 413. [CrossRef] [PubMed]
7. Chou, P.-Y.; Chou, Y.-C.; Lai, Y.-H.; Lin, Y.-T.; Lu, C.-J.; Liu, S.-J. Fabrication of Drug-Eluting Nano-Hydroxylapatite Filled Polycaprolactone Nanocomposites Using Solution-Extrusion 3D Printing Technique. *Polymers* **2021**, *13*, 318. [CrossRef] [PubMed]
8. Park, G.; Yun, J.; Lee, C.; Jung, H.W. Effect of Material Parameter of Viscoelastic Giesekus Fluids on Extensional Properties in Spinline and Draw Resonance Instability in Isothermal Melt Spinning Process. *Polymers* **2021**, *13*, 139. [CrossRef] [PubMed]
9. Parres, F.; Peydro, M.A.; Juarez, D.; Arrieta, M.P.; Aldas, M. Study of the Properties of a Biodegradable Polymer Filled with Different Wood Flour Particles. *Polymers* **2020**, *12*, 2974. [CrossRef] [PubMed]
10. Erchiqui, F.; Zaafrane, K.; Baatti, A.; Kaddami, H.; Imad, A. Reliability of Free Inflation and Dynamic Mechanics Tests on the Prediction of the Behavior of the Polymethylsilsesquioxane–High-Density Polyethylene Nanocomposite for Thermoforming Applications. *Polymers* **2020**, *12*, 2753. [CrossRef] [PubMed]
11. Wang, J.; Adami, D.; Lu, B.; Liu, C.; Maazouz, A.; Lamnawar, K. Multiscale Structural Evolution and Its Relationship to Dielectric Properties of Micro-/Nano-Layer Coextruded PVDF-HFP/PC Films. *Polymers* **2020**, *12*, 2596. [CrossRef] [PubMed]
12. Fang, M.; Zhang, N.; Huang, M.; Lu, B.; Lamnawar, K.; Liu, C.; Shen, C. Effects of Hydrothermal Aging of Carbon Fiber Reinforced Polycarbonate Composites on Mechanical Performance and Sand Erosion Resistance. *Polymers* **2020**, *12*, 2453. [CrossRef] [PubMed]
13. Hel, C.L.; Bounor-Legaré, V.; Catherin, M.; Lucas, A.; Thèvenon, A.; Cassagnau, P. TPV: A New Insight on the Rubber Morphology and Mechanic/Elastic Properties. *Polymers* **2020**, *12*, 2315. [CrossRef] [PubMed]
14. Guan, X.; Cao, B.; Cai, J.; Ye, Z.; Lu, X.; Huang, H.; Liu, S.; Zhao, J. Design and Synthesis of Polysiloxane Based Side Chain Liquid Crystal Polymer for Improving the Processability and Toughness of Magnesium Hydrate/Linear Low-Density Polyethylene Composites. *Polymers* **2020**, *12*, 911. [CrossRef] [PubMed]
15. Weingart, N.; Raps, D.; Lu, M.; Endner, L.; Altstädt, V. Comparison of the Foamability of Linear and Long-Chain Branched Polypropylene—The Legend of Strain-Hardening as a Requirement for Good Foamability. *Polymers* **2020**, *12*, 725. [CrossRef] [PubMed]

Review

Interfacial Phenomena in Multi-Micro-/Nanolayered Polymer Coextrusion: A Review of Fundamental and Engineering Aspects

Bo Lu [1,2], Huagui Zhang [3], Abderrahim Maazouz [2,4] and Khalid Lamnawar [2,3,*]

[1] Key Laboratory of Materials Processing and Mold (Ministry of Education), National Engineering Research Center for Advanced Polymer Processing Technology, Zhengzhou University, Zhengzhou 450002, China; bolu@zzu.edu.cn

[2] CNRS, UMR 5223, Ingénierie des Matériaux Polymères, INSA Lyon, Université de Lyon, F-69621 Villeurbanne, France; abderrahim.maazouz@insa-lyon.fr

[3] Fujian Key Laboratory of Polymer Science, College of Chemistry and Materials Science, Fujian Normal University, Fuzhou 350007, China; huagui.zhang@fjnu.edu.cn

[4] Hassan II Academy of Science and Technology, Rabat 10100, Morocco

* Correspondence: khalid.lamnawar@insa-lyon.fr

Citation: Lu, B.; Zhang, H.; Maazouz, A.; Lamnawar, K. Interfacial Phenomena in Multi-Micro-/Nanolayered Polymer Coextrusion: A Review of Fundamental and Engineering Aspects. *Polymers* **2021**, *13*, 417. https://doi.org/10.3390/polym13030417

Academic Editor: Eduardo Guzmán
Received: 31 December 2020
Accepted: 25 January 2021
Published: 28 January 2021

Publisher's Note: MDPI stays neutral with regard to jurisdictional claims in published maps and institutional affiliations.

Copyright: © 2021 by the authors. Licensee MDPI, Basel, Switzerland. This article is an open access article distributed under the terms and conditions of the Creative Commons Attribution (CC BY) license (https://creativecommons.org/licenses/by/4.0/).

Abstract: The multilayer coextrusion process is known to be a reliable technique for the continuous fabrication of high-performance micro-/nanolayered polymeric products. Using laminar flow conditions to combine polymer pairs, one can produce multilayer films and composites with a large number of interfaces at the polymer-polymer boundary. Interfacial phenomena, including interlayer diffusion, interlayer reaction, interfacial instabilities, and interfacial geometrical confinement, are always present during multilayer coextrusion depending on the processed polymers. They are critical in defining the microstructural development and resulting macroscopic properties of multilayered products. This paper, therefore, presents a comprehensive review of these interfacial phenomena and illustrates systematically how these phenomena develop and influence the resulting physicochemical properties. This review will promote the understanding of interfacial evolution in the micro-/nanolayer coextrusion process while enabling the better control of the microstructure and end use properties.

Keywords: micro-/nanolayered polymers; interfacial phenomena; multilayer coextrusion

1. Introduction

Polymer blends and multilayers account for a very large fraction of high-performance materials used in industry today, because they are easy and cheap to prepare. In these multiphase polymer systems, interfacial phenomena such as interdiffusion, interfacial slippage, and/or interfacial reactions occurring at the polymer–polymer interface are of high interest for researchers in polymer science [1–3]. For instance, the interdiffusion process that occurs in the region between polymers plays a very important role in mixing and homogenizing composition gradients during processing [4,5]. Interfacial slippage may also occur in incompatible polymer systems, a phenomenon that has been suggested to explain the negative deviation in viscosity [3]. Nevertheless, it is undesirable due to its detrimental effect on the final properties of products, such as reduced adhesion between polymers. To improve this weak polymer–polymer adhesion in an incompatible system, a compatibilization effect is often introduced by adding premade copolymers as compatibilizers or by locally incorporating chemical reactions at the polymer–polymer interface [6,7]. Hence, an interphase that is a nonzero thickness 3D zone, as opposed to a purely geometrical plane interface, could be triggered either by interdiffusion or by chemical reaction at the polymer–polymer interfacial zone in a multiphase system. So far, the rheology, physics, morphological, and geometrical properties associated with such an interphase, as well its contribution to the multiphase blends and to multilayer systems, persist as a hot issue among polymer scientists and engineers.

More specifically, for multilayered polymers that are widely used multiphase materials, polymer–polymer interfaces play a crucial role in the resulting properties. Generally, there are many techniques for combining different polymers to create multilayered films, including layer-by-layer assembly (LbL) [8], multilayer coextrusion [9], lamination, solvent casting, and spin coating. One of the most appealing techniques for industry is the coextrusion process, which is widely used to form multilayered sheets or films that are suitable for various products ranging from food packaging materials to dielectric capacitors to reflective polarizers [10–12]. In contrast to the LbL method, multilayer coextrusion is known as a forced-assembly concept in which two or more polymers are extruded by two or more extruders and combined in a feedblock or die to form a finished product with multiple layers (Figure 1) [5,13]. Notably, as opposed to traditional approaches using LbL assembly and spin coating that suffer from low productivity, this technology is a top-down approach that enables the industrial manufacture of multilayered films. Multilayer coextrusion is also a good tool for fundamental studies, especially for studying interfacial dynamics in polymer blending, since the morphology of coextruded multilayers is well defined by layer number and layer thicknesses, making this a suitable model system for multiphase systems [3,14]. When it comes to the processing problems posed by coextrusion, the subject is very vast. Researchers have dealt with many aspects ranging from equipment designs [15], fluid mechanics, and analysis of multilayer flow [16,17] to interfacial defects and optimization of processing conditions [18,19], etc. It is known that under certain operating conditions, interfacial defects can be observed inside the die, especially in the case of coextruded polymers with a large rheological contrast. These defects can be divided into two common types: one featuring a waviness and/or rugged shape at the polymer–polymer interface (i.e., interfacial flow instability) and another characterized by nonuniformity in layer thicknesses (i.e., encapsulation) [12]. The important theoretical and experimental advances made in the last few decades with regard to such interfacial defects have been mainly limited to mechanical and numerical approaches.

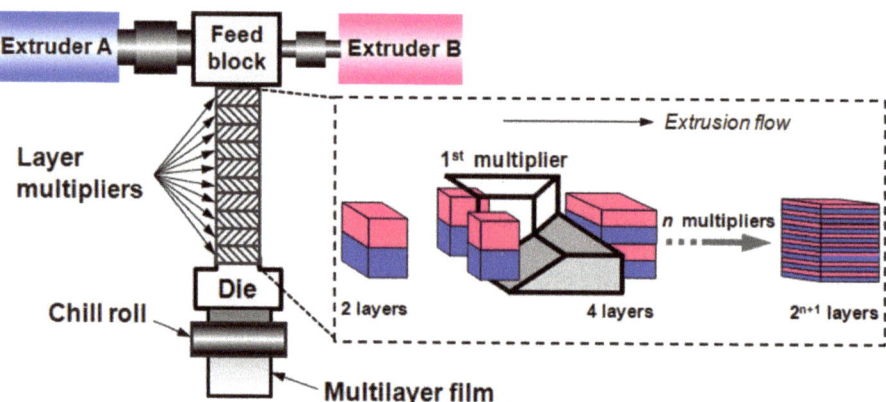

Figure 1. Schematic illustration of the multilayer coextrusion process. Reproduced with permission from Ref. [13]. Copyright (2020) John Wiley and Sons.

In addition, with the rapid development of coextrusion processing technology, the nanolayer coextrusion technique invented by incorporating the concept of layer multiplication has made much progress in the past two decades [20,21]. Using layer multipliers or layer-multiplying elements, products with thousands of layers can be produced, in which the layer thickness can be reduced to the nanometer scale (Figure 1) [22,23]. Interfacial spatial confinement always dominates when layer thickness is decreased, which greatly alters the microstructure and dynamics of the multilayer polymers [24,25]. Macroscopic properties, including mechanical, electric, and gas/liquid barrier properties, are also dramatically affected. In addition, for the nanolayer coextrusion process, the laminar flow conditions combine polymers in the layer multipliers by producing a large number of interlayer interfaces without complete mixing [5]. Interfacial behaviors involving interlayer diffusion and reactions are also critical in defining the structure and properties [5,13]. The measurement of the interfacial properties is important for understanding the interdependence of processing, structure, and properties.

Therefore, this paper provides an overview of the research work carried out in academia in recent years regarding interfacial phenomena, notably interlayer diffusion, interlayer reaction, interfacial instabilities, and interfacial geometrical confinement. The effects of these phenomena on microstructural development and microscopic properties have been summarized. Relevant basic theories and experimental techniques are also included. This review will enable a better understanding of the interfacial phenomena involved in multilayer coextrusion, in view of establishing the processing-structure-property relationship of multi-micro-/nanolayered polymer products.

2. Interfacial Phenomena at Polymer–Polymer Interfaces
2.1. Interfacial Diffusion

When two polymers are brought into contact at a temperature above the glass transition temperature, particularly at the melt state, the movement of polymer chains across the interface results in an interdiffusion process driven by the entropic advantage between components [26–29]. Interdiffusion has been an important issue in multicomponent polymer blends and multilayered composites. The presence of interdiffusion enhances polymer–polymer adhesion and stabilizes the interfaces [26,30,31]. This phenomenon is also important in controlling the glass transition temperatures [32,33], microstructure development [5,34], gas permeability [35], and mechanical [30], optical [4], and electric properties [36] of multicomponent polymer products.

2.1.1. Basic Interdiffusion Theories

During the diffusion process, the rate of disappearance of the concentration gradient across the interface, referred to as the mutual diffusion coefficient (D_m), is dependent on the composition of the system, as well as the excess enthalpy and entropy of segment–segment mixing:

$$D_m = 2(\chi_s - \chi)\varphi_A \varphi_B D_T = \left(\frac{\varphi_B}{N_A} + \frac{\varphi_A}{N_B} - 2\chi \varphi_A \varphi_B\right) D_T, \qquad (1)$$

where χ_s is the interaction parameter at the spinodal of the A/B mixture:

$$\chi_s = \frac{1}{2}\left(\frac{1}{\varphi_A N_A} + \frac{1}{\varphi_B N_B}\right). \qquad (2)$$

D_T is the transport coefficient related to the mobility of segments or the monomeric friction coefficient of the components involved [37]. Two different theories concerning D_T have been proposed, including the fast-mode theory with a faster moving species dominating the diffusion mechanism [38] and the slow-mode theory with a slower moving species [39]. Both theories were originated from the Flory–Huggins Lattice theory via Onsager formalism. The driving force for interdiffusion is the chemical potential gradient. The flux J_i is related to the chemical potential μ_i as follows:

$$J_A = -\Lambda_A \nabla(\mu_A - \mu_V) \qquad (3)$$

$$J_B = -\Lambda_B \nabla(\mu_B - \mu_V) \qquad (4)$$

$$J_V = \Lambda_A \nabla(\mu_A - \mu_V) + \Lambda_B \nabla(\mu_B - \mu_V). \qquad (5)$$

where Λ_i is the Onsager coefficient of lattice i and subscripts A, B, and V denote molecule A, molecule B, and vacancy, respectively.

The chemical potential gradient is obtained from the Flory–Huggins theory:

$$\nabla \mu_i = \frac{k_B T}{\varphi_i} \left(\frac{\varphi_B}{N_A} + \frac{\varphi_A}{N_B} - 2\chi \varphi_A \varphi_B \right) \nabla \varphi_i, \tag{6}$$

where k_B is the Boltzmann constant and T is the temperature.

The slow-mode theory assumes no vacancy flux ($J_V = 0$), leading to:

$$J_A = -J_B = \frac{-\Lambda_A \Lambda_B}{\Lambda_A + \Lambda_B} \nabla (\mu_A - \mu_B). \tag{7}$$

Combining Equations (5)–(7) of continuity for component A:

$$\frac{1}{\Omega} \frac{\partial \varphi}{\partial t} = \nabla(-J_A) \tag{8}$$

yields:

$$\frac{\partial \varphi}{\partial t} = \nabla(D_m \nabla \varphi) = \nabla(-\Omega J_A) = \nabla \left[\frac{\Omega k_B T}{\varphi_A \varphi_B} \left(\frac{\Lambda_A \Lambda_B}{\Lambda_A + \Lambda_B} \right) \left(\frac{\varphi_B}{N_A} + \frac{\varphi_A}{N_B} - 2\chi \varphi_A \varphi_B \right) \nabla \varphi_i \right] \tag{9}$$

and D_m can be thus obtained:

$$D_m = \frac{\Omega k_B T}{\varphi_A \varphi_B} \left(\frac{\Lambda_A \Lambda_B}{\Lambda_A + \Lambda_B} \right) \left(\frac{\varphi_B}{N_A} + \frac{\varphi_A}{N_B} - 2\chi \varphi_A \varphi_B \right), \tag{10}$$

where Ω is the volume of a quasi-lattice site.

On the contrary, the fast-mode theory assumes $J_V \neq 0$ but rather $\nabla \mu_V = 0$, which leads to the total flux J_A^T of A:

$$J_A^T = -\Lambda_A \varphi_B \nabla \mu_A + \Lambda_B \varphi_A \nabla \mu_B. \tag{11}$$

Similarly, an expression of D_m can be given:

$$D_m = \Omega k_B T \left(\frac{\varphi_B}{\varphi_A} \Lambda_A + \frac{\varphi_A}{\varphi_B} \Lambda_B \right) \left(\frac{\varphi_B}{N_A} + \frac{\varphi_A}{N_B} - 2\chi \varphi_A \varphi_B \right). \tag{12}$$

Comparing Equation (13) with Equation (1) and making some rearrangements, we can obtain:

$$D_T = \varphi_B N_A D_A^* + \varphi_A N_B D_B^*, \tag{13}$$

with the tracer diffusion coefficient of component i, D_i^*, as:

$$D_i^* = \frac{\Omega k_B T}{N_i \varphi_i} \Lambda_i. \tag{14}$$

The Onsager coefficient (Λ_i) is then extracted in terms of the curvilinear Rouse mobility ($B_{0,i}$) or in terms of the monomeric friction coefficient (ζ_i) of the segment i:

$$\Lambda_i = \frac{B_{0,i} N_i^e}{\Omega N_i} \varphi_i = \frac{N_i^e}{\zeta_i \Omega N_i} \varphi_i, \tag{15}$$

where N_i^e is the number of repeat units per entanglement length of component i, respectively.

The slow-mode theory assumes the equal and opposite fluxes of the two polymers, indicating that the interface remains symmetrical as interdiffusion proceeds. By contrast, the fast-mode theory describes the interdiffusion with a moving interface by unequal fluxes that are balanced by a net flux of vacancies across the interface. It suggests a movement of the interface toward the more quickly diffusing component and a broadening of the concentration profile on the slower component side.

2.1.2. Interdiffusion in Micro-/Nanolayer Coextrusion

Interdiffusion also occurs in multilayer coextrusion, in which a resulting interphase is generated by localized interlayer mixing [4,35,36]. This interdiffusion behavior in multilayer coextrusion can be viewed in Figure 2, which illustrates the coextrusion of nylon with two different grades of ethylene vinyl alcohol (EVOH) with different ethylene contents: EVOH44 (44 mol% ethylene) and EVOH24 (24 mol% ethylene) [40]. Obviously, well-defined layer boundaries with sharp interfaces can be observed in 17-layered films, and 129-layered films show a diffused interphase, but the two polymer components are still visible in this sample. When the number of layers increases to 1025, however, completely interdiffused nylon/EVOH layers are observed therein. This clearly indicates the existence of interdiffusion during coextrusion due to the intermixing of component layers. Miscibility between the nylon and EVOH melts causes the layers to diffuse from the time that they are combined in the feedblock until the melt is quenched after exiting the die. Coextrusion through an increasing number of layer-multiplying dies increases the number of layer interfaces and prolongs the melt contact time while decreasing the layer thicknesses. The degree of interlayer interdiffusion rapidly increases and, after several multiplications, forces the entire melt toward a homogeneous blend.

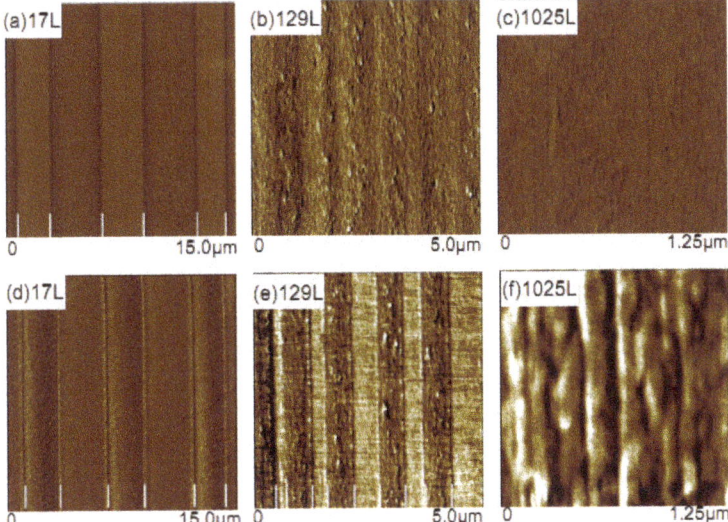

Figure 2. Atomic force microscope (AFM) phase images showing the layer morphology in multilayer films with a varying number of layers for (**a**–**c**) nylon/ethylene vinyl alcohol (44 mol% ethylene) (EVOH44) multilayers and (**d**–**f**) nylon/24 mol% ethylene (EVOH24) multilayers. (**a**,**d**) It shows 17 layers with nominal layer thickness of 3.2 µm; (**b**,**e**) 129 layers with nominal layer thickness of 400 nm; and (**c**,**f**) 1025 layers with nominal layer thickness of 50 nm. Reproduced with permission from Ref. [40]. Copyright (2016) Elsevier.

A few studies have addressed theories describing the interdiffusion phenomenon during practical coextrusion processing. In a pioneering work, Baer and his coworkers studied interdiffusion in a microlayered system of polycarbonate (PC) and copolyester (KODAR) and proposed an interdiffusion model to explain the diffusion occurring during the coextrusion process [41]. An interfacial composition profile can be calculated via Fick's equation as follows:

$$\frac{\partial W_i}{\partial t} = D_m \frac{\partial^2 W_i}{\partial x^2}, \tag{16}$$

where W_i is the weight fraction of species i, t is the diffusion time, x is the position, and D_m is the mutual diffusion coefficient. Only one-half layers of the two components were considered due to the symmetry of the layered structure. Many assumptions equivalent to the case of weakly interacting polymer pairs were used in this analysis. There are some assumptions: (i) the mutual diffusion coefficient is dependent on temperature but independent of the composition; (ii) the position of the interface between adjacent layers is constant; and (iii) the initial interface is sharp, and the composition gradient at the boundaries of the interdiffusion element is equal to zero:

$$\left(\frac{\partial W_{PC}}{\partial t}\right)_{x=0} = \left(\frac{\partial W_{PC}}{\partial t}\right)_{x=(L_{PC}+L_k)/2} = 0. \tag{17}$$

On the basis of these assumptions, this equation can be solved by the method of separation of variables as follows:

$$W_{PC}(x,t) = \frac{L_{PC}}{L_{PC}+L_K} + \sum_{n=1}^{\infty} \frac{2}{n\pi} \sin\left(\frac{n\pi L_{PC}}{L_{PC}+L_K}\right) \\ \times \cos\left(\frac{2\pi n x}{L_{PC}+L_K}\right) \exp\left(-\frac{4\pi^2 n^2 D_m t}{(L_{PC}+L_K)^2}\right) \tag{18}$$

where L_{PC} and L_K are the layer thicknesses of PC and KODAR, respectively. By this equation, the concentration profiles across the layers can be mapped by relating the mutual diffusion coefficient to the gas permeability.

Note that the above interdiffusion model proposed by Baer and his coworkers follows a slow-model diffusion mechanism that assumes the mutual diffusion coefficient to be independent of the composition. It, therefore, might be unable to describe fast-controlled diffusion, especially for cases of diffusion in which the mutual diffusion coefficient is strongly dependent on the composition [42]. Very recently, Lamnawar and coworkers proposed a modified rheological model from a primitive Qiu–Bousmina model [42] to determine the mutual diffusion coefficient (D_m) with composition dependence and subsequently mapped the interfacial diffusion profile in coextruded layers based on polymer dynamics theory and fast-controlled mode theory [43–46]. Assuming an apparent friction coefficient (ζ_b) for the chain mixture at the diffuse interphase, the Onsager coefficient (Λ_i) takes the following form:

$$\Lambda_i = \frac{N_b^e}{\zeta_b \Omega N_i} \varphi_i. \tag{19}$$

where N_b^e is the average number of repeat units between entanglements for the polymer pair, and ζ_b is strongly composition dependent [43,47]. Thus, the mutual diffusion coefficient (D_m) can be related to the structural properties of the mixture of A and B:

$$D_m = \frac{k_B T N_b^e}{\zeta_b} \left(\frac{\varphi_B}{N_A} + \frac{\varphi_A}{N_B}\right) \left(\frac{\varphi_B}{N_A} + \frac{\varphi_A}{N_B} - 2\chi \varphi_A \varphi_B\right). \tag{20}$$

The mutual diffusion coefficient (D_m) can be experimentally quantified by its relationship with the rheological behavior of the interphase using a planar polymer A/B sandwich with healing time:

$$\frac{1}{G_I^*(t)} = \frac{H}{2(D_m t)^{1/2}} \left(\frac{1}{G_{s,t}^*} - \frac{1}{G_{s,0}^*}\right) + \left(\frac{\varphi_A}{G_{A,0}^*} + \frac{\varphi_B}{G_{B,0}^*}\right), \tag{21}$$

where $G_I^*(t)$ is the complex modulus of the interphase at the healing time t; H is the total thickness of sandwich assembly; $G_{s,t}^*$ and $G_{s,0}^*$ are the overall complex moduli of sandwich assembly at healing times of t and 0, respectively; and $G_{A,0}^*$ and $G_{B,0}^*$ are the complex moduli of polymers A and B at healing time $t = 0$, respectively. D_m can be related to the monomeric friction coefficient (ζ_b) for the mixture as follows:

$$\zeta_b = \frac{\pi^2 k_B e_b^2 T}{N_b^3 b^4} \frac{1}{\omega} \left[\left(\frac{8 G_{N,b}^0}{\pi^2 G_I^*(t)} \right)^2 - 1 \right]^{-1/2}, \qquad (22)$$

where e_b represents the stem length on the order of the gyration radius of entanglements, N_b is the number of repeat units of the blend, b is the effective bond length, ω is the angular frequency, and $G_{N,b}^0$ represents the average plateau modulus of the interphase/blend. The mutual diffusion coefficient can be calculated using:

$$D_m = \left[\frac{(2/3)^{1/3} p}{\left(9q + \sqrt{3} \times \sqrt{-4p^3 + 27q^2}\right)^{1/3}} + \frac{\left(9q + \sqrt{3} \times \sqrt{-4p^3 + 27q^2}\right)^{1/3}}{2^{1/3} \times 3^{2/3}} \right]^2 \qquad (23)$$

with

$$p = \frac{8\delta\omega G_{N,b}^0}{\pi^2} \left(\frac{\varphi_A}{G_{A,0}^*} + \frac{\varphi_B}{G_{B,0}^*} \right) \qquad (24)$$

$$q = \frac{8\delta\omega G_{N,b}^0}{\pi^2} \frac{H}{2t^{1/2}} \left(\frac{1}{G_{s,t}^*} - \frac{1}{G_{s,0}^*} \right) \qquad (25)$$

$$\delta = \frac{N_b^e N_b^3 b_b^4}{\pi^2 e_b^2} \left(\frac{\varphi_B}{N_A} + \frac{\varphi_A}{N_B} \right) \left(\frac{\varphi_B}{N_A} + \frac{\varphi_A}{N_B} - 2\chi \varphi_A \varphi_B \right). \qquad (26)$$

The resultant interlayer interphase thickness can be also estimated according to:

$$h_I' = 2(D_m t)^{1/2}. \qquad (27)$$

The concentration profile of diffusing species across the interlayer interface can be also approximated by the simple Fickian solution:

$$\varphi(z,t) = \frac{1}{2} \left[erf\left(\frac{h-z}{2(D_m t)^{1/2}} \right) + erf\left(\frac{h+z}{2(D_m t)^{1/2}} \right) \right], \qquad (28)$$

where erf is the error function, h is the layer thickness, and z is the spatial axis along the diffusion direction with the boundary of layers as $z = 0$.

The interdiffusion model above makes it possible to map the time evolution of the interphase profile including the mutual diffusion coefficient, interphase thickness, and concentration profile of diffusing species across the interface. Figure 3 shows an example of the quantified evolution of the mutual diffusion coefficient and the interphase thickness determined from rheological modeling for a compatible poly(vinylidene fluoride) (PVDF)/poly(methyl methacrylate) (PMMA) bilayered system, as well as the concentration profile determined from dispersive X-ray analysis (EDX) in the cross-section of the healed bilayer and the coextruded bilayer [45]. Notably, the calculated interphase thicknesses are in quantitative agreement with that determined by energy-dispersive X-ray analysis (EDX), indicating the validation of the interdiffusion model. In addition, as well documented in a recent study, the development of multiple interlayer interphases from interdiffusion and relevant length scales in nanolayered coextruded polymer films could be further quantitatively determined by the combination of dielectric relaxation spectroscopy and energy-dispersive X-ray analysis (Figure 4) [5].

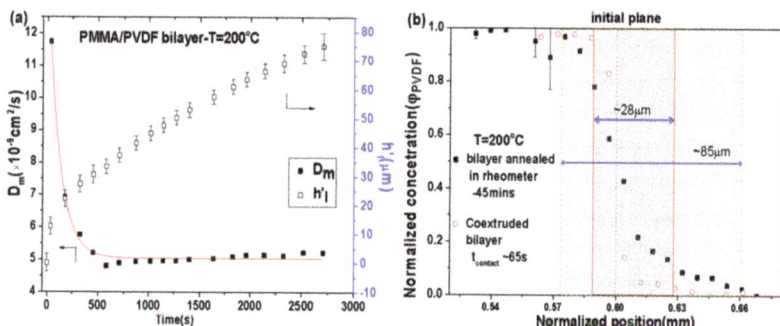

Figure 3. (**a**) Time evolution of the mutual diffusion coefficient and the interphase thickness determined from rheology with a poly(methyl methacrylate) (PMMA)/poly(vinylidene fluoride) (PVDF) bilayer healed at 200 °C, (**b**) normalized PVDF concentration profile versus normalized position determined from SEM-energy-dispersive X-ray analysis (EDX) in the cross-section of the bilayer after healing for 45 min (black solid squares) and the coextruded bilayer (red open circles). The sparse and dense shadow zones designate the interphase scale in the healed and coextruded bilayer, respectively. Reproduced with permission from Ref. [45]. Copyright (2016) AIP Publishing.

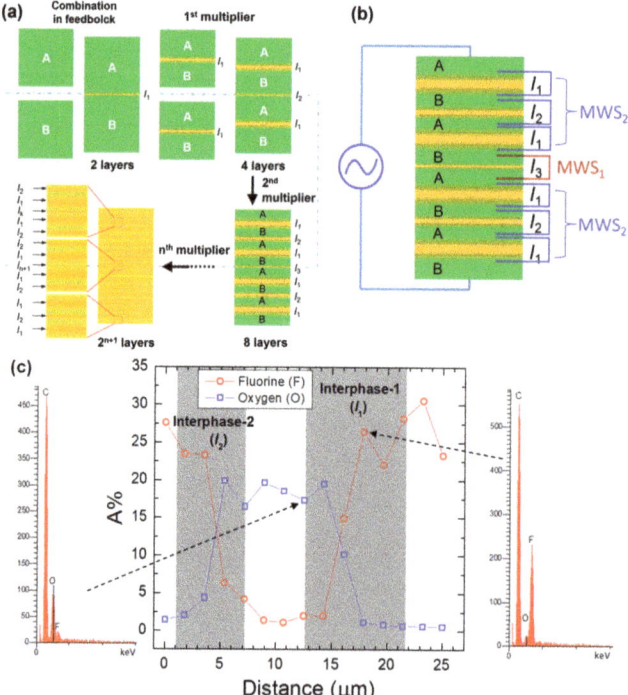

Figure 4. (**a**) A schematic illustration showing the interdiffusion and development of diffuse interphases during the coextrusion process for a compatible multilayered A/B system. "I" stands for the interphase shown by the bright regions in the schematic layered structures. (**b**) Scheme explaining the bimodal interfacial Maxwell–Wagner–Sillars (MWS) relaxations measured by dielectric relaxation spectroscopy for multilayered films subjected to an alternating electric field. (**c**) EDX measured concentration profile of F and O (in atomic fraction) versus measured positions for an eight-layer PVDF/PMMA film. Reproduced with permission from Ref. [5]. Copyright (2018) American Chemical Society.

2.2. Interfacial Instabilities

2.2.1. Interfacial Flow Instabilities

Interfacial instability is an unstable process in which the interface between neighboring layers changes locally or loses its continuity. Interfacial distortion by flow instability can cause nonuniform layer thicknesses, irregular interfaces, and even nonuniform film thickness. Other interfacial flow instabilities include zigzag and wave-type in the extrudate [48,49]. A zigzag is observed along the flow direction, which is triggered in the die land above the critical interfacial shear stress [50]. In addition, a layer breakup phenomenon appears during coextrusion when the thicknesses of layers are reduced to beneath a critical value [51]. Generally, these interfacial instabilities can be reduced or eliminated by controlling the shear stress, extrusion rates, die gap, feedblock design, polymer viscoelasticity, etc.

2.2.2. Interfacial Slip

Most commercial polymers are immiscible, and the interfacial region between them may be of a lower density than the constituent bulk materials. A significant slip can, therefore, occur during the flow due to reduced entanglements at their interface [52]. Especially for the cases in which components in a layered material have little compatibility or adhesion, densitometry variations within layered films with high numbers of layer interfaces can result in nontrivial interphase volumes. Interphases with reduced polymer chain entanglements have been demonstrated to induce interfacial instability, when films are processed under high shear rates through a mechanism assigned to the interfacial slip.

Interfacial slip in multilayer systems was ever studied using coextruded polypropylene (PP)/polystyrene (PS) multilayers with closely matched viscosities [3,53,54]. An obvious deviation was found in the measured pressure drops of multilayers from those of the homopolymers as well as in the nominal viscosity, especially for larger numbers of layers (Figure 5) [53]. These deviations increased with the flow rate. These data are strong evidence for apparent slip, indicating a thin, low-viscosity layer between the PP and PS interfaces. Significant layer instability, characterized by layer breakup and/or delamination, was found in nanolayered samples, which could result from interfacial slip (Figure 5). Besides, with another multilayered system of PS/PMMA (smaller Flory–Huggins parameter χ than PP/PS), a slight pressure drop and slip velocity decreased with the interfacial width [3]. In addition, the incorporation of a premade polystyrene-block-ethyl-ethylene-diblock copolymer P(S-b-EE) to the PP/PS system could suppress the interfacial slip [3]. The interfacial slip phenomenon suggests the importance of interfacial interactions and adequate adhesion in multilayered polymers during processing, beyond the common expectation of matching viscosities. Eliminating the possibility of interfacial slip is of remarkable importance if highly regular and continuous layer architecture is required for the target properties.

2.2.3. Layer Breakup in Nanolayer Coextrusion

The layer breakup phenomenon has been observed in many nanolayered systems during coextrusion, with layers breaking spontaneously during the process and losing their integrity [33,51,55–57]. This breakup phenomenon has been observed in different polymer pairs, and the layer-continuity limit appeared to be system dependent. Hiltner and coworkers found that efforts to obtain PMMA nanolayers thinner than 5 nm resulted in layer breakup and instability in polycarbonate (PC)/PMMA multilayers [33]. Meanwhile, in a PP/PS system, the layers break up when the layer thickness is thinner than 25 nm PP/PS [56]. Layer breakup alters the layered structure and can even adversely affect the final properties. In the presence of layer breakup, Hiltner and coworkers observed a reduction in the barrier property of PP/poly(ethylene oxide) (PEO) nanolayer film [58]. For the moment, the mechanism governing layer breakup in nanolayers is still lacking in the open literature. Interfacial distortions (viscous encapsulation or secondary flows) during coextrusion might be generally responsible for the layer breakup. In addition, interfacial instabilities due to an initial disturbance at the interfaces may be amplified along the flow in the die, which can also induce layer ruptures [59]. In addition, material characteristics have also been demonstrated to contribute to the layer breakup. With a PS/PMMA multilayer system, Bironeau et al. recently reported the existence of a critical layer thickness of around 10 nm, below which the layers break up [51]. They attributed this breakup phenomenon to small interfacial disturbances that are amplified by van der Waals disjoining forces. The critical layer thickness is independent of the processing parameters, but presumably dependent only on material characteristics. In another system of poly(vinylidene fluoride-co-hexafluoropropylene) (PVDF-HFP)/polycarbonate (PC) with substantial rheological mismatch, at a nominal layer thickness below 160 nm, layers break up into microsheets and droplets triggered by viscoelastic differences between component melts (Figure 6), which dramatically alters the resulting dielectric properties [60]. Nevertheless, a clear mechanism responsible for layer breakup in the multilayered polymers still remains as an open question. More experimental and theoretical studies still need to be performed in order to better understand this phenomenon in the multilayer coextrusion process.

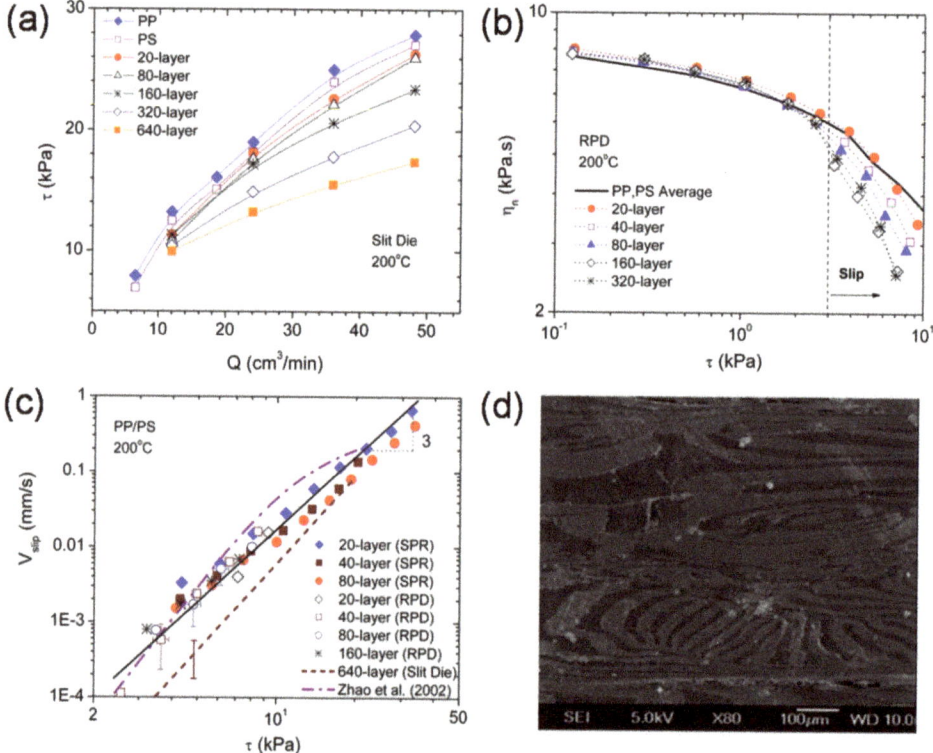

Figure 5. (**a**) Pressure drop of multilayered polypropylene (PP)/polystyrene (PS) samples and neat components in an in-line slit rheometer; (**b**) nominal viscosity of multilayer samples and harmonic average (solid lines) of the neat components, as measured by steady shear experiments with a rotational parallel-disk rheometer; (**c**) interfacial slip velocities of multilayers; and (**d**) SEM micrographs of 320-layer PP/PS multilayer samples after steady shear. Reproduced with permission from Ref. [53]. Copyright (2009) AIP Publishing.

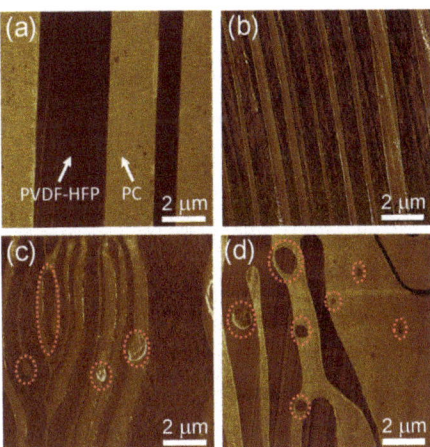

Figure 6. AFM phase images for cross-sections of poly(vinylidene fluoride-*co*-hexafluoropropylene) (PVDF-HFP)/polycarbonate (PC) films with various numbers of layers: (**a**) 32 L, (**b**) 256 L, (**c**) 2048 L, and (**d**) 16384 L.

2.3. Interfacial Reaction

2.3.1. Basic Theories

Considerable efforts have been dedicated to understanding the fundamental kinetics and mechanisms of interfacial reaction over the past 20 years. Theoretically, many studies are based on a planar interface in a simple model bilayer under static conditions, and the reaction kinetics reported can be generally categorized under two different mechanisms: a reaction-controlled mechanism and a diffusion-controlled mechanism. Fredrickson and Milner [61,62] studied the coupling reactions occurring at the interface between two symmetrical polymers (with the same degree of polymerization, $N_A = N_B$). The formation of copolymers, i.e., reactive coupling kinetics at polymer–polymer interfaces, was predicted to take place in three stages as follows. (i) At early times, the kinetics of the coupling reaction at the interface between reactive moieties controls the amount of copolymers at the interface, which (interfacial coverage) grows linearly with time. (ii) At intermediate times, the saturation of copolymers at the interface remains negligible, but a depletion hole of reactants builds up around the interface; the growth is dominated by the diffusion of the more dilute reactive species to the interface. (iii) For longer times, copolymers gradually form a copolymer layer at the interface. This interfacial copolymer layer generates a significant chemical potential barrier for unreacted homopolymers that strongly suppresses and controls the kinetics of the reaction by limiting the diffusion of reactive moieties to the interface for further reaction. O'Shaughnessy et al. [63,64] argued that the reaction over the first two steps should depend on the reactivity of the functional groups, especially for the reaction with weakly reactive pairs at the very beginning of the process. Nevertheless, they agreed on the third regime regarding the formation of an energy barrier resulting from the accumulation of copolymers at the interface. Generally speaking, the reaction-controlled mechanism is believed to be more appropriate in describing the short-time reaction for polymers with low reactivity. On the other hand, the diffusion-controlled mechanism is suitable in situations in which a reaction with high reactivity takes place over a longer time period. The transition from reaction-controlled kinetics to diffusion-controlled kinetics can be expected for reactions with high reactivity [65].

2.3.2. Interfacial Morphology under Reaction

Theories on interfacial reaction are focused mainly on the reaction kinetics at very short times. Thus, researchers have shown that the accumulation of copolymers at the interface slows down the kinetics, because of the creation of an energy barrier preventing unreacted species from reaching the interface for further reaction. These studies were completed by experimental studies showing that this slower kinetics was spontaneously compensated, even in static conditions, by an acceleration due to the roughening of the interface. These interfacial fluctuations allow new reactive moieties to reach the interface and, therefore, the conversion rate to increase under certain conditions. In addition to an acceleration of the reaction kinetics, this interfacial roughening can generate small micelles in blend phases.

Reaction under Equilibrium Conditions

Macosko and coworkers performed a considerable number of studies on the interfacial reactions and morphological development at polymer–polymer interfaces under equilibrium conditions (without flows) [2]. For example, they examined the changes in the interfacial morphology upon the coupling reaction between aliphatic amine-terminated polystyrene (PS-NH$_2$) and anhydride-terminated poly(methyl methacrylate) (PMMA-ah) [66]. A roughening interface can be observed after 20 min annealing as a result of the coupling reaction (Figure 7). Another interesting observation is that a lamellar microstructure appears at the interface of the thin domains after 1 h annealing, which indicates the formation of a PS-PMMA block copolymer. The interfacial roughening is explained by the large decrease in interfacial tension due to the creation of copolymers at the interface. The authors also supposed that local fluctuations might be induced by thermal fluctuations when block copolymer coverage at the interface decreases below the saturation level. This creates a new interfacial area that allows new reactive moieties to reach the interface and new copolymers to be formed. Interfacial instabilities continuously create a new interface in such a way that the interface becomes rough.

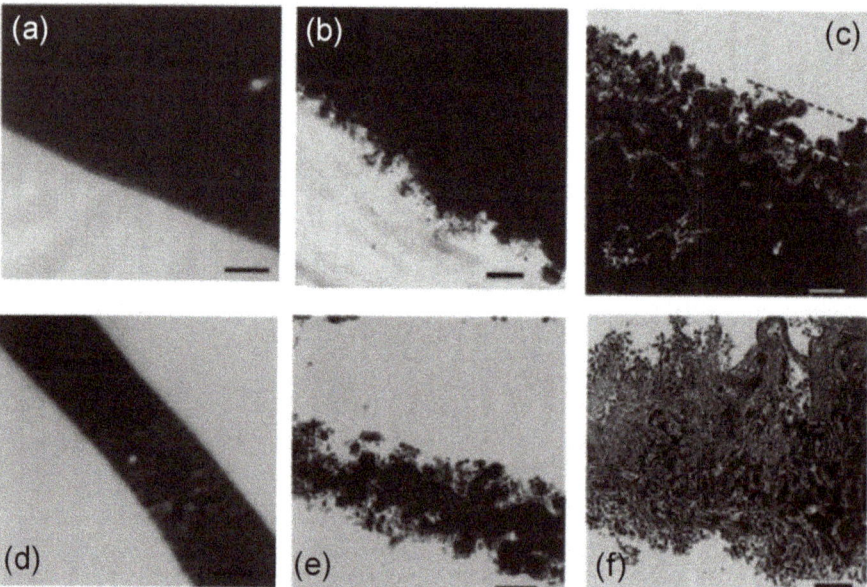

Figure 7. Representative morphologies of aliphatic amine-terminated polystyrene (PS-NH$_2$)/anhydride-terminated poly(methyl methacrylate) (PMMA-ah) blends after static reaction times of (**a**) 0, (**b**) 20, and (**c**) 60 min at a large domain interface and after static reaction times of (**d**) 0, (**e**) 20, and (**f**) 60 min at a thin sheet interface at 200 °C. All scale bars are 500 nm. Dash lines in (**c**) approximately indicate the roughening zone. The magnitude of its width is roughly 500 nm. Reprinted with permission from Ref. [66]. Copyright (1999) American Chemical Society.

Morphology Development Revealed by Rheometry

Using a dynamic rheological measurement (small-amplitude oscillatory shear), the relationship between interfacial reactions and interfacial morphology development at a planar interface could be monitored the rheological responses [67–69]. Figure 8 presents the time evolution of the complex viscosity of a bilayer system based on end-functionalized monocarboxylated polystyrene (PS-mCOOH) and poly(methyl methacrylate-ran-glycidyl methacrylate) (PMMA-GMA) with reaction time measured at a strain of 0.005 and an angular frequency of 0.1 rad/s [67]. Based on the rheological responses and interfacial morphology captured by electron microscopes, the authors proposed three stages for the reaction: (i) in stage I, the sharp interface begins to undulate due to the creation of copolymers from the reaction at the interface; (ii) in stage II, the reactive moieties around the interface are totally consumed. New reactive moieties have to diffuse thorough the brush-like copolymer layer formed during stage I. The interface becomes corrugated; and (iii) in the final stage III, the reactive polymer chains penetrate again into the densely packed copolymer layers and allow the reaction to proceed further. The interfacial thickness and roughness increase. Further reaction even leads to the formation of some micelles (10–20 nm) and microemulsions (~100 nm) (which are micelles swollen with homopolymers) at the interfacial region. Apart from rheometry, the interfacial coupling reactions and morphology development at the planar interfaces of layered polymers could be also probed in situ by dielectric relaxation spectroscopy [13,70]. Using the dielectric molecular relaxation spectrum as a probe, the interfacial copolymer accumulation and resulting roughness could be captured and quantified, which agrees well with the rheological and morphological characterizations [70].

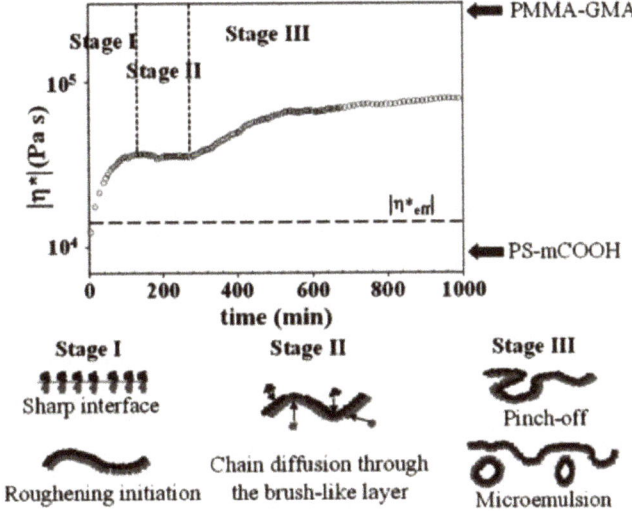

Figure 8. Complex viscosity as a function of reaction time for a monocarboxylated polystyrene (PS-mCOOH)/poly(methyl methacrylate-ran-glycidyl methacrylate) (PMMA-GMA) bilayer and schematic illustration describing variations of interfacial morphology. The dashed line is predicted by the reciprocal rule. Reprinted with permission from Ref. [67]. Copyright (2003) American Chemical Society.

Effects of Oscillatory Shear Flow

Oscillatory shear amplitude and angular frequency also have a substantial influence on the interfacial reaction kinetics and morphology development [71]. A smaller strain amplitude (γ_0) and angular frequency (ω) were found to enhance the extent of reaction and the generation of an interphase. By contrast, at larger strains and angular frequencies, oscillatory shearing inhibits the diffusion of polymer chains to the interface and thus also inhibits reactions therein. In addition, a large strain can generate alternating layers of PS-mCOOH/PMMA-*graft*-PS copolymer, which act as a barrier for the diffusion of reactive chains to the interface, thereby restricting interfacial reactions. On the other hand, a higher ω can break the interface and generate a multilayer of graft copolymer. In this situation, even though a lower ω is applied again, further reaction does not occur; this inhibition, therefore, becomes a permanent obstacle to further interfacial reactions. Additionally, it is observed that the perpendicular shear force greatly enhances the roughness of the interface as compared with parallel shear force applied to a reactive bilayer [72]. Here, the perpendicular shearing applied to the reactive bilayer also leads to the formation of microemulsions that could markedly stabilize the interfacial morphology.

2.3.3. Interfacial Reaction in Multilayer Coextrusion

Macosko and coworkers [73] investigated the interfacial coupling reaction during coextrusion with a 640-layer PS-NH$_2$/anthracene-labeled anhydride-terminal PMMA (PMMA-anh-anth) system prepared by multilayer coextrusion. They found that a significant amount of PS-b-PMMA copolymers are formed during coextrusion, such that the interfaces were almost saturated with the block copolymers formed in situ. The interface became roughened, and interfacial emulsification could be observed due to high block copolymer coverage. Coextrusion processing significantly accelerated the coupling reaction of PS-NH$_2$/PMMA-anh and reduced the reaction time by more than three orders of magnitude as compared with that under quiescent annealing. The external flow under coextrusion could overcome the combined effects of high surface energy of the functional groups and slow diffusion, increasing the functional group concentration in the interfaces. In particular, the extensional deformation could be more important than shear in accelerating coupling. Flow-accelerated interfacial reaction kinetics were also noticed in other layered systems [74,75]. In a recent study, the compressive flow in coextrusion was demonstrated to play a crucial role in accelerating the coupling reaction [76]. Strikingly, coextrusion with compressive flow resulted in a reaction rate two orders of magnitude faster than that without compressive flow and enabled stronger adhesion. Besides, mechanical properties of coextruded multilayers with in-situ compatibilization reactions are significantly enhanced [7,77,78]. As illustrated by Figure 9, coextruded layers of polyurethane (PU) and functional polyethylene (PE) with amine groups (NHR) exhibit dramatically the higher peel strength at a shorter reaction time in comparison with these produced from lamination process [79]. This acceleration was attributed to the combined extensional and compressive flows in coextrusion that overcome the diffusion barrier at the interface and forcing reactive species to penetrate the interface. The interfacial reinforcement in the real micro-/nanolayer coextrusion show a strong dependence on the reaction time and the number of layers [13,80]. The resulting large enhancement in interfacial stress with the number of layers, characterized by an extensional strain hardening behavior, is attributed to the increased density of copolymers at the interfaces.

Lamnawar and coworkers [81–83] prepared multilayer films of PA6/PP and EVOH/PP systems with maleic anhydride-grafted polypropylene (PP-*g*-AM) as a tie layer using reactive coextrusion. The macromolecular architecture of the copolymers formed at the interfaces critically influences the interfacial morphology evolution, viscoelasticity and processability of layered systems. Specifically, the interface in the pure grafting case of PA6/PP-*g*-AM bilayer is relatively flat, while the interface in EVOH/PP-*g*-AM bilayer is irregular and rough due to the presence of mixed grafting and cross-linking reactions. Particularly, under real coextrusion flows, with these copolymers of complex architecture, interfacial roughness and irregularities are further enlarged in the die exit as compared to those in the feedblock due to the accelerated reaction kinetics (Figure 10). Additionally, the higher surface density of the EVOH-co-PP-*g*-MA copolymers generated at the interfaces further leads to an optical grainy defect and a loss of transparency in the resulting multilayer films [82,84].

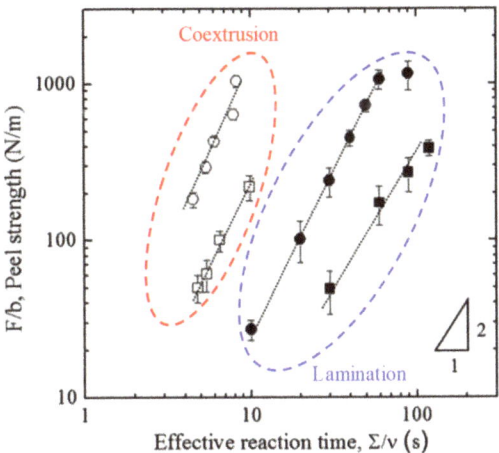

Figure 9. Comparison for peel strength of polyurethane (PU) and PE-amine groups (NHR) with different NHR functionalities produced from coextrusion and lamination. The circles indicate NHR functionality of 3 wt%, and squares represent NHR functionality of 1 wt%. Reproduced with permission from Ref. [79]. Copyright (2011) John Wiley and Sons.

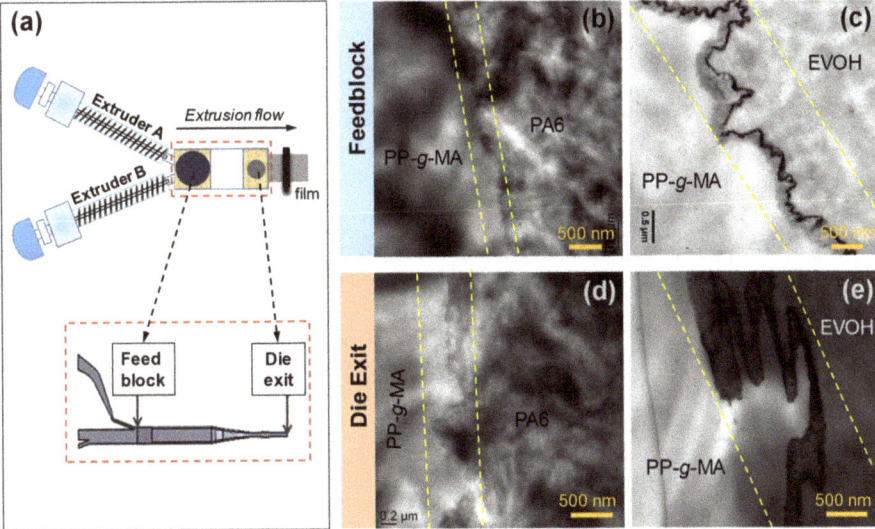

Figure 10. (a) Schematic of the coextrusion device with feedblock and die exit system. TEM micrographs of interfacial morphology for (b) PP-g-MA/PA6 and (c) PP-g-MA/EVOH bilayers quenched after leaving the feedblock and for (d) PP-g-MA/PA6 and (e) PP-g-MA/EVOH bilayers quenched after leaving the die exit. The area sandwiched between parallel dashed lines in (b–e) indicates the interfacial regions. Reprinted with permission from Ref. [83]. Copyright (2020) American Chemical Society.

2.4. Interfacial Confinement in Micro-/Nanolayer Coextrusion

Confinement of polymeric systems at the nanometer scale has revealed that under constraint, many properties including crystallization, physical aging, permittivity, and glass transition deviate from bulk material characteristics [85–87]. Interestingly, confinement phenomena also occur in the multilayer coextruded polymers, especially for nanolayered films with thousands of layers in which all layers have nanometric thicknesses. The nanoconfinement effect plays an important role in tailoring macroscopic properties of as-coextruded films [24,88,89]. In this section, we introduce the effects of interfacial geometrical confinement on crystallization and glass transitions in multilayered films produced by coextrusion.

2.4.1. Confined Crystallization

Geometrical/spatial confinement in multilayer coextrusion remarkably alters the crystallization behaviors and morphology within the confined crystalline polymer layers [25]. When layer thickness decreases, the crystalline morphology is gradually changed from a three-dimensional (3D) spherulite morphology into one-dimensional (1D) lamellar morphology. The confined crystallization in multilayer coextrusion has been observed in many systems with semicrystalline polymers as confined materials, including polypropylene (PP) [90,91], polyethylene (PE) [92,93], poly(ε-caprolactone) (PCL) [94], and poly(ethylene oxide) (PEO) [23,95], typically confined by amorphous rigid polymers with the higher glass transition temperatures (e.g., PS) (i.e., confining materials). When the layer thickness decreases below the diameter of spherulites (100 μm), the lamellae organize into flattened/compressed spherulites as discoids, and the lamellae usually have a preferred in-plane orientation, such as edge-on morphology. For example, PEO confined by PS forms in-plane crystals with an improved orientation by reducing the layer thickness and even crystallizes into a single, in-plane, high aspect ratio lamellae crystal when layer thickness reaches the scale of lamellae thickness (Figure 11) [25,96]. With further reduction in the layer thickness, the lamellae display edge-on morphology. In addition, a substantial reduction in permeability (by more than two orders of magnitude) has been noticed under confinement due to increased tortuosity of the diffusion pathway through the highly oriented lamellae [96]. With other crystalline polymers such as PVDF, geometrical confinement can induce other types of crystal morphologies and lamellar orientation [25,97,98]. It has been generally accepted that this confinement in the PVDF-containing multilayers greatly affects the resulting dielectric properties when used as capacitors [99–101].

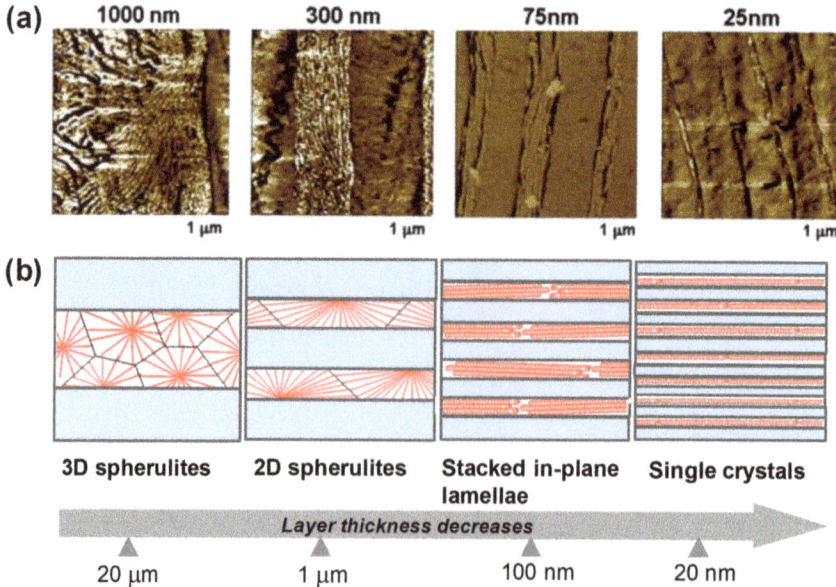

Figure 11. (a) AFM phase images of cross-sections of PS/poly(ethylene oxide) (PEO) layered films with 1000, 300, 75, and 25 nm thick PEO layers from left to right. (b) Schematic morphology evolution of PEO confined layers as the layer thickness is reduced from the microscale to the nanoscale. Reprinted with permission from Ref. [25]. Copyright (2012) Cambridge University Press.

2.4.2. Confined Glass Transition Dynamics

Confinement at the layer–layer interfaces also influences the glass transitions of as-coextruded multilayer films [102]. For instance, a merging in glass transition temperatures (T_g) with increasing the number of layers was ever observed in PC/PMMA multilayer films, especially when the individual layer thicknesses were below 10 nm [32,33]. In particular, the degree of confinement depends on the layer thickness and the correction length scale of the confined layer materials. Delbreilh and coworkers [103] investigated molecular mobility at the glass transition evolution in PC/PMMA multilayered films from micro- to nanoscale, including the glass transition temperature (T_g) and cooperatively rearranging region (CRR) size. The molecular mobility in each polymer is found to be altered in entirely different ways as the layer thickness becomes thinner than 125 nm, whereby the constituent polymers exist as two-dimensional layers under these conditions. PC exhibits a drastic decrease in cooperativity volume at the glass transition due to the confinement effect associated with conformational changes in the macromolecular chain. By contrast, slight modifications were observed for PMMA, due to the weaker intermolecular interactions in the main chains compared to those of PC. However, in a recent study with PC/PMMA multilayered films with layer thickness as thin as 4 nm, Casalini et al. recently reported an absence of geometric confinement effects on dynamic correlation (CCR size) of PMMA (i.e., the same as the bulk), especially when the confinement length scale (layer thickness) approaches or is less than the correlation length scale (ξ) (Figure 12a) [104]. Neither the fragility nor the breadth of the relaxation dispersion was affected by the geometric confinement therein (Figure 12b,c). More significantly, the dynamic correlation volume/length representing the cooperativity of the dynamics was also unaffected. Instead, the increase in the local segmental relaxation time and glass transition temperature of PMMA with decreasing layer thickness was primarily attributed to its intermixing with the high-T_g component (PC) at the extended interfacial region.

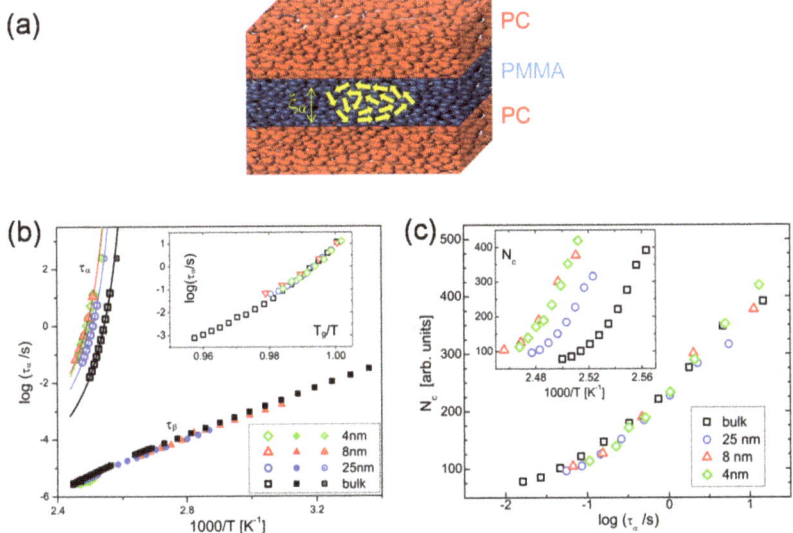

Figure 12. (a) Comparison of confinement length scale with correction length of confining PMMA in PC/PMMA multilayers. (b) Activation plots of segmental α-relaxation time (τ_α) for PC/PMMA multilayers and bulk PMMA. (c) Number of correlated units (N_c) versus the α-relaxation time for bulk and nanoconfined PMMA samples. Notwithstanding the large changes in N_c due to confinement (inset), the dependence of N_c on τ_α is the same as for the bulk PMMA. Reproduced with permission from Ref. [104]. Copyright (2016) Elsevier.

3. Conclusions

In conclusion, this paper presents a systematic review of the interfacial phenomena involved in multi-micro-/nanolayered polymer coextrusion, from the dual viewpoints of fundamental science and engineering. These phenomena, including interlayer diffusion, interlayer reaction, interfacial instabilities, and interfacial geometrical confinement, frequently occur at the layer–layer interfaces of multilayered polymers produced by multilayer coextrusion processing. The origin and basic theories of these interfacial phenomena are explained, along with the way in which they can affect microstructural development and the resulting macroscopic properties. In particular, these interfacial phenomena play a significant role in determining the interlayer adhesive strength, glass transitions, mechanical, transport, and other physicochemical properties of coextruded multilayer polymers. These phenomena occurring in multilayer systems generally show strong dependences on the layer compositions, layer thicknesses, the number of layers, processing conditions (i.e., flow fields and temperature), and the inherent characteristics of constituent polymers as well. Different approaches to investigating such effects have also been summarized, including numerical methods, microscopy, rheology, dielectric spectroscopy, energy-dispersive X-ray analysis, thermal and dynamic analyses, etc. More intensive work should be carried out in both the academia and industry to further clarify exactly what occurs at layer–layer interfaces with a view toward the scaling-up of multilayer coextrusion. Hopefully, this paper will promote a better understanding of the interfacial phenomena of multicomponent polymers and facilitate the establishment of a processing-structure-property relationship. In addition, this review provides some guidelines for controlling the interfaces and microstructure in multilayered polymer systems intended for use in advanced applications.

Author Contributions: Writing—original draft preparation, B.L.; validation, K.L. and A.M.; visualization, B.L.; writing—review and editing, H.Z., K.L. and A.M. All authors have read and agreed to the published version of the manuscript.

Funding: B.L. acknowledges the financial support of the National Key R&D Program of China (2019YFA0706801), the China Postdoctoral Science Foundation (2019M650174), and the China 111 project (D18023). H.Z. thanks the National Natural Science Foundation of China (21903015), the Natural Science Foundation of Fujian Province of China (2020J01145), and the Award Program of Fujian Minjiang Scholar Professorship (2018). K.L. is grateful for the support of the French National Research Agency (ANR, grants no. ANR-11-RMNP-0002 and ANR-20-CE06-0003), the Région Auvergne-Rhône-Alpes (ARC AURA 2017–2020), and the "100 Foreign Experts" of Fujian Province of China. A.M. also thanks the "National Foreign Experts" of China (G20200220017).

Institutional Review Board Statement: Not applicable.

Informed Consent Statement: Not applicable.

Data Availability Statement: Not applicable.

Acknowledgments: This manuscript is written in honor of the 50th anniversary of the French Polymer Group (Groupe Français des Polymères-GFP).

Conflicts of Interest: The authors declare no conflict of interest.

References

1. Qiu, H.; Bousmina, M. New technique allowing the quantification of diffusion at polymer/polymer interfaces using rheological analysis: Theoretical and experimental results. *J. Rheol.* **1999**, *43*, 551–568. [CrossRef]
2. Macosko, C.W.; Jeon, H.K.; Hoye, T.R. Reactions at polymer–polymer interfaces for blend compatibilization. *Prog. Polym. Sci.* **2005**, *30*, 939–947. [CrossRef]
3. Zhao, R.; Macosko, C.W. Slip at polymer–polymer interfaces: Rheological measurements on coextruded multilayers. *J. Rheol.* **2002**, *46*, 145–167. [CrossRef]
4. Lai, C.-Y.; Ponting, M.T.; Baer, E. Influence of interdiffusion on multilayered gradient refractive index (GRIN) lens materials. *Polymer* **2012**, *53*, 1393–1403. [CrossRef]

5. Lu, B.; Lamnawar, K.; Maazouz, A.; Sudre, G. Critical Role of Interfacial Diffusion and Diffuse Interphases Formed in Multi-Micro-/Nanolayered Polymer Films Based on Poly(vinylidene fluoride) and Poly(methyl methacrylate). *ACS Appl. Mater. Interfaces* **2018**, *10*, 29019–29037. [CrossRef]
6. Sun, Y.-J.; Hu, G.-H.; Lambla, M.; Kotlar, H.K. In situ compatibilization of polypropylene and poly(butylene terephthalate) polymer blends by one-step reactive extrusion. *Polymer* **1996**, *37*, 4119–4127. [CrossRef]
7. Jiang, G.; Wu, H.; Guo, S. Reinforcement of adhesion and development of morphology at polymer–polymer interface via reactive compatibilization: A review. *Polym. Eng. Sci.* **2010**, *50*, 2273–2286. [CrossRef]
8. Richardson, J.J.; Cui, J.; Björnmalm, M.; Braunger, J.A.; Ejima, H.; Caruso, F. Innovation in Layer-by-Layer Assembly. *Chem. Rev.* **2016**, *116*, 14828–14867. [CrossRef] [PubMed]
9. Schrenk, W. Method for Multilayer Coextrusion. U.S. Patent 3,773,882 A, 20 November 1973.
10. Dooley, J. Viscoelastic Flow Effects in Multilayer Polymer Coextrusion. Ph.D. Thesis, Eindhoven University of Technology, Eindhoven, The Netherlands, 2002.
11. Han, C.D. Multiphase Flow in Polymer Processing. In *Rheology: Volume 3: Applications*; Astarita, G., Marrucci, G., Nicolais, L., Eds.; Springer US: Boston, MA, USA, 1980; pp. 121–128.
12. Langhe, D.; Ponting, M. *Manufacturing and Novel Applications of Multilayer Polymer Films*; Elsevier: Oxford, UK, 2016.
13. Lu, B.; Alcouffe, P.; Sudre, G.; Pruvost, S.; Serghei, A.; Liu, C.; Maazouz, A.; Lamnawar, K. Unveiling the Effects of In Situ Layer–Layer Interfacial Reaction in Multilayer Polymer Films via Multilayered Assembly: From Microlayers to Nanolayers. *Macromol. Mater. Eng.* **2020**, *305*, 2000076. [CrossRef]
14. Mueller, C.D.; Nazarenko, S.; Ebeling, T.; Schuman, T.L.; Hiltner, A.; Baer, E. Novel structures by microlayer coextrusion—talc-filled PP, PC/SAN, and HDPE/LLDPE. *Polym. Eng. Sci.* **1997**, *37*, 355–362. [CrossRef]
15. Mitsoulis, E.; Wagner, R.; Heng, F.L. Numerical simulation of wire-coating low-density polyethylene: Theory and experiments. *Polym. Eng. Sci.* **1988**, *28*, 291–310. [CrossRef]
16. Mitsoulis, E. Multilayer Film Coextrusion of Polymer Melts: Analysis of Industrial Lines with the Finite Element Method. *J. Polym. Eng.* **2005**, *25*, 393–410. [CrossRef]
17. Valette, R.; Laure, P.; Demay, Y.; Agassant, J.F. Convective linear stability analysis of two-layer coextrusion flow for molten polymers. *J. Non-Newton. Fluid Mech.* **2004**, *121*, 41–53. [CrossRef]
18. Lamnawar, K.; Maazouz, A. Role of the interphase in the flow stability of reactive coextruded multilayer polymers. *Polym. Eng. Sci.* **2009**, *49*, 727–739. [CrossRef]
19. Wilson, G.M.; Khomami, B. An experimental investigation of interfacial instabilities in multilayer flow of viscoelastic fluids. III. Compatible polymer systems. *J. Rheol.* **1993**, *37*, 341–354. [CrossRef]
20. Ponting, M.; Hiltner, A.; Baer, E. Polymer Nanostructures by Forced Assembly: Process, Structure, and Properties. *Macromol. Symp.* **2010**, *294*, 19–32. [CrossRef]
21. Burt, T.M.; Jordan, A.M.; Korley, L.T.J. Toward Anisotropic Materials via Forced Assembly Coextrusion. *ACS Appl. Mater. Interfaces* **2012**, *4*, 5155–5161. [CrossRef]
22. Burt, T.M.; Keum, J.; Hiltner, A.; Baer, E.; Korley, L.T.J. Confinement of Elastomeric Block Copolymers via Forced Assembly Coextrusion. *ACS Appl. Mater. Interfaces* **2011**, *3*, 4804–4811. [CrossRef]
23. Lai, C.-Y.; Hiltner, A.; Baer, E.; Korley, L.T.J. Deformation of Confined Poly(ethylene oxide) in Multilayer Films. *ACS Appl. Mater. Interfaces* **2012**, *4*, 2218–2227. [CrossRef]
24. Baer, E.; Zhu, L. 50th Anniversary Perspective: Dielectric Phenomena in Polymers and Multilayered Dielectric Films. *Macromolecules* **2017**, *50*, 2239–2256. [CrossRef]
25. Carr, J.M.; Langhe, D.S.; Ponting, M.T.; Hiltner, A.; Baer, E. Confined Crystallization in Polymer Nanolayered Films: A review. *J. Mater. Res.* **2012**, *27*, 1326–1350. [CrossRef]
26. Aradian, A.; Raphaël, E.; de Gennes, P.G. Strengthening of a Polymer Interface: Interdiffusion and Cross-Linking. *Macromolecules* **2000**, *33*, 9444–9451. [CrossRef]
27. Du, W.; Yuan, Y.; Wang, M.; Han, C.C.; Satija, S.K.; Akgun, B. Initial Stages of Interdiffusion between Asymmetrical Polymeric Layers: Glassy Polycarbonate and Melt Poly(methyl methacrylate) Interface Studied by Neutron Reflectometry. *Macromolecules* **2014**, *47*, 713–720. [CrossRef]
28. Kim, J.K.; Han, C.D. Polymer-polymer interdiffusion during coextrusion. *Polym. Eng. Sci.* **1991**, *31*, 258–269. [CrossRef]
29. Yang, L.; Suo, T.; Niu, Y.; Wang, Z.; Yan, D.; Wang, H. Effects of phase behavior on mutual diffusion at polymer layers interface. *Polymer* **2010**, *51*, 5276–5281. [CrossRef]
30. Shi, M.; Zhang, Y.; Cheng, L.; Jiao, Z.; Yang, W.; Tan, J.; Ding, Y. Interfacial Diffusion and Bonding in Multilayer Polymer Films: A Molecular Dynamics Simulation. *J. Phys. Chem. B* **2016**, *120*, 10018–10029. [CrossRef]
31. Zhao, R.; Macosko, C.W. Polymer–polymer mutual diffusion via rheology of coextruded multilayers. *AIChE J.* **2007**, *53*, 978–985. [CrossRef]
32. Liu, R.Y.F.; Jin, Y.; Hiltner, A.; Baer, E. Probing Nanoscale Polymer Interactions by Forced-Assembly. *Macromol. Rapid Commun.* **2003**, *24*, 943–948. [CrossRef]
33. Liu, R.Y.F.; Ranade, A.P.; Wang, H.P.; Bernal-Lara, T.E.; Hiltner, A.; Baer, E. Forced Assembly of Polymer Nanolayers Thinner Than the Interphase. *Macromolecules* **2005**, *38*, 10721–10727. [CrossRef]

34. Ji, X.; Chen, D.; Zheng, Y.; Shen, J.; Guo, S.; Harkin-Jones, E. Multilayered assembly of poly(vinylidene fluoride) and poly(methyl methacrylate) for achieving multi-shape memory effects. *Chem. Eng. J.* **2019**, *362*, 190–198. [CrossRef]
35. Decker, J.J.; Meyers, K.P.; Paul, D.R.; Schiraldi, D.A.; Hiltner, A.; Nazarenko, S. Polyethylene-based nanocomposites containing organoclay: A new approach to enhance gas barrier via multilayer coextrusion and interdiffusion. *Polymer* **2015**, *61*, 42–54. [CrossRef]
36. Yin, K.; Zhou, Z.; Schuele, D.E.; Wolak, M.; Zhu, L.; Baer, E. Effects of Interphase Modification and Biaxial Orientation on Dielectric Properties of Poly(ethylene terephthalate)/Poly(vinylidene fluoride-co-hexafluoropropylene) Multilayer Films. *ACS Appl. Mater. Interfaces* **2016**, *8*, 13555–13566. [CrossRef] [PubMed]
37. Composto, R.J.; Kramer, E.J.; White, D.M. Mutual diffusion in the miscible polymer blend polystyrene/poly(xylenyl ether). *Macromolecules* **1988**, *21*, 2580–2588. [CrossRef]
38. Kramer, E.J.; Green, P.; Palmstrøm, C.J. Interdiffusion and Marker Movements in Concentrated Polymer-Polymer Diffusion Couples. *Polymer* **1984**, *25*, 473–480. [CrossRef]
39. Brochard, F.; Jouffroy, J.; Levinson, P. Polymer-polymer diffusion in melts. *Macromolecules* **1983**, *16*, 1638–1641. [CrossRef]
40. Langhe, D.; Ponting, M. Gas Transport, Mechanical, Interphase, and Interdiffusion Properties in Coextruded-Multilayered Films. In *Manufacturing and Novel Applications of Multilayer Polymer Films*; Langhe, D., Ponting, M., Eds.; William Andrew Publishing: Boston, MA, USA, 2016; pp. 46–116.
41. Pollock, G.; Nazarenko, S.; Hiltner, A.; Baer, E. Interdiffusion in Microlayered Polymer Composites of Polycarbonate and a Copolyester. *J. Appl. Polym. Sci.* **1994**, *52*, 163–176. [CrossRef]
42. Qiu, H.; Bousmina, M. Determination of Mutual Diffusion Coefficients at Nonsymmetric Polymer/Polymer Interfaces from Rheometry. *Macromolecules* **2000**, *33*, 6588–6594. [CrossRef]
43. Zhang, H.; Lamnawar, K.; Maazouz, A. Rheological Modeling of the Mutual Diffusion and the Interphase Development for an Asymmetrical Bilayer Based on PMMA and PVDF Model Compatible Polymers. *Macromolecules* **2013**, *46*, 276–299. [CrossRef]
44. Zhang, H.; Lamnawar, K.; Maazouz, A. Fundamental Understanding and Modeling of Diffuse Interphase Properties and Its Role in Interfacial Flow Stability of Multilayer Polymers. *Polym. Eng. Sci.* **2015**, *55*, 771–791. [CrossRef]
45. Zhang, H.; Lamnawar, K.; Maazouz, A.; Maia, J.M. A Nonlinear Shear and Elongation Rheological Study of Interfacial Failure in Compatible Bilayer Systems. *J. Rheol.* **2016**, *60*, 1–23. [CrossRef]
46. Zhang, H.; Lamnawar, K.; Maazouz, A. Understanding of Transient Rheology in Step Shear and Its Implication to Explore Nonlinear Relaxation Dynamics of Interphase in Compatible Polymer Multi-microlayered Systems. *Ind. Eng. Chem. Res.* **2018**, *57*, 8093–8104. [CrossRef]
47. Arendt, B.H.; Krishnamoorti, R.; Kornfield, J.A.; Smith, S.D. Component Dynamics in Miscible Blends: Equally and Unequally Entangled Polyisoprene/Polyvinylethylene. *Macromolecules* **1997**, *30*, 1127–1137. [CrossRef]
48. Schrenk, W.J.; Bradley, N.L.; Alfrey, T.; Maack, H. Interfacial flow instability in multilayer coextrusion. *Polym. Eng. Sci.* **1978**, *18*, 620–623. [CrossRef]
49. Han, C.D.; Shetty, R. Studies on multilayer film coextrusion II. Interfacial instability in flat film coextrusion. *Polym. Eng. Sci.* **1978**, *18*, 180–186. [CrossRef]
50. Ramanathan, R.; Shanker, R.; Rehg, T.; Jons, S.; Headley, D.; Schrenk, W. *"Wave" Pattern Instability in Multilayer Coextrusion-An Experimental Investigation*; SPE ANTEC Proc; Society of Plastics Engineers Inc.: Indianapolis, IN, USA, 1996; pp. 224–228.
51. Bironeau, A.; Salez, T.; Miquelard-Garnier, G.; Sollogoub, C. Existence of a Critical Layer Thickness in PS/PMMA Nanolayered Films. *Macromolecules* **2017**, *50*, 4064–4073. [CrossRef]
52. Jordan, A.M.; Lee, B.; Kim, K.; Ludtke, E.; Lhost, O.; Jaffer, S.A.; Bates, F.S.; Macosko, C.W. Rheology of polymer multilayers: Slip in shear, hardening in extension. *J. Rheol.* **2019**, *63*, 751–761. [CrossRef]
53. Lee, P.C.; Park, H.E.; Morse, D.C.; Macosko, C.W. Polymer-polymer interfacial slip in multilayered films. *J. Rheol.* **2009**, *53*, 893–915. [CrossRef]
54. Lee, P.C.; Macosko, C.W. Polymer-polymer interfacial slip by direct visualization and by stress reduction. *J. Rheol.* **2010**, *54*, 1207–1218. [CrossRef]
55. Bernal-Lara, T.E.; Liu, R.Y.F.; Hiltner, A.; Baer, E. Structure and thermal stability of polyethylene nanolayers. *Polymer* **2005**, *46*, 3043–3055. [CrossRef]
56. Scholtyssek, S.; Adhikari, R.; Seydewitz, V.; Michler, G.H.; Baer, E.; Hiltner, A. Evaluation of Morphology and Deformation Micromechanisms in Multilayered PP/PS Films: An Electron Microscopy Study. *Macromol. Symp.* **2010**, *294*, 33–44. [CrossRef]
57. Zhu, Y.; Bironeau, A.; Restagno, F.; Sollogoub, C.; Miquelard-Garnier, G. Kinetics of thin polymer film rupture: Model experiments for a better understanding of layer breakups in the multilayer coextrusion process. *Polymer* **2016**, *90* (Suppl. C), 156–164. [CrossRef]
58. Lin, Y.; Hiltner, A.; Baer, E. A new method for achieving nanoscale reinforcement of biaxially oriented polypropylene film. *Polymer* **2010**, *51*, 4218–4224. [CrossRef]
59. Feng, J.; Zhang, Z.; Bironeau, A.; Guinault, A.; Miquelard-Garnier, G.; Sollogoub, C.; Olah, A.; Baer, E. Breakup behavior of nanolayers in polymeric multilayer systems—Creation of nanosheets and nanodroplets. *Polymer* **2018**, *143*, 19–27. [CrossRef]
60. Wang, J.; Adami, D.; Lu, B.; Liu, C.; Maazouz, A. Multiscale Structural Evolution and Its Relationship to Dielectric Properties of Micro-/Nano-Layer Coextruded PVDF-HFP/PC Films. *Polymers* **2020**, *12*, 2596. [CrossRef] [PubMed]
61. Fredrickson, G.H.; Milner, S.T. Time-Dependent Reactive Coupling at Polymer–Polymer Interfaces. *Macromolecules* **1996**, *29*, 7386–7390. [CrossRef]

62. Fredrickson, G.H. Diffusion-controlled reactions at polymer-polymer interfaces. *Phys. Rev. Lett.* **1996**, *76*, 3440. [CrossRef]
63. O'shaughnessy, B.; Sawhney, U. Polymer reaction kinetics at interfaces. *Phys. Rev. Lett.* **1996**, *76*, 3444. [CrossRef]
64. O'Shaughnessy, B.; Sawhney, U. Reaction Kinetics at Polymer−Polymer Interfaces. *Macromolecules* **1996**, *29*, 7230–7239. [CrossRef]
65. Wang, M.; Yuan, G.; Han, C.C. Reaction process in polycarbonate/polyamide bilayer film and blend. *Polymer* **2013**, *54*, 3612–3619. [CrossRef]
66. Lyu, S.-P.; Cernohous, J.J.; Bates, F.S.; Macosko, C.W. Interfacial Reaction Induced Roughening in Polymer Blends. *Macromolecules* **1999**, *32*, 106–110. [CrossRef]
67. Kim, H.Y.; Jeong, U.; Kim, J.K. Reaction Kinetics and Morphological Changes of Reactive Polymer−Polymer Interface. *Macromolecules* **2003**, *36*, 1594–1602. [CrossRef]
68. Kim, H.Y.; Kim, H.J.; Kim, J.K. Effect of interfacial reaction and morphology on rheological properties of reactive bilayer. *Polym. J.* **2006**, *38*, 1165–1172. [CrossRef]
69. Lamnawar, K.; Baudouin, A.; Maazouz, A. Interdiffusion/reaction at the polymer/polymer interface in multilayer systems probed by linear viscoelasticity coupled to FTIR and NMR measurements. *Eur. Polym. J.* **2010**, *46*, 1604–1622. [CrossRef]
70. Lu, B.; Lamnawar, K.; Maazouz, A. Rheological and dynamic insights into an in situ reactive interphase with graft copolymers in multilayered polymer systems. *Soft Matter* **2017**, *13*, 2523–2535. [CrossRef]
71. Kim, H.Y.; Lee, D.H.; Kim, J.K. Effect of oscillatory shear on the interfacial morphology of a reactive bilayer polymer system. *Polymer* **2006**, *47*, 5108–5116. [CrossRef]
72. Kim, H.Y.; Joo, W.; Kim, J.K. Effect of Perpendicular Shear Force on the Interfacial Morphology of Reactive Polymer Bilayer. *Macromol. Chem. Phys.* **2008**, *209*, 746–753. [CrossRef]
73. Zhang, J.; Ji, S.; Song, J.; Lodge, T.P.; Macosko, C.W. Flow Accelerates Interfacial Coupling Reactions. *Macromolecules* **2010**, *43*, 7617–7624. [CrossRef]
74. Kiparissoff-Bondil, H.; Devisme, S.; Rauline, D.; Chopinez, F.; Restagno, F.; Léger, L. Evidences for flow-assisted interfacial reaction in coextruded PA6/PP/PA6 films. *Polym. Eng. Sci.* **2019**, *59*, E44–E50. [CrossRef]
75. Barraud, T.; Devisme, S.; Hervet, H.; Brunello, D.; Klein, V.; Poulard, C.; Restagno, F.; Léger, L. Convection and diffusion assisted reactive coupling at incompatible semi-crystalline polymer interfaces. *J. Phys. Mater.* **2020**, *3*, 035001. [CrossRef]
76. Song, J.; Baker, A.M.; Macosko, C.W.; Ewoldt, R.H. Reactive Coupling between Immiscible Polymer Chains: Acceleration by Compressive Flow. *AIChE J.* **2013**, *59*, 3391–3402. [CrossRef]
77. Du, Q.; Jiang, Z.; Li, J.; Guo, S. Adhesion and delamination failure mechanisms in alternating layered polyamide and polyethylene with compatibilizer. *Polym. Eng. Sci.* **2010**, *50*, 1111–1121. [CrossRef]
78. Lamnawar, K.; Maazouz, A. Rheological study of multilayer functionalized polymers: Characterization of interdiffusion and reaction at polymer/polymer interface. *Rheol. Acta* **2006**, *45*, 411–424. [CrossRef]
79. Song, J.; Ewoldt, R.H.; Hu, W.; Craig Silvis, H.; Macosko, C.W. Flow accelerates adhesion between functional polyethylene and polyurethane. *AIChE J.* **2011**, *57*, 3496–3506. [CrossRef]
80. Lu, B.; Lamnawar, K.; Maazouz, A. Influence of in situ reactive interphase with graft copolymer on shear and extensional rheology in a model bilayered polymer system. *Polym. Test.* **2017**, *61*, 289–299. [CrossRef]
81. Bondon, A.; Lamnawar, K.; Maazouz, A. Influence of copolymer architecture on generation of defects in reactive multilayer coextrusion. *Key Eng. Mater.* **2015**, *651*, 836–841. [CrossRef]
82. Bondon, A.; Lamnawar, K.; Maazouz, A. Experimental investigation of a new type of interfacial instability in a reactive coextrusion process. *Polym. Eng. Sci.* **2015**, *55*, 2542–2552. [CrossRef]
83. Lu, B.; Bondon, A.; Touil, I.; Zhang, H.; Alcouffe, P.; Pruvost, S.; Liu, C.; Maazouz, A.; Lamnawar, K. Role of the Macromolecular Architecture of Copolymers at Layer–Layer Interfaces of Multilayered Polymer Films: A Combined Morphological and Rheological Investigation. *Ind. Eng. Chem. Res.* **2020**, *59*, 22144–22154. [CrossRef]
84. Vuong, S.; Léger, L.; Restagno, F. Controlling interfacial instabilities in PP/EVOH coextruded multilayer films through the surface density of interfacial copolymers. *Polym. Eng. Sci.* **2020**, *60*, 1420–1429. [CrossRef]
85. Michell, R.M.; Müller, A.J. Confined crystallization of polymeric materials. *Prog. Polym. Sci.* **2016**, *54* (Suppl. C), 183–213. [CrossRef]
86. Mijangos, C.; Hernández, R.; Martín, J. A review on the progress of polymer nanostructures with modulated morphologies and properties, using nanoporous AAO templates. *Prog. Polym. Sci.* **2016**, *54* (Suppl. C), 148–182. [CrossRef]
87. Cangialosi, D.; Alegría, A.; Colmenero, J. Effect of nanostructure on the thermal glass transition and physical aging in polymer materials. *Prog. Polym. Sci.* **2016**, *54* (Suppl. C), 128–147. [CrossRef]
88. Wang, H.; Keum, J.K.; Hiltner, A.; Baer, E.; Freeman, B.; Rozanski, A.; Galeski, A. Confined Crystallization of Polyethylene Oxide in Nanolayer Assemblies. *Science* **2009**, *323*, 757–760. [CrossRef] [PubMed]
89. Beadie, G.; Shirk, J.S.; Rosenberg, A.; Lane, P.A.; Fleet, E.; Kamdar, A.R.; Jin, Y.; Ponting, M.; Kazmierczak, T.; Yang, Y.; et al. Optical properties of a bio-inspired gradient refractive index polymer lens. *Opt. Express* **2008**, *16*, 11540–11547. [CrossRef]
90. Jin, Y.; Rogunova, M.; Hiltner, A.; Baer, E.; Nowacki, R.; Galeski, A.; Piorkowska, E. Structure of polypropylene crystallized in confined nanolayers. *J. Polym. Sci. Part. B Polym. Phys.* **2004**, *42*, 3380–3396. [CrossRef]
91. Luo, S.; Yi, L.; Zheng, Y.; Shen, J.; Guo, S. Crystallization of polypropylene in multilayered spaces: Controllable morphologies and properties. *Eur. Polym. J.* **2017**, *89*, 138–149. [CrossRef]

92. Bernal-Lara, T.E.; Masirek, R.; Hiltner, A.; Baer, E.; Piorkowska, E.; Galeski, A. Morphology studies of multilayered HDPE/PS systems. *J. Appl. Polym. Sci.* **2006**, *99*, 597–612. [CrossRef]
93. Cheng, J.; Pu, H. Orientation of LDPE crystals from microscale to nanoscale via microlayer or nanolayer coextrusion. *Chin. J. Polym. Sci.* **2016**, *34*, 1411–1422. [CrossRef]
94. Ponting, M.; Lin, Y.; Keum, J.K.; Hiltner, A.; Baer, E. Effect of Substrate on the Isothermal Crystallization Kinetics of Confined Poly(ε-caprolactone) Nanolayers. *Macromolecules* **2010**, *43*, 8619–8627. [CrossRef]
95. Wang, H.; Keum, J.K.; Hiltner, A.; Baer, E. Crystallization Kinetics of Poly(ethylene oxide) in Confined Nanolayers. *Macromolecules* **2010**, *43*, 3359–3364. [CrossRef]
96. Wang, H.; Keum, J.K.; Hiltner, A.; Baer, E. Confined Crystallization of PEO in Nanolayered Films Impacting Structure and Oxygen Permeability. *Macromolecules* **2009**, *42*, 7055–7066. [CrossRef]
97. Chen, X.; Tseng, J.-K.; Treufeld, I.; Mackey, M.; Schuele, D.E.; Li, R.; Fukuto, M.; Baer, E.; Zhu, L. Enhanced Dielectric Properties due to Space Charge-induced Interfacial Polarization in Multilayer Polymer Films. *J. Mater. Chem. C* **2017**, *5*, 10417–10426. [CrossRef]
98. Huang, H.; Chen, X.; Li, R.; Fukuto, M.; Schuele, D.E.; Ponting, M.; Langhe, D.; Baer, E.; Zhu, L. Flat-On Secondary Crystals as Effective Blocks To Reduce Ionic Conduction Loss in Polysulfone/Poly(vinylidene fluoride) Multilayer Dielectric Films. *Macromolecules* **2018**, *51*, 5019–5026. [CrossRef]
99. Mackey, M.; Schuele, D.E.; Zhu, L.; Baer, E. Layer confinement effect on charge migration in polycarbonate/poly(vinylidene fluorid-co-hexafluoropropylene) multilayered films. *J. Appl. Phys.* **2012**, *111*, 113702. [CrossRef]
100. Mackey, M.; Schuele, D.E.; Zhu, L.; Flandin, L.; Wolak, M.A.; Shirk, J.S.; Hiltner, A.; Baer, E. Reduction of Dielectric Hysteresis in Multilayered Films via Nanoconfinement. *Macromolecules* **2012**, *45*, 1954–1962. [CrossRef]
101. Chen, X.; Li, Q.; Langhe, D.; Ponting, M.; Li, R.; Fukuto, M.; Baer, E.; Zhu, L. Achieving Flat-on Primary Crystals by Nanoconfined Crystallization in High-Temperature Polycarbonate/Poly(vinylidene fluoride) Multilayer Films and Its Effect on Dielectric Insulation. *ACS Appl. Mater. Interfaces* **2020**, *12*, 44892–44901. [CrossRef] [PubMed]
102. Langhe, D.S.; Murphy, T.M.; Shaver, A.; LaPorte, C.; Freeman, B.D.; Paul, D.R.; Baer, E. Structural relaxation of polystyrene in nanolayer confinement. *Polymer* **2012**, *53*, 1925–1931. [CrossRef]
103. Arabeche, K.; Delbreilh, L.; Adhikari, R.; Michler, G.H.; Hiltner, A.; Baer, E.; Saiter, J.-M. Study of the Cooperativity at the Glass Transition Temperature in PC/PMMA Multilayered Films: Influence of Thickness Reduction from Macro- to Nanoscale. *Polymer* **2012**, *53*, 1355–1361. [CrossRef]
104. Casalini, R.; Zhu, L.; Baer, E.; Roland, C.M. Segmental Dynamics and the Correlation Length in Nanoconfined PMMA. *Polymer* **2016**, *88*, 133–136. [CrossRef]

Article

Comparison of the Foamability of Linear and Long-Chain Branched Polypropylene—The Legend of Strain-Hardening as a Requirement for Good Foamability

Nick Weingart [1,†], Daniel Raps [1,†], Mingfu Lu [2], Lukas Endner [1] and Volker Altstädt [1,*]

1. Department of Polymer Engineering, University of Bayreuth, 95447 Bayreuth, Germany;
nick.weingart@uni-bayreuth.de (N.W.); daniel.raps@gmx.net (D.R.); endner.lukas@gmx.de (L.E.)
2. SINOPEC Beijing Research Institute of Chemical Industry, Beijing, 100013, China; lumf.bjhy@sinopec.com
* Correspondence: altstaedt@uni-bayreuth.de; Tel.: +49-921-557471
† Equally contributed.

Received: 9 March 2020; Accepted: 21 March 2020; Published: 24 March 2020

Abstract: Polypropylene (PP) is an outstanding material for polymeric foams due to its favorable mechanical and chemical properties. However, its low melt strength and fast crystallization result in unfavorable foaming properties. Long-chain branching of PP is regarded as a game changer in foaming due to the introduction of strain hardening, which stabilizes the foam morphology. In this work, a thorough characterization with respect to rheology and crystallization characteristics of a linear PP, a PP/PE-block co-polymer, and a long-chain branched PP are conducted. Using these results, the processing window in foam-extrusion trials with CO_2 and finally the foam properties are explained. Although only LCB-PP exhibits strain hardening, it neither provide the broadest foaming window nor the best foam quality. Therefore, multiwave experiments were conducted to study the gelation due to crystallization and its influence on foaming. Here, linear PP exhibited a gel-like behavior over a broad time frame, whereas the other two froze quickly. Thus, apart from strain hardening, the crystallization behavior/crystallization kinetics is of utmost importance for foaming in terms of a broad processing window, low-density, and good morphology. Therefore, the question arises, whether strain hardening is really essential for low density foams with a good cellular morphology.

Keywords: polypropylene; foam-extrusion; morphology; foaming; crystallization kinetics

1. Introduction

Polypropylene (PP) is an outstanding choice as matrix material for polymeric foams due to its favorable mechanical properties and chemical resistance. However, its low melt strength and fast crystallization result in unfavorable foaming properties. It is difficult to obtain homogeneous lightweight foams due to their low melt elasticity, viscosity, and low temperature dependence of melt viscosity in fully molten state, which makes the control of viscosity by temperature variation highly challenging. Furthermore, crystallization can occur during foaming, which can lead to either shutdowns of the process (freezing of the die) or unfavorable foam structures due to cell rupture. It is concluded, that a good knowledge on the rheology and crystallization properties is essential to adjust the morphology and thereby the elastic modulus, strength [1], impact behavior [2], and thermal conductivity [3] of foamed products.

During foam extrusion, temperature, pressure, and blowing agent affect the flow behavior of the melt and, ultimately, the foam properties. In addition, the melt behaves differently under shear and elongational flow. The flow patterns in the extruder are dominated by shear deformation. However, the elongational properties become relevant after the melt leaves the die, as in the foaming stage, the melt is subjected to elongational deformation during bubble growth, namely, equi-biaxial extension. The rheology of polypropylene is a major challenge in achieving low-density foams. Generally, PP has a comparatively low viscosity in fully molten state compared to amorphous polymers such as polystyrene or polycarbonate.

For foaming semicrystalline polymers like PP, the crystallization behavior is one of the key factors for good and controllable foamability. In the state-of-the-art for continuous foaming, the polymer is first molten, mixed with the blowing agent, followed by subsequent cooling of the melt/gas solution along with increasing pressure towards the die. Thus, the temperature of the melt at the die (relative to the onset of crystallization) determines how long the polymer remains in the molten state until it solidifies. For semicrystalline polymers, this threshold value is the crystallization temperature, which can be shifted towards lower values due to the plasticizing effect of the blowing agent (concentration-dependent) [4]. As soon as the expanded polymer melt reaches this temperature, the foam starts solidifying. The outer shell solidifies first as it is exposed to the cold surrounding. In the foam core, the higher temperature is retained longer, which can cause cell/foam coalescence if the favorable foam structure is not stabilized (frozen) in time. Therefore, it is a necessity to cool down the melt to an optimal temperature window (processing window). A temperature that is too low causes a very high viscosity (freezing of the die for semicrystalline polymers) and inhibits the growth of the foam cells, whereas a high temperature makes it difficult to stabilize the foam (low viscosity and high solidification time) to keep the blowing agent dissolved, and finally leads to foam collapse. Conclusively, the processing window represents a compromise between remaining temperature (required for elasticity, expansion, and stabilization) and foaming potential/time (crystallization speed and blowing agent concentration) [5–13]. Therefore, for standard grades of polypropylene, the temperature window for extrusion foaming is generally very narrow.

Furthermore, the elongational properties of the melt are of utmost importance, as they govern the cell growth and influence the rupture of the forming cell walls. The viscoelastic behavior of the melt in elongational deformation is crucial for the formation of the foam morphology [14], as mostly low average cell size, a narrow cell size distribution, and low density are required. Polypropylene is well known to possess very disadvantageous elongational properties; both melt strength and drawability are very low for standard grades.

Fortunately, chain topology can be modified in order to improve its foamability. In particular, the effect of strain hardening at large elongational strains is desired to support cell stabilization. In this frame, strain hardening is highly beneficial for foaming due to the effect of self-healing of the cell walls during the expansion of foam cells. Strain hardening means the rise of the elongational viscosity above the zero-rate elongational viscosity. The strain hardening helps to prevent cell coalescence and to widen the processing window [13], i.e., for the foaming of PP [15]. In terms of chain topology, strain hardening is caused by long-chain branching (LCB) [7,8]. Furthermore, LCB often causes thermo-rheological complexity, as shown by several authors [16–18]. Although the determination of thermo-rheological complexity is not straightforward, it can help to widen the temperature processing window. For example, Raps at al. [19] found that long-chain branching leads to an increased temperature sensitivity of the melt viscosity at strain rates relevant for foaming, which should allow a better process control during foaming. However, LCB-PP has the disadvantage of a lower solubility of CO_2 compared to linear PP [20]. This arises due to the fact that LCB-PP has a lower specific volume compared to linear PP as well as a more pronounced resistance against swelling, which is caused by dissolution of CO_2 [21].

The crystallization kinetics and crystalline morphology of ready available semicrystalline polymers are well understood because of their high importance for the manufacturing of polymer products by techniques like fiber-spinning, blow molding, and injection molding [22]. In the case of foam and cellular polymeric materials, although to a lesser extent, it is well known that crystallization phenomena play a major role in foaming [23]. Crystallization phenomena at the relevant conditions for gas loaded polymers intended for foaming must be understood to obtain good foam morphologies, to reduce development time and set the lower limit of the processing window of temperature. Many factors change the crystallization behavior of polymer melts. Those factors will be discussed subsequently.

The effect of pressure on the crystallization of various polymers has been discussed in many publications in the past. Elevated pressure leads to higher crystallization temperatures as the driving force for chain alignment is increased. Besides pressure and cooling rate, also the molecular structure determines the crystallization behavior. Especially important is long-chain branching (LCB), as discussed before. One effect of LCB on crystallization is an increase in crystallization temperature [24] and the amount of γ-phase [11].

As gas is dissolved into the polymer for the purpose of foaming, its effect on crystallization must be considered as well. Takada et al. studied the effect of CO_2 on crystallization of iPP [25]. Dissolved gas increases the molecular mobility and the free volume, thus leading a decrease of chain–chain interactions. Because of this the motion of chains into the crystal-amorphous boundary is improved, leading to a higher crystallization rate. It can be summarized that, if crystallization is nucleation-controlled, the overall rate is decreased by the incorporation of CO_2; otherwise, it is increased. Takada et al. also investigated the effect of dissolved CO_2 on the isothermal crystallization of PET in yet another work [26]. They found that if the reduction of the glass transition temperature T_g is higher than the melting temperature T_m, the crystallization rate is increased. If their reduction is fairly similar, the crystallization rate is decreased (nucleation-controlled region). They concluded that this rule may be applicable to all semicrystalline polymers. Moreover, Oda and Saito [27] found that spherulite growth rate reached an optimum at moderate CO_2 pressure, thus emphasizing the competition between plasticization effects, which dominate at low concentrations, and exclusion effects pushing CO_2 away from the crystallization front at high CO_2 concentrations. However, there is also a back side in crystallization: the occurrence of crystals leads to a significant reduction in gas solubility as well as diffusivity [23].

Furthermore, the timing and kinetics of crystallization are crucial for a well-defined and fine cellular morphology. If crystallization proceeds too quickly and in a very narrow window, the danger of the die freezing in foam extrusion is more pronounced. Also, if crystallization takes place during cell growth, crystals contribute as a nucleus for bubble growth, but it might also result in a collapse of the cellular structure, become very inhomogeneous or partially open-cellular. Moreover, effects like nucleation due to shear and elongational flow must be taken into account [28].

It can be seen that both rheology and crystallization play an important role in foaming. Especially, long-chain branching is regarded as a major contributor for good foam morphology, i.e., low density and small cells with a narrow distribution by causing strain hardening. Currently, strain hardening is generally regarded as a prerequisite for good foamability of semicrystalline thermoplastics. However, is strain hardening really essential for low density foams with a good cellular morphology? This paper aims to shed some light on this question and whether other factors might also come into play. Therefore, we study the rheology and crystallization properties of three PPs (linear PP with a broad molecular weight distribution, a PP-PE block-co-polymer, and a LCB-PP as benchmark material).

2. Materials and Methods

In the scope of this work, three polypropylene grades were studied: two PP grades of HMS20Z (linear, homopolymer) and E02ES (linear, PP-PE copolymer) were provided by Sinopec, 100013 Beijing, China, (trade names) as high melt strength PP. They will be referred to as Sinopec HMS20Z and Sinopec E02ES CoPo, respectively. Sinopec HMS20Z has a M_W of 474,000 g/mol (PDI of 11.05) and a MVR of 2.1 g/10 min (230 °C with 2.16 kg). The higher melt strength of the materials is achieved through a broad molecular weight distribution. Sinopec E02ES CoPo has a M_W of 456,000 g/mol (PD of 9.16) and a MVR of 1.5 g/10 min (230 °C with 2.16 kg). The third HMS-PP grade, Borealis Daploy WB140HMS with a M_W of 350,000 g/mol (PDI of 4.6) and a MVR of 2.1 g/10 min (230 °C with 2.16 kg), was used as a reference in this work (referred to as Borealis WB140 HMS). The higher melt strength of this grade is generated through long-chain branching. The Sinopec materials are chemically pure. Borealis WB140 HMS is a commercial grade, thus additives cannot be excluded.

The determination of the crystallization kinetics was performed with a Mettler Toledo DSC 1, 43085 Columbus, OH, USA (cooling realized by a compressor) in the temperature range of 0 to 220 °C with a heating rate of 10 K/min and cooling rates of 2, 4, 8, 10, and 16 K/min under nitrogen atmosphere. The evaluation was carried out with the STARe-software with a ΔH_m^0 of 207.1 J/g [29] according to Khanna et al. [30] with a straight base line.

The analysis of the non-isothermal crystallization kinetics for the DSC results was performed according to Jeziorny modified Avrami theory [31]. The DSC thermograms are often inconclusive where the crystallization process begins and where it ends. Avrami plots strongly depend on the shape and position of the borders of the crystallization process. As no clear reproducible proceeding was reported in literature, and to guarantee reproducibility, the heat flow was differentiated (1st order) and plotted over temperature to identify the start and the end of the crystallization window for the Avrami evaluation.

The rheological investigation in shear deformation was performed with an Anton Paar MCR 702 TwinDrive, 804x Graz, Austria rotational rheometer under nitrogen atmosphere (50 mL/min) with specimen geometry of 25 mm diameter and 2 mm thickness. Strain sweeps were carried out in the deformation range of 0 to 100% with an angular frequency of 1 rad/s at 180, 190, and 200 °C. Frequency sweeps were performed at constant temperature of 180, 190, and 200 °C with a decreasing angular frequency of 200 to 0.1 rad/s and amplitude of 5 %. Non-isothermal multiwave (NiMW) measurements were performed with an Anton Paar MCR 702 rotational rheometer under nitrogen atmosphere and cooling rates of 0.5, 1, 2, and 4 K/min from 200 to 100 °C to determine the gel point due to crystallization and the starting point of the crystallization. The harmonics were chosen as factor 5, 25, and 125 of the fundamental sinus wave (1 rad/s).

Isothermal multiwave analysis was analogously performed with an Anton Paar MCR 702 TwinDrive device under nitrogen atmosphere. The measurement procedure for this work was specifically designed to prevent the specimen from premature crystallization. The measurement procedure consisted of three steps: **(I)** Cooling the polymer melt to a temperature 1 °C above the investigation temperature (previously determined by DSC), while the shear stress is increased from 25 to 40 Pa. **(II)** Cooling down from the onset temperature to the investigation temperature with constant shear stress of 50 Pa. **(III)** Measurement at investigation temperature with 1 rad/s and a shear stress of 50 Pa. Sinopec E02ES CoPo was investigated at 130, 129, 128, and 127 °C; Sinopec HMS20Z at 136.5, 135.5, 134.5, and 133.5 °C; and Borealis WB140 HMS at 148.5, 147.5, 146.5, and 145.5 °C. The normal force was kept constant at 0 N during the measurements as the specimen volume decreases due to crystallization. In Table 1, the measurement procedures are exemplarily displayed for Sinopec HMS20Z.

Table 1. Detailed isothermal multiwave measuring procedure/parameters for Sinopec HMS20Z.

Procedure Measuring Time	Step (I) 13 min	Step (II) 2,4 min	Step (III) 180 min
τ	25–40 Pa	50 Pa	50 Pa
ω	1 rad/s	1 rad/s	1 rad/s
T	200 → $T_{measure}$ + 1	$T_{measure}$ + 1 → $T_{measure}$	$T_{measure}$
F_N	0 N	0 N	0 N

The investigation of the melt strength of the PP-grades was performed with a Rheotens 71.97 mounted on a Göttfert high-pressure capillary rheometer 6000, 74722 Buchen, Germany. Previous to testing, the material was dried overnight at 70 °C under vacuum. The measurement was carried out at a melt temperature of 220 °C and a shear rate of 30 s^{-1}. The temperature was carefully chosen to avoid strand expansion and sagging of the melt after leaving the die, hence preventing internal stress.

The nozzle diameter was 2 mm with a length of 30 mm and a distance of 95 mm between the nozzle and the upper wheel. The maximum wheel speed (measuring range) was set to 1000 mm/s, with a starting speed of 7.5 mm/s.

Furthermore, the elongational viscosity was studied using a universal extensional fixture (UXF). This is a filament stretching tool for rotational rheometers. The device was manufactured by Anton Paar, Austria and used with a MCR 702 rheometer. A strip-like sample of $10 \times 18 \times 0.6$ mm^3 is uniaxially stretched by means of two cylinders, which rotate around their own axis ($\omega 1 = \omega 2$). The tensile stress is obtained from the measured torque and the strain rate from the rotational speed. The maximum Hencky strain $\varepsilon = \ln\left(\frac{l_{max}}{l_0}\right)$ is 5 for one revolution of the drums. To reduce the sagging of the sample, it was subjected to a pre-stretch. Therefore, a constant torque of 2.5 µNm at 180 °C, which equals a stress of 81 Pa in the sample, and 1.25 µNm at 200 °C, which equals a stress of 47 Pa in the sample, was applied. As a drawback, this can cause measuring errors due to a certain degree of orientation of the polymer chains. However, the error due to this initial tension on the results of the elongational experiments is markedly smaller than that of a deformed sample due to gravity. The final stretching step was carried out at strain rates of 1, 3, and 10 s^{-1} up to a Hencky strain of 4.7. The higher elongation rates were deliberately chosen to simulate the rates occurring during foaming [13].

The foam extrusion was carried out on a tandem extrusion line from Dr. Collin GmbH. The line consists of a twin-screw kneader ZK 25 P x 42 L/D, maximum throughput of 15 kg/h and (A-extruder) with co-rotating screws for compounding and gas injection and a single-screw extruder E 45 M x 30 D (B extruder) for pressure build-up and cooling. The foam extrusion line was operated at a throughput of 5.5 kg/h, 2–6 wt.% CO_2 as blowing agent (supercritical dosing by a Maximator) and a nozzle temperature of 190 °C (156 °C for Sinopec). The nozzle diameter was 3 mm and the length 100 mm. The speeds of rotation for the screws were 135 rpm in the A-extruder and 12 rpm (9 rpm for WB140HMS) in the B-extruder. The single-screw in the B-Extruder has a constant flight depth over the whole screw. The sealing applied to prevent CO_2-loss works via a melt buffer upstream of the transfer pipe between A- and B-Extruder.

The aim was to determine a processing window for foam extrusion, defined through best foam properties in regard to density and morphology, for the investigated PP-grades. Foaming Borealis WB140 HMS was rather challenging, as high pressures and viscosities narrowed the operating window. As well, this grade was not able to contain more than 4 wt.% of CO_2 (more blowing agent resulted in strong fluctuations and gaseous leakage at the die). For general comparison, the materials with the best result were chosen. Detailed processing parameters and the found processing windows are summarized in Results. No additives were used for comparable neat material characterization.

The foam morphology was studied with a scanning electron microscope JEOL JSM-6510, 196-8558 Tokyo, Japan with an acceleration voltage of 10 kV and a secondary electrons (SE) detector. The samples were fractured in liquid nitrogen and sputtered with a 13 nm thick gold layer. Evaluation of cell diameters and cell size distribution was performed with the program ImageJ on an average of at least 80 cells.

3. Results and Discussion

In the following the characterization of the samples is laid out. Thereafter, the results of the foam extrusion trials (processing window and the final foam morphology) are presented and explained with the characteristic properties of the PPs.

3.1. Characterization of the Base-PPs

3.1.1. Characterization of the Crystallization Behavior

For the characterization of the thermal properties, heating and cooling rates of maximum 16 K/min were used. During the actual foaming process, the melt leaves the die at the temperature between T_{die} and T_{melt} and is cooled to ~50 °C usually in less than one minute. Those high cooling rates cannot be realized with standard DSC measurements and therefore have to be assumed, based on lower possible cooling/heating rates analog to literature. Nevertheless, it gives good insight into the crystallization behavior and how the crystallization behavior progresses with increasing cooling rates [31,32].

Non-Isothermal Crystallization Kinetics

First, thermal characterization of the different PP-grades is presented. The three samples differ in their total crystallinity (Figure 1; Table 2) as well as the melting/crystallization peak temperature. The Sinopec HMS20Z has the highest degree of crystallinity with 54%, followed by Borealis WB10 HMS with 44% and finally Sinopec E02ES CoPo with 40%, showing that a small amount of PE in the backbone reduces crystallinity more than long chain branches.

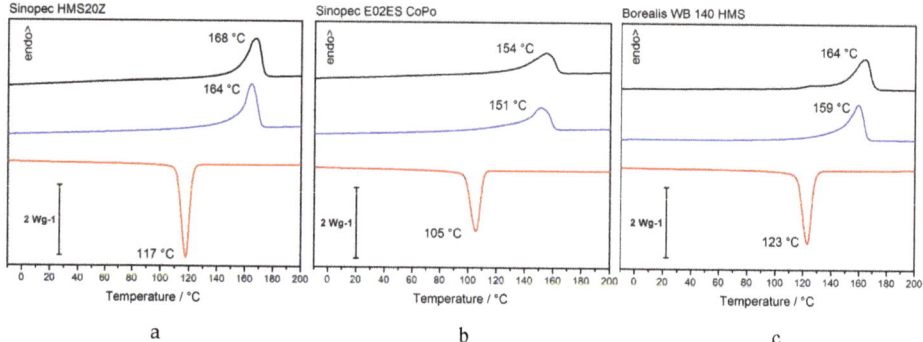

Figure 1. DSC-analysis of Sinopec HMS20Z (**a**), E02ES CoPo (**b**), and Borealis WB140 HMS (**c**) at 10 K/min.

Table 2. Crystallinity, half-time crystallization, and Avrami-temperature range for PP grades.

Sinopec HMS20Z	Crystallinity χ	$t_{1/2}$	$(t_{1/2})^{-1}$	Avrami Range
−2 K/min	55 %	241.8 s	4.14×10^{-3} s^{-1}	134–106 °C
−4 K/min	54 %	159 s	6.29×10^{-3} s^{-1}	133–98 °C
−8 K/min	54 %	103.8 s	9.63×10^{-3} s^{-1}	132–90 °C
−16 K/min	52 %	83.4 s	11.99×10^{-3} s^{-1}	136–74 °C
Sinopec E02ES CoPo				
−2 K/min	39 %	235.2 s	4.25×10^{-3} s^{-1}	124–99 °C
−4 K/min	40 %	148.8 s	6.72×10^{-3} s^{-1}	122–92 °C
−8 K/min	39 %	103.2 s	9.69×10^{-3} s^{-1}	122–86 °C
−16 K/min	41 %	83.4 s	11.99×10^{-3} s^{-1}	123–75 °C
Borealis WB140 HMS				
−2 K/min	44 %	354 s	2.82×10^{-3} s^{-1}	143–112 °C
−4 K/min	44 %	215.4 s	4.64×10^{-3} s^{-1}	142–99 °C
−8 K/min	44 %	137.4 s	7.28×10^{-3} s^{-1}	143–94 °C
−16 K/min	39 %	87.6 s	11.42×10^{-3} s^{-1}	145–85 °C

The determination of the half time point for crystallization and the kinetic study samples was measured at different cooling rates. As depicted in Figure 2, the temperature range for crystallization expectedly shifts towards lower temperatures for higher cooling rates and the higher supercooling in all samples.

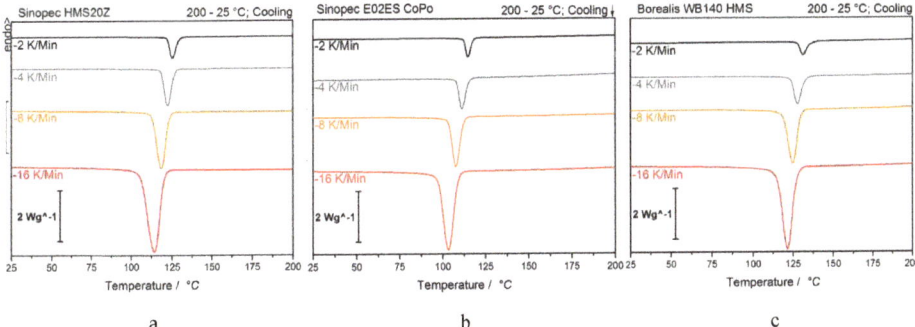

Figure 2. 2ND cooling curves of Sinopec HMS20Z (**a**), E02ES CoPo (**b**), and Borealis WB 140 HMS (**c**) at −2, −4, −8, and −16 K/min used for Avrami evaluation.

To compare the non-isothermal crystallization kinetics (in dependence on the cooling rate) of the PP grades with multiwave measurements, an evaluation with the Avrami approach was performed:

$$X_t = 1 - \exp(-k\,t^n) \tag{1}$$

$$\ln[-\ln(1 - X_t)] = \ln k + n \ln t \tag{2}$$

The Avrami evaluation is used to describe isothermal crystallization kinetics with X_t being the time-dependent relative crystallinity, k is a constant representing the crystallization rate, and n is the Avrami-exponent describing the crystal growth and the crystalline superstructures. However, the Avrami-approach ignores the influence of cooling rate and thermal gradient on the sample. Thus, to describe the non-isothermal crystallization kinetic, a modified Avrami theory by Jeziorny [31] was used. This approach considers the heating rates φ, as they influence the nucleation and growth of the crystalline polymer phase. The constant for the kinetic crystallization rate k_c is described as follows,

$$\ln k_c = \ln \frac{k}{\varphi} \tag{3}$$

In Figure 3, the conversion plots of each PP grade over time are shown beside the corresponding Avrami–Jeziorny plots. Both Sinopec grades show a similar course of conversion with comparable times for the half-time crystallization $\tau_{1/2}$, whereas Borealis WB 140 HMS takes longer times (at lower cooling rates) to reach 50% crystallinity. At higher cooling rates, the differences become more and more negligible. Additionally, it is evident that the whole crystallization process for Sinopec E02ES CoPo is very fast and finishes after ~7 min, whereas Sinopec HMS20Z takes twice as long and Borealis WB140 HMS takes even longer. Detailed values are listed in Table 2.

For low cooling rates, all Avrami-plots indicate a multi-stage crystallization with an Avrami-exponent n of ~3 (platelets/spherulites) at the beginning (Region I). Furthermore, a change in the slope with an exponent around 5 to 6 (complicated nucleation and spherulite form) occurs (Region II) until a high degree of crystallinity (80–95% of total crystallinity) is reached, after which the perfection of the crystalline structures takes place (n around 1, Region III) [33]. With increasing cooling rates, Avrami-exponents increase in all three regions and both of the first spherulite growing states become less distinctive until 8 K/min (Figure 3). At 8 K/min, Sinopec E02ES CoPo and Borealis WB140 HMS show similar values in region I and II, while this effect is less pronounced for Sinopec HMS20Z. Considering the Avrami-exponent n as well as the nucleation and growth rate constant k at the highest cooling rate, the following trends can be observed for the crystallization process of the PP grades.

- Sinopec E02ES CoPo and Borealis WB140 HMS crystallize (main crystallization area) in a very similar manner and crystallize faster than Sinopec HMS20Z at low cooling rates.

- At higher cooling rates, the most rapid main crystallization occurs for Sinopec E02ES CoPo, followed by Sinopec HMS20Z, and the slowest is Borealis WB140 HMS.
- The perfection of crystalline structures (Region III) takes significantly longer for Sinopec HMS20Z, than for Borealis WB140 HMS or Sinopec E02ES CoPo (fastest).

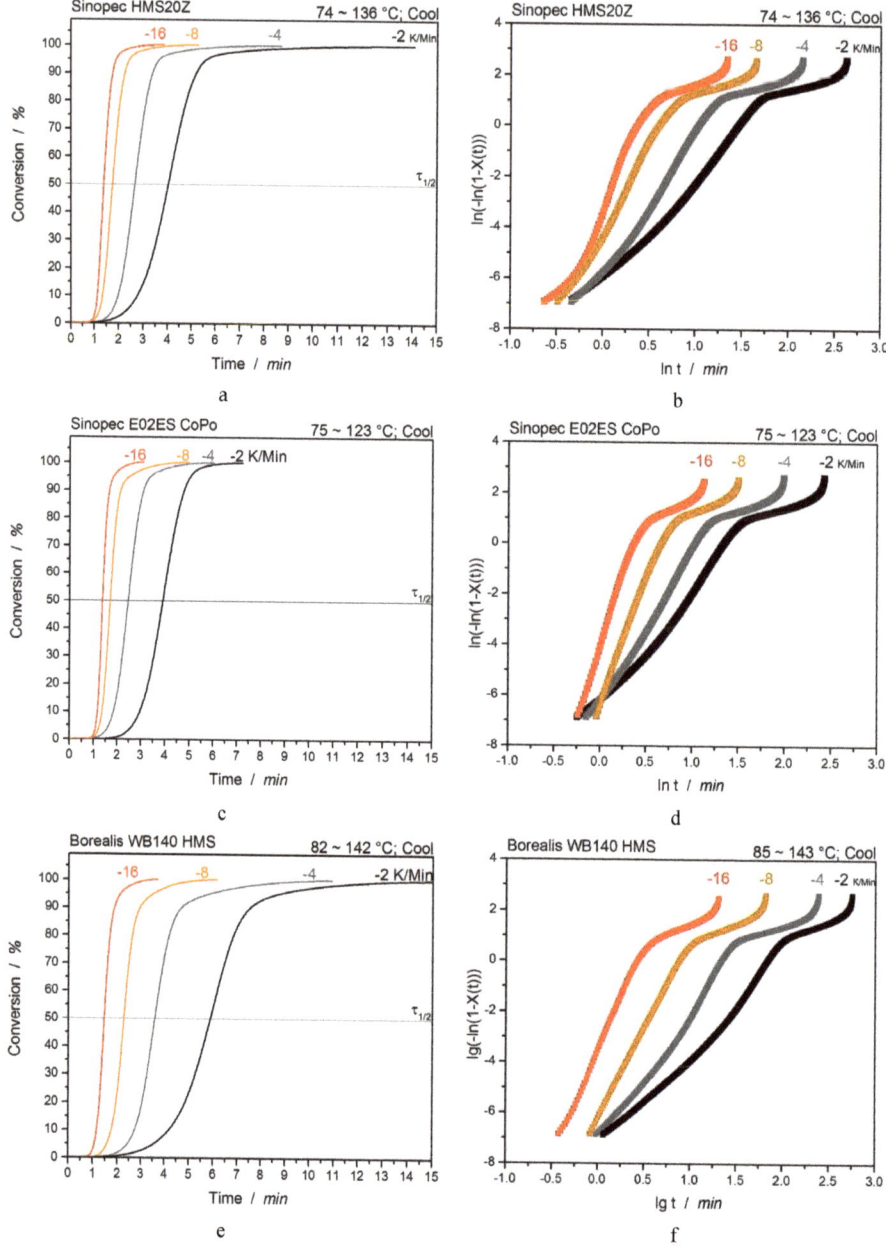

Figure 3. Conversion- and Avrami-plots of Sinopec HMS20Z (**a**,**b**), Sinopec E02ES CoPo (**c**,**d**), and Borealis WB 140 HMS (**e**,**f**).

The complete data is summarized in Table 3.

Table 3. Avrami-exponent *n*, nucleation, and growth rate constant *k* at the different cooling rates.

	n Region I	n Region II	n Region III	k I / k II / k III
Sinopec HMS20Z				
−2 K/min	2.9	5.0	0.9	2.0 / 1.5 / 0.5
−4 K/min	3.5	6.4	1.0	1.7 / 1.0 / 0.1
−8 K/min	4.6	8.1	1.1	1.0 / 0.6 / −0.2
−16 K/min	6.4	10.7	1.2	0.6 / 0.3 / −0.3
Sinopec E02ES CoPo				
−2 K/min	3.2	5.9	1.1	1.9 / 1.3 / 0.7
−4 K/min	4.9	6.8	1.2	1.3 / 1.0 / 0.4
−8 K/min	10.7	9.8	1.4	0.6 / 0.6 / 0.2
−16 K/min	11.7	11.3	1.6	0.4 / 0.3 / −0.1
Borealis WB140 HMS				
−2 K/min	3.1	5.8	1.3	2.3 / 1.8 / 1.3
−4 K/min	4.1	6.9	1.1	1.7 / 1.3 / 0.8
−8 K/min	7.3	7.1	1.5	0.9 / 0.9 / 0.5
−16 K/min	9.2	8.4	1.5	0.4 / 0.4 / 0.1

3.2. Rheological Characterization

3.2.1. Shear-Rheological Non-Isothermal Multiwave Measurements

Usually, the determination of cross-linking due to crystallization is qualified by the gel point. As the crystallization of polymers is a sol–gel transition of polymeric materials, a network point-like behavior of the crystals can be assumed. For the standard evaluation, the gel point is defined as crossover point of storage and loss modulus at a fixed angular frequency. The main disadvantage of this method is the dependency of the crossover on the frequency, meaning this method is only an approximation. A multiwave test, on the other hand, is performed with many frequencies simultaneously in superposition (consisting of multiple harmonics based on the fundamental wave). The material answers for each frequency is separated by a Fourier transformation. Compared to the previously mentioned method, here the loss factor tan (δ) is frequency independent at the gel point. It is then a point of intersection of all frequencies [34–36].

First, the Borealis WB140 HMS benchmark material was investigated. Starting from the molten polymer state, the material was cooled down at 0.5, 1, 2, and 4 K/min with different frequencies. As shown in Figure 4, the crystallization starts at ~146 °C at a cooling rate of 0.5 K/min (onset of curve drop) and hits the gel point at ~142 °C. This temperature range (cooling rate dependent) represents the foam processing window for a semicrystalline polymer, as the crystallization kinetics can strongly influence the foaming process, e.g., melt freezing in the die, cell rupture due to local freezing in a cell wall, and inhibited foam expansion (decreased elasticity).

Compared to the benchmark material, the Sinopec E02ES CoPo behaves clearly different. The crystallization onset lies, cooling rate dependent, at lower temperatures (~12 °C) due to the PE-content, which also hinders the crystallization. When the gel point is reached the copolymer freezes immediately, showing a very narrow temperature window for processing (Figure 5). Cooling rates above 4 K/min illustrate the fast crystallization, as the valid measurement area is very limited (beginning of background noise).

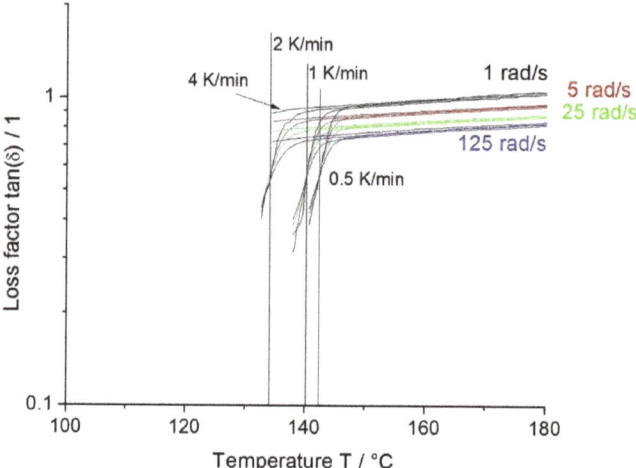

Figure 4. Non-isothermal crystallization multiwave measurement of Borealis WB140 HMS.

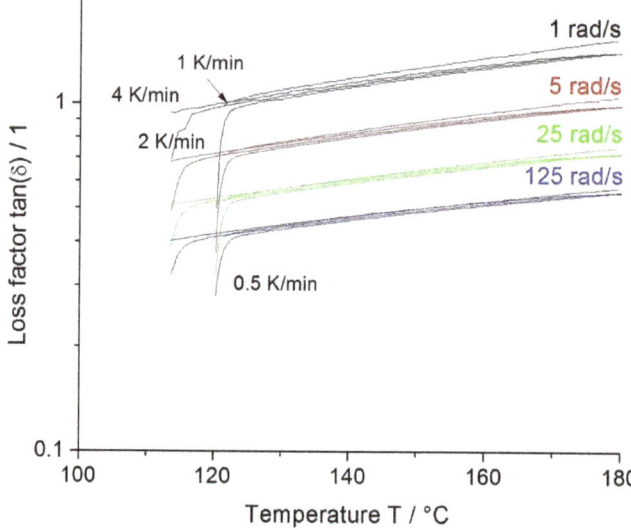

Figure 5. Non-isothermal crystallization multiwave measurement of Sinopec E02ES CoPo.

The Sinopec HMS20Z PP also exhibits an onset point at lower temperatures, compared to the Borealis long-chain branched material, namely, at ~136 °C (Figure 6). The most noteworthy observation is that Sinopec HMS20Z solidifies gradually with a comparably broad temperature window to LCB-PP (Borealis WB140 HMS).

Therefore, it is expected that this PP-grade will exhibit lower susceptibility to process fluctuations (more stable process) and show the most favorable crystallization behavior for foaming.

The multiwave results allow for a better understanding of the process as well as help narrow down the processing window and estimate the process stability (temperature difference between onset and gel point of the crystallization). In this frame, the Sinopec HMS20Z-PP and the Borealis WB140 HMS LCB-PP are expected to perform better in foam extrusion in terms of processing window and foam quality.

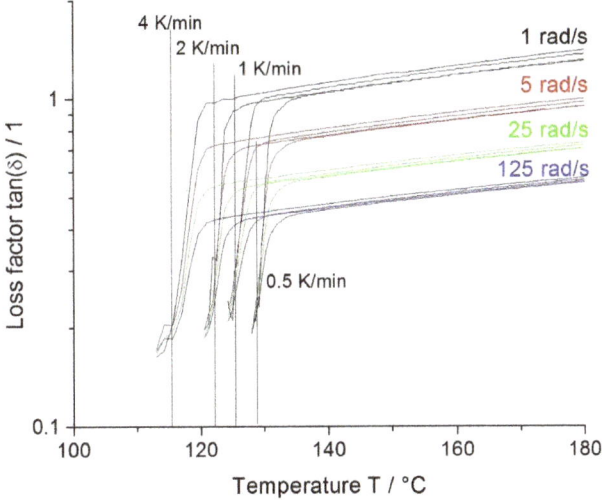

Figure 6. Non-isothermal crystallization multiwave measurement of Sinopec HMS20Z. The vertical line indicates the gel point.

3.2.2. Isothermal Multiwave Measurements

The crystallization behavior was further analyzed with isothermal multiwave measurements. The required temperatures for this measurement were obtained around the DSC-Avrami evaluation range. For better comparability, only the temperatures closest to the Avrami range at 1 rad/s (due to similar temperature ranges for crystallization) will be discussed. As shown in Figure 7, the Borealis WB140 HMS-PP starts to crystallize after 432 s at 145.5 °C (onset after 5 % loss in tan (δ)) and remains elastic after passing through the gel point for a longer period of time (984 s) until measurement becomes invalid due to noise (after 1884 s). This indicates that even after the immobilization through the gelation, some of the elastic behavior is retained for additional 984 s (1 rad/s), making it an advantageous feature for foaming. At higher measured temperatures, Borealis WB140 HMS crystallizes significantly slower. Crystallization starts after 864 s, hits the gel point at 2213 s measuring time, and retains elasticity for further 979 s (1 rad/s). Interestingly, the times for retained elasticity are more or less similar at all investigated temperatures for this material grade.

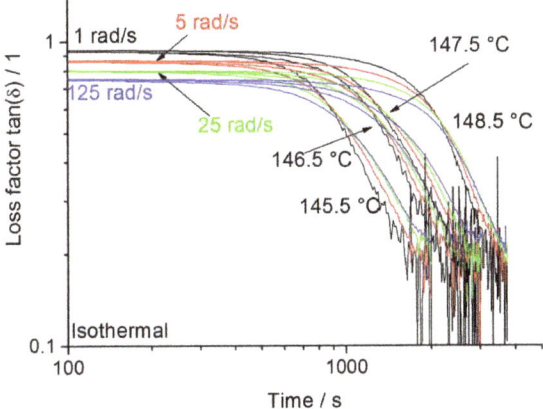

Figure 7. Isothermal crystallization multiwave measurement of Borealis WB140HMS.

Compared to that, the crystallization process for the Sinopec E02ES CoPo PP (Figure 8) starts later after 846 s (5% loss in tan δ) at a temperature of 127 °C and reaches the gel point at 1908 s (1 rad/s). The observed times for further elastic behavior after the gel point are very short, compared to other grades. Thus, the fast crystallization process (abrupt change) can be validated again as well as the high material rigidity, as short remaining elasticity (unusual termination) is shown and noise starts almost directly after the gel point due to low deformability.

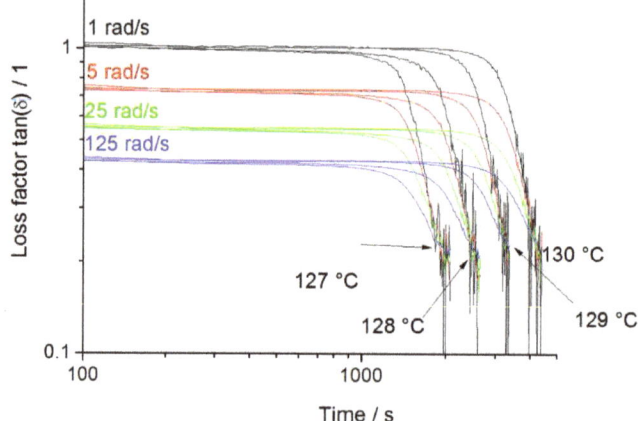

Figure 8. Isothermal crystallization multiwave measurement of Sinopec E02ES CoPo.

In Figure 9, isothermal multiwave results for the Sinopec HMS20Z PP are shown. It can be seen that this grade starts to crystallize after 756 s (onset), pacing the gel point at 2088 s and showing measurable values even for 2106 s after gelation. Consequentially, this PP-grade shows the longest retained overall elasticity of 4194 s, compared to Sinopec E02ES CoPo (elasticity for 2052 s) and Borealis WB140 HMS-PP (1884 s) and can be correlated to the crystallization process. Further data is summarized in Table 4. The difference in isothermal crystallization process between Borealis WB 140 HMS and Sinopec HMS 20Z arises from the different response of the chain topology to shear thinning, making it easier for linear chains to be disentangled (more susceptible to shearing). Retaining elasticity after the gel point, combined with slow crystallization, enables a broader processing window for foaming, thus it is possible to estimate foaming performance again.

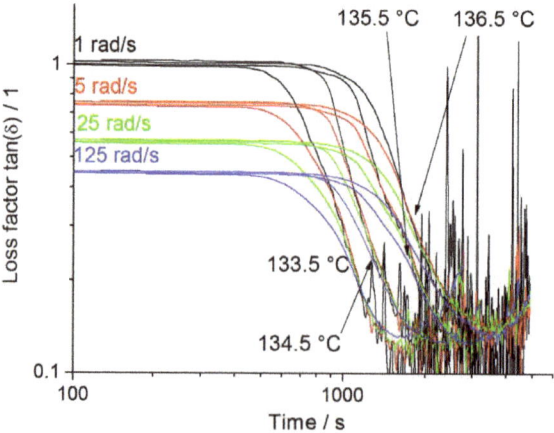

Figure 9. Isothermal crystallization multiwave measurement of Sinopec HMS20Z.

Table 4. Supplementary data for isothermal multiwave investigation of PP grades at 1 rad/s.

	Cryst. Start at	Gel-Point at	Onset => Gel-Point	Elastic after GP for
Sinopec HMS20Z				
133.5 °C	486	1170 s	684 s	1206 s
134.5 °C	648 s	1548 s	900 s	1548 s
135.5 °C	**756 s**	**2088 s**	**1332 s**	**2106 s**
136.5 °C	846 s	2592 s	1746 s	2034 s
Sinopec E02ES CoPo				
127 °C	**846 s**	**1908 s**	**1062 s**	**144 s**
128 °C	990 s	2412 s	1422 s	234 s
129 °C	1584 s	3289 s	1705 s	76 s
130 °C	2214 s	4176 s	1962 s	219 s
Borealis WB140 HMS				
145.5 °C	**432 s**	**900 s**	**468**	**984 s**
146.5 °C	522 s	1310 s	788	979 s
147.5 °C	684 s	1377 s	693	1112 s
148.5 °C	864 s	2213 s	1349	979 s

Note: highlighted temperatures resemble nearest data sets of multiwave to avrami-ranges.

The findings and tendencies of the isothermal and non-isothermal multiwave measurements can be correlated very well with the Avrami–Jeziorny analysis. The fast crystallization process of Sinopec E02ES CoPo can be observed in the conversion and Jeziorny plots (fast overall crystallization also represented in n and k-values), as well as in the non-isothermal multiwave showing an abrupt termination of the measured curves after reaching the T_C. Additionally, same tendencies are confirmed by isothermal multiwave.

According to the DSC-evaluation, Borealis WB140 HMS has a slower crystallization process compared to Sinopec HMS 20Z-PP. The results of the multiwave, non-isothermal as well as isothermal, contradict that, as Borealis WB140 HMS crystallizes faster in isothermal measurements compared to Sinopec HMS20Z. The reason for this can be the chosen temperatures from the DSC-evaluations and should be determined through rheology in future. In general, Borealis WB140 HMS and Sinopec HMS20Z show similar behavior, and therefore a similar foaming potential can be expected.

3.2.3. Elongational Rheology

As already mentioned, the elongational properties of the polymer are very important for foaming and controlling the foam morphology [14]. One of the main contributing factors for good foams usually stated in literature is strain hardening [13,37,38]. Strain hardening enables self-healing (thinner regions of cells require more energy to be stretched than thick ones, thus preventing cell coalescence and to widening processing window), which is the easiest way to ensure low-density foams [13,15–19]. However, this research shows that strain hardening is not a necessity to obtain good foams in continuous foaming and can be compensated material-wise by control of crystallization behavior [39].

Rheotens Test Result

The drawability and melt strength of the investigated PP-grades were analyzed to get an estimate for the material's behavior during bubble expansion. As shown in Figure 10, Borealis WB140 HMS has the highest melt strength (MS) of all tested PP grades with 0.25 N, but a rather low drawability of the melt. The strong increase in MS is caused by the branched structure in the polymer backbone, which enables short-term physical network points due to entanglements during stress. For Sinopec HMS20Z-PP, high melt strength is achieved by a different approach of combining linear polymer chains of different lengths into a grade (broad PD) according to the manufacturer. Therefore, the short-term networking, caused by branching, is missing, and both Sinopec-grades exhibit a lower melt strength at ~0.035 N but a significantly higher drawability compared to the Borealis grade.

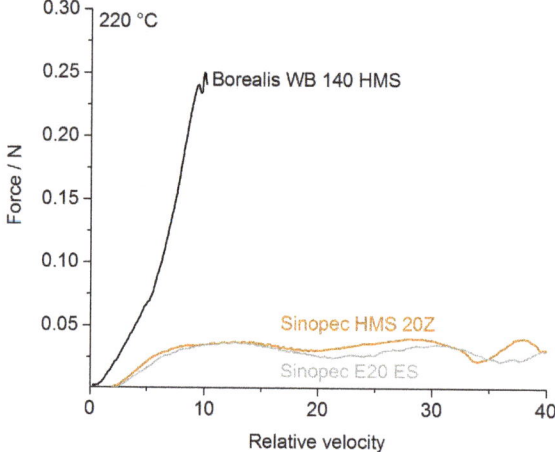

Figure 10. Melt drawability and strength for Borealis WB140 HMS, Sinopec E02ES CoPo, and Sinopec HMS20Z.

Universal Extensional Fixture

For a better illustration of the transient elongational viscosities, the investigated PP-grades are compared in Figures 11 and 12. The Borealis WB140 HMS-PP shows a clear strain hardening behavior at all tested strain rates. This effect is caused by the long-chain branching in the polymer structure and enables the "self-healing effect".

This means that thin cell walls being exposed to higher strain rates, and thereby a higher chain-stretching degree, undergo an increase in elongational viscosity making thick sections easier to extend. This contributes to a more homogeneous morphology. The advantage for foaming is the preventing of cell rupture as well as coalescence through viscosity increase [6]. Sinopec E02ES CoPo expectedly exhibits no strain hardening at all strain rates investigated due to the linear nature of the polymer chains.

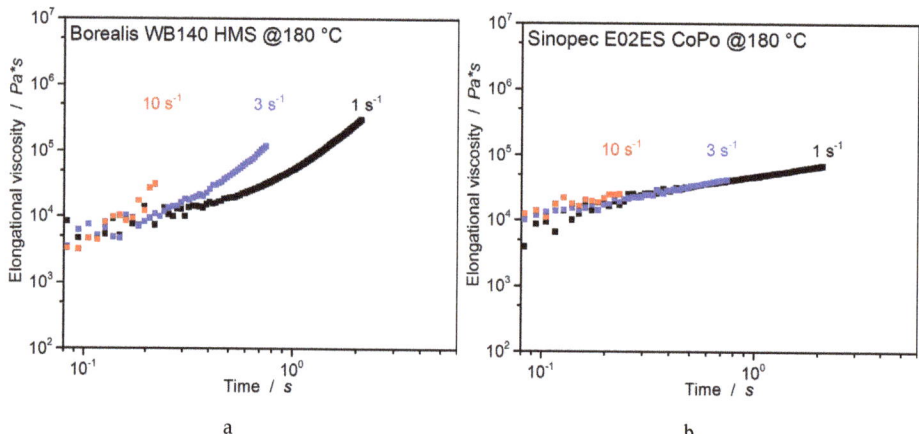

Figure 11. Elongation viscosity of Borealis WB140 HMS (**a**) and Sinopec E02ES CoPo (**b**) at 180 °C.

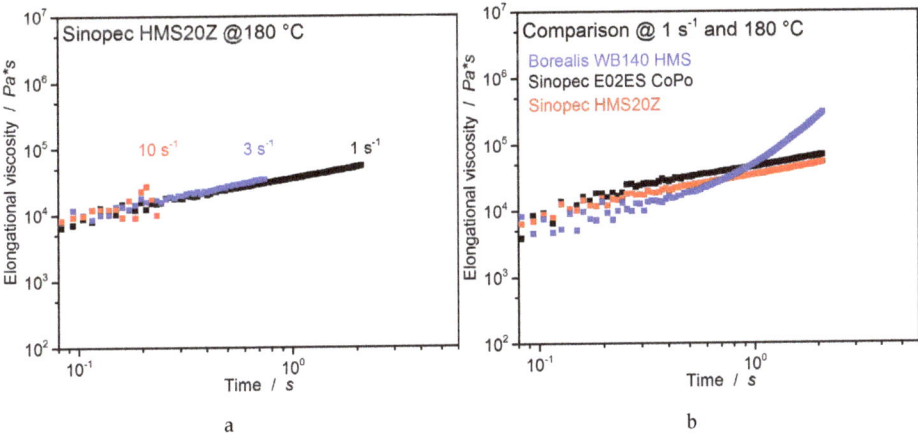

Figure 12. Elongation viscosity of Sinopec HMS20Z (**a**) and a comparison of all PP (**b**) at 180 °C.

Sinopec HMS20Z homo-PP behaves similarly to Sinopec E02ES CoPo, showing no strain hardening at all rates. The course of the curves and the values are very similar (Figure 12), indicating similar polymer basis. The comparison of elongational properties at a strain rate of 1 s^{-1} illustrates the similarity in performance of both Sinopec PP-grades at elongational rheology, while only Borealis WB140 HMS exhibits strain hardening (Figure 12b).

As the Sinopec materials lack strain hardening at all strain rates, they are expected (based on these results) to perform worse during extrusion foaming, resulting in higher densities and inhomogeneous morphology. On the contrary, Borealis WB140 HMS is expected to outperform both Sinopec grades in density and morphology. This conclusion is based on an established and well-known fact in literature, that strain hardening is considered advantageous and contributing significantly to good foamability.

3.3. Foam Extrusion

A proper foaming window for a polymer usually constitutes a compromise between the temperature (energy for expansion/elastic properties) and pressure (drop rate as driving force for gas expansion). Most characterizing methods for this difficult process aim for a thorough material characterization to extract material characteristics and thus obtain a qualitative understanding of the processing window and resulting foam morphology, as many influencing factors cannot be taken into account (e.g., high shearing rates in extruder, which cause (locally) increased melt temperatures, cannot be reproduced in a plate-plate setup in a rotational rheometer). In this frame, the obtained multiwave results (material characteristics) help narrow down the actual foaming window, as the estimations from the non-isothermal and isothermal multiwave are transferable to the actual process. For foam extrusion, the three parameters, CO_2 content, cCO_2; die temperature, T_{Die}; and, to a certain degree, the temperature of the polymer melt, T_{Melt}, were systematically studied. The zone temperatures in the A-extruder were adjusted to 180–200 °C with a pressure of 45–140 bar and in the B-extruder to 155–160 °C with an average pressure at the die around 81–95 bar. In some cases, when no satisfactory foams yielded (as for example for the Borealis WB140 HMS Material with 6% CO_2), no results are reported.

3.3.1. Variation of Blowing Agent Concentration

Three different CO_2 concentrations (2, 4, and 6 wt.%) were studied regarding their foaming potential. The resulting foam densities of the materials from these trials are shown in Figure 13. Generally, it is observed, that with increasing CO_2 concentration, the potential for expansion increases. From the variation in density, especially for the Sinopec E02ES CoPo and the Borealis WB140 HMS polymer, it can be deducted that all processing conditions have to be well controlled to achieve a high expansion ratio. In contrast, the Sinope HMS20Z-PP delivers consistently low densities over a wide range of CO_2 concentrations. Unfortunately, the same conditions as for the Sinopec material, were not possible for Borealis WB140 HMS because the die pressure was too high (140 bar) and fluctuating. Thus, the screw speed in the B-extruder was reduced to 9 rpm. Furthermore, it was not possible to dissolve and keep more than 4% CO_2 at these conditions in WB140HMS; otherwise, gas leakage at the die resulted.

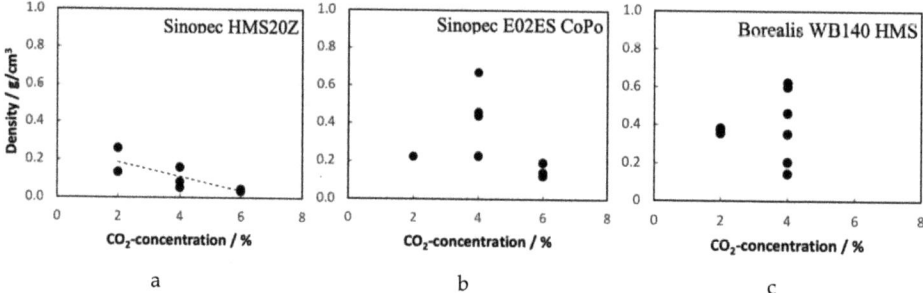

Figure 13. Summary of the foaming trials in terms of density as function of CO_2 concentration for (**a**) Sinopec HMS20Z, (**b**) Sinopec E02ES CoPo, and (**c**) Borealis WB140 HMS. Please note that not only the CO_2 concentration was varied, but also die temperature, melt temperature, and throughput.

Densities ranging between 120 and 600 kg/m^3 are observed for Sinopec E02ES CoPo, whereas the Sinopec HMS20Z-PP shows densities ranging between 35 and 260 kg/m^3. Interestingly, and unexpectedly, the Borealis WB140 HMS-PP exhibits the worst foaming behavior with densities between 140 and 620 kg/m^3. Possible explanations are given below in the section for the summary of the extrusion trials.

3.3.2. Variation of Die Temperature

The die temperature is a suitable parameter to control the foam expansion and obtain information about the width of the processing window. To test the variability in foam density with the die temperature, it was varied in a range between 150 and 190 °C (when possible). The results are plotted in Figure 14. It is striking, that both Sinopec E02ES CoPo and HMS20Z exhibit a rather broad processing window as the density remains almost constant as function of die temperature. In contrast, the foamability of the Borealis WB140 HMS PP strongly depends on the die temperature. Only within a narrow range of ~10 °C (165–175 °C), good foamability is observed for this material. The Sinopec HMS20Z-PP shows only minor variations with temperature, consistently low densities between 45 and 32 kg/m^3 are observed over the studied temperature range. When comparing the linear PPs and the LCB-PP, it can be observed that a much lower melt temperature is required for a low-density PP-foam. It is hypothesized, that the introduction of long-chain branching (as found for the Borealis type) leads to higher melt temperatures during processing due to additional viscous dissipation. Therefore, the branched polymer structure makes it difficult to achieve the required lower foaming temperatures and cooling efficiency.

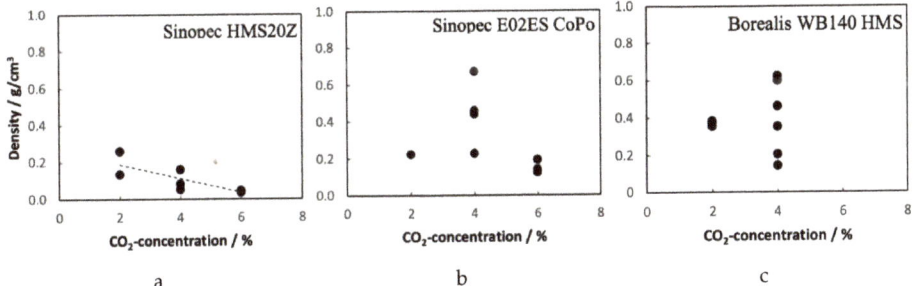

Figure 14. Effect of the die temperature on the foamability/density of the three PP grades for (**a**) Sinopec HMS20Z @ 6 wt.% CO_2 & T_{Melt} = 162 °C; (**b**) Sinopec E02ES CoPo @ 6 wt.% CO_2 & T_{Melt} = 156 °C; and (**c**) Borealis WB140 HMS @ 4 wt.% CO_2 & T_{Melt} = 160 °C.

3.3.3. Resulting Morphology

The morphology of the foamed strands (produced at best foaming conditions) was further investigated with SEM. The resulting SEM-micrographs are shown in Figure 15. The Borealis WB140 HMS PP-grade reaches a density of 140 kg/m³ with an overall coarse foam morphology (Figure 15a). It shows only a few foamed cells (cell density of $2.3*10^6$ cells/cm² calculated according to work in [39]) with a bimodal cell size distribution, containing large (> 1 mm) and small cells (no homogeneous foam structure) with an average cell diameter of 0.9 mm. Borealis WB140 HMS exhibits inferior properties during foam extrusion and has not the desired foam structure, but is rather a porous solid with randomly and inhomogeneously distributed cells. By contrast, lower densities as well as smaller and more uniform cells can be achieved with Sinopec E02ES CoPo. At the lowest density of 120 kg/m³, a cell density of $1.7*10^7$ cells/cm² and a mean cell diameter of 0.4 mm were achieved with clearly narrower size distribution (Figure 15b). Sinopec E02ES CoPo consists of mainly PP with a low percentage of PE units in the polymer backbone, making crystallization and nucleation easier due to the linear character, but is also negatively influenced by the PE-units. Therefore, a lot of cell rupture can be observed, as well as cracks in the cell walls (open cellular structure) at higher magnification. Sinopec HMS20Z-PP shows the best foamability with the lowest density of 40 kg/m³, a cell density of $1.6*10^8$ cells/cm² and a rather fine cellular morphology, compared to other two grades, with an average cell diameter of 0.19 mm (Figure 15c). Nevertheless, a bimodal cell distribution can be observed (with very small cells in the core), which could be controlled by adding nucleating agents. As Sinopec HMS20Z only contains linear PP chains of different lengths, it exhibits the highest cell nucleation density, compared to Borealis WB140 HMS. One of the contributing factors is the presence of spherulites (correlates with degree of crystallinity) that can enhance the amount of cell nucleation sites during foaming and contributes to a finer foam morphology. During a continuous process, the melt is usually undercooled due to plasticizing effect of the blowing agent, thus the melt nearest to the barrel start to crystallize earlier. Through shearing, these crystals are mixed into the melt phase increasing viscosity and contributing as nucleating sites. When the amorphous phase crystallizes it expels the CO_2 from crystalline regions. Therefore, locally increased blowing agent concentrations yield around spherulites (heterogeneous nucleation at amorphous/crystalline interfaces), leading to an increase in bubble nucleation [5,28,40]. This relationship was already reported and discussed in literature. Sinopec E02ES CoPo also has an increased cell nucleation density due to possible nucleation at the PP-PE interface, despite having a lower crystallinity than Borealis WB140 HMS, confirming that several factors influence the overall foam nucleation process. This can be further improved by adding nucleating agents. Once again, Borealis WB140 HMS proved to be unsuitable for continuous foaming due to the resulting bad morphology, despite expressing a strain hardening behavior.

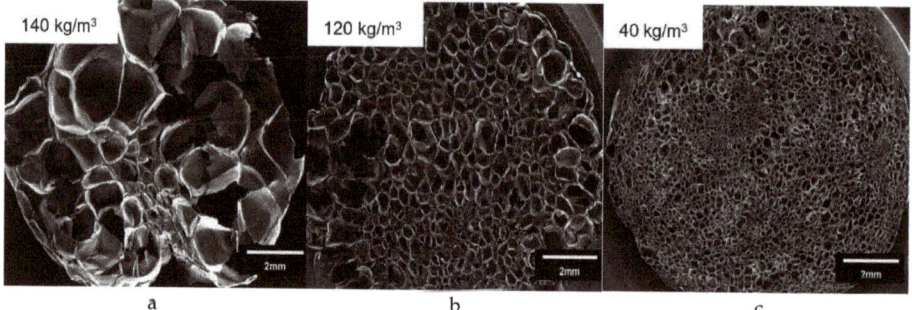

Figure 15. SEM-Images of Borealis WB140 HMS (**a**), Sinopec E02ES CoPo (**b**), and HMS20Z (**c**).

3.3.4. Correlation of Previous Findings to Foam Extrusion Trials

When comparing the three studied materials, the Sinopec HMS PP shows superior foaming characteristics, as can be seen from the lowest achieved density in Table 5 with the corresponding foaming conditions. The material also had the broadest foaming window thus guaranteeing a stable and reproducible process.

Table 5. Best performing settings for foam extrusion with lowest densities.

	Sinopec E02ES CoPo	**Sinopec HMS20Z**	**Borealis WB140 HMS**
CO_2 / %	6	6	4
T_{melt} / °C	156	165	164
T_{die} / °C	156	190	170
P_{die} / bar	95	81	43
Density / kg/m^3	121	34	143

Now the question arises, where this behavior comes from. To shed some light into this fact, the Sinopec HMS-PP is compared with its competitors regarding their properties. The bubble growth at the exit of the die is mainly controlled by the viscosity, making the rheological and crystallization analysis a powerful tool for estimating foaming performance.

The crystallization behavior of Borealis WB140 HMS is predestined for discontinuous foaming and is therefore quite different compared with Sinopec HMS20Z and E02ES CoPo, with tacticity not being taken into account. The Borealis WB140 HMS-PP has a significantly higher crystallization temperature (caused by long-chain branching), which leads to lower melt expansion after the die due to earlier freezing (linear Sinopec grades do not freeze that fast). Thus, no high expansion ratios can be achieved and lower crystallization temperatures (like Sinopec Material) appear beneficial.

However, the isothermal multiwave measurements are more revealing than non-isothermal measurements, therefore making the assumptions on the material foam performance more accurate and fitting. Here, the times between start of the crystallization and the gel-point are considered. At temperatures, closest to the Avrami temperature range (DSC), the Borealis WB140 HMS PP-grade exhibits the shortest time until gelation is reached, which can be correlated to difficult processability as well as premature freezing (lower density reduction) and supports the DSC results. Thus, the Borealis WB140 HMS is rather unsuitable for continuous foaming as pure material. Blending with a linear PP-grade should be considered to counter the disadvantages as described in [29,38,39], also taking into account that this grade was designed for discontinuous foaming in the first place.

On the other hand, the linear Sinopec PP-grades have lower crystallization temperatures, which enable higher expansion ration and less premature die freezing. Additionally, the time from the onset to the gelation point of the linear PPs is 2–3-fold higher compared to Borealis WB140 HMS, which results in increased foaming performance. The enhanced foamability of Sinopec HMS20Z (compared to Sinopec E02ES) is attributed to the significantly prolonged elasticity after the gel point, which is enabled by the high PDI of 11.

In summary, a combination of low crystallization temperature and longer period of time until polymer-gelation appears to be one major key for good foamability. In contrast, "high melt strength" is only one part of the puzzle of a good foaming material and contributes significantly less in continuous foaming.

4. Conclusions

Polypropylene (PP) is an outstanding choice as matrix material for polymeric foams due to its favorable mechanical properties and chemical resistance. However, its semicrystalline nature and low melt elasticity result in unfavorable foaming properties. Long-chain branching of polypropylene is regarded as a game changer in foaming this material due to the introduction of strain hardening, which stabilizes the foam morphology. However, the question arises, whether it is the singular factor for obtaining low-density foams? Therefore, three PPs dedicated for foaming are studied regarding the dependency of their foaming behavior on the rheological and crystallization properties. In this frame, a linear PP Sinopec HMS20Z, a PP/PE-block co-polymer Sinopec E02ES CoPo and a long-chain branched PP Borealis WB140 HMS are the polymers of interest.

Although only the LCB-PP exhibits strain hardening and has five times the melt strength of the other grades, it does not provide the broadest foaming window or the best foam quality in terms of density (140 g/L) and cellular morphology. On the contrary, the linear Sinopec HMS20Z delivers low densities (<40 g/L) and superior foam morphology. The Sinopec E02ES CoPo performs rather negatively in terms of density and mediocre in terms of cellular morphology.

The beneficial foaming behavior of the Sinopec HMS20Z-PP is attributed to slower crystallization and a low crystallization temperature compared to the other two materials. Multiwave experiments were conducted to study the gelation due to crystallization. In isothermal multiwave experiments, the Sinopec HMS20Z exhibited a gel-like behavior over a broad time frame, whereas the other two PPs froze quickly. Non-isothermal multiwave tests also underline the finding of a broad processing window for the linear PP (Sinopec HMS20Z) as the temperature (and time) difference between the onset of crystallization and the gel point is significantly broader compared to the other PP-grades. These findings were also confirmed in DSC experiments, as especially the crystal-perfection occurs notably slower for Sinopec HMS20Z, which in turn leads to a longer gel-like state before solidification. Again, it is noted that this PP-grade exhibits no strain hardening. Thus, it is concluded that, besides sufficient rheological properties, the crystallization behavior is of utmost importance for foaming in terms of a broad processing window, when aiming to obtain low-densities and good foam morphology. In detail, a broad temperature window between the onset of crystallization and gelation are preferred for cooling, for isothermal processes an extended time in gel-like state appears beneficial.

Author Contributions: Conceptualization, N.W. and D.R; methodology, D.R. and L.E.; software, D.R. and N.W.; validation, D.R., N.W., and L.E.; formal analysis, D.R.; investigation, N.W. and D.R.; resources, M.L.; data curation, N.W.; writing—original draft preparation, N.W. and D.R.; writing—review and editing, D.R. and N.W.; visualization, N.W.; supervision, V.A.; project administration, V.A.; funding acquisition, V.A. All authors have read and agreed to the published version of the manuscript.

Funding: This research was funded by the National Key Research and Development Program of China (2016YFB0302000) and Research and Development Program of SINOPEC (No. 2018013-3). Open access charges were funded by the German Research Foundation (DFG) and the University of Bayreuth in the funding program Open Access Publishing.

Acknowledgments: The authors would like to thank the SINOPEC Beijing Research Institute of Chemical Industry, as this work was supported by the National Key Research and Development Program of China (2016YFB0302000) and Research and Development Program of SINOPEC (No. 2018013-3). Special thanks goes to Martin Demleitner for the fruitful discussions about multiwave rheology.

Conflicts of Interest: The authors declare no conflicts of interest.

References

1. Doroudiani, S.; Kortschot, M.T. Polystyrene foams. III. Structure-tensile properties relationships. *J. Appl. Polym. Sci.* **2003**, *90*, 1427–1434. [CrossRef]
2. Rachtanapun, P.; Selke, S.E.M.; Matuana, L.M. Relationship between cell morphology and impact strength of microcellular foamed high-density polyethylene/polypropylene blends. *Polym. Eng. Sci.* **2004**, *44*, 1551–1560. [CrossRef]
3. Ferkl, P.; Pokorný, R.; Bobák, M.; Kosek, J. Heat transfer in one-dimensional micro- and nano-cellular foams. *Chem. Eng. Sci.* **2013**, *97*, 50–58. [CrossRef]
4. Raps, D.; Köppl, T.; De Anda, A.R.; Altstädt, V. Rheological and crystallisation behaviour of high melt strength polypropylene under gas-loading. *Polymer* **2014**, *55*, 1537–1545. [CrossRef]
5. Baldwin, D.F.; Park, C.B.; Suh, N.P. A microcellular processing study of poly(ethylene terephthalate) in the amorphous and semicrystalline states. Part II: Cell growth and process design. *Polym. Eng. Sci.* **1996**, *36*, 1446–1453. [CrossRef]
6. Raps, D.; Hossieny, N.; Park, C.B.; Altstädt, V. Past and present developments in polymer bead foams and bead foaming technology. *Polymer* **2015**, *56*, 5–19. [CrossRef]
7. Varma-Nair, M.; Handa, P.Y.; Mehta, A.K.; Agarwal, P. Effect of compressed CO_2 on crystallization and melting behavior of isotactic polypropylene. *Thermochim. Acta* **2003**, *396*, 57–65. [CrossRef]
8. Kao, N.; Chandra, A.; Bhattacharya, S. Melt strength of calcium carbonate filled polypropylene melts. *Polym. Int.* **2002**, *51*, 1385–1389. [CrossRef]
9. Kaewmesri, W.; Lee, P.C.; Park, C.B.; Pumchusak, J. Effects of CO_2 and talc contents on foaming behavior of recyclable high-melt-strength PP. *J. Cell. Plast.* **2006**, *42*, 405–428. [CrossRef]
10. Wang, M.; Wang, Z.; Zhou, N. Extrusion foaming of polypropylene with supercritical carbon dioxide. *Hecheng Shuzhi Ji Suliao/China Synth. Resin Plast.* **2011**, *28*, 13–16.
11. Liao, R.; Yu, W.; Zhou, C.; Yu, F.; Tian, J. The formation of γ-crystal in long-chain branched polypropylene under supercritical carbon dioxide. *J. Polym. Sci. Part B Polym. Phys.* **2008**, *46*, 441–451. [CrossRef]
12. Kaewmesri, W.; Rachtanapun, P.; Pumchusak, J. Effect of solvent plasticization on polypropylene microcellular foaming process and foam characteristics. *J. Appl. Polym. Sci.* **2008**, *107*, 63–70. [CrossRef]
13. Spitael, P.; Macosko, C.W. Strain hardening in polypropylenes and its role in extrusion foaming. *Polym. Eng. Sci.* **2004**, *44*, 2090–2100. [CrossRef]
14. Koppl, T.; Raps, D.; Altstadt, V. E-PBT–Bead foaming of poly(butylene terephthalate) by underwater pelletizing. *J. Cell. Plast.* **2014**, *50*, 475–487. [CrossRef]
15. *Borealis Daploy WB140HMS, Techn. Report*; Borealis AG: Vienna, Austria, 2010.
16. Carella, J.M.; Gotro, J.T.; Graessley, W.W. Thermorheological effects of long-chain branching in entangled polymer melts. *Macromolecules* **1986**, *19*, 659–667. [CrossRef]
17. Malmberg, A.; Liimatta, J.; Lehtinen, A.; Löfgren, B. Characteristics of Long Chain Branching in Ethene Polymerization with Single Site Catalysts. *Macromolecules* **1999**, *32*, 6687–6696. [CrossRef]
18. Wood-Adams, P.; Costeux, S. Thermorheological Behavior of Polyethylene: Effects of Microstructure and Long Chain Branching. *Macromolecules* **2001**, *34*, 6281–6290. [CrossRef]
19. Raps, D.; Köppl, T.; Heymann, L.; Altstädt, V. Rheological behaviour of a high-melt-strength polypropylene at elevated pressure and gas loading for foaming purposes. *Rheol. Acta* **2017**, *56*. [CrossRef]
20. Hasan, M.M.; Li, Y.G.; Li, G.; Park, C.B.; Chen, P. Determination of Solubilities of CO_2 in Linear and Branched Polypropylene Using a Magnetic Suspension Balance and a PVT Apparatus. *J. Chem. Eng. Data* **2010**, *55*, 4885–4895. [CrossRef]
21. Li, Y.G.; Park, C.B. Effects of Branching on the Pressure–Volume–Temperature Behaviors of PP/CO_2 Solutions. *Ind. Eng. Chem. Res.* **2009**, *48*, 6633–6640. [CrossRef]
22. Watanabe, K.; Suzuki, T.; Masubuchi, Y.; Taniguchi, T.; Takimoto, J.; Koyama, K. Crystallization kinetics of polypropylene under high pressure and steady shear flow. *Polymer* **2003**, *44*, 5843–5849. [CrossRef]
23. Doroudiani, S.; Park, C.B.; Kortschot, M.T. Effect of the crystallinity and morphology on the microcellular foam structure of semicrystalline polymers. *Polym. Eng. Sci.* **1996**, *36*, 2645–2662. [CrossRef]
24. Naguib, H.E.; Park, C.B.; Song, S.-W. Effect of Supercritical Gas on Crystallization of Linear and Branched Polypropylene Resins with Foaming Additives. *Ind. Eng. Chem. Res.* **2005**, *44*, 6685–6691. [CrossRef]

25. Takada, M.; Tanigaki, M.; Ohshima, M. Effects of CO_2 on crystallization kinetics of polypropylene. *Polym. Eng. Sci.* **2001**, *41*, 1938–1946. [CrossRef]
26. Takada, M.; Ohshima, M. Effect of CO_2 on crystallization kinetics of poly(ethylene terephthalate). *Polym. Eng. Sci.* **2003**, *43*, 479–489. [CrossRef]
27. Oda, T.; Saito, H. Exclusion effect of carbon dioxide on the crystallization of polypropylene. *J. Polym. Sci. Part B Polym. Phys.* **2004**, *42*, 1565–1572. [CrossRef]
28. Lee, S.-T.; Park, C.B. *Foam Extrusion: Principles and Practice, Second Edition (Polymeric Foams)*, 2nd ed.; CRC Press: Boca Raton, FL, USA, 2014; ISBN 9781439898598.
29. Setiawan, A.H. Determination of Crystallization and Melting Behaviour of Poly-lactic Acid and Polypropyleneblends as a Food Packaging Materials by Differential Scanning Calorimeter. *Procedia Chem.* **2015**, *16*, 489–494. [CrossRef]
30. Khanna, Y.P.; Kuhn, W.P. Measurement of crystalline index in nylons by DSC: Complexities and recommendations. *J. Polym. Sci. Part B Polym. Phys.* **1997**, *35*, 2219–2231. [CrossRef]
31. Jeziorny, A. Parameters characterizing the kinetics of the non-isothermal crystallization of poly(ethylene terephthalate) determined by d.s.c. *Polymer* **1978**, *19*, 1142–1144. [CrossRef]
32. Hay, J.N. Application of the modified avrami equations to polymer crystallisation kinetics. *Br. Polym. J.* **1971**, *3*, 74. [CrossRef]
33. Rimdusit, S.; Ishida, H. Gelation study of high processability and high reliability ternary systems based on benzoxazine, epoxy, and phenolic resins for an application as electronic packaging materials. *Rheol. Acta* **2002**, *41*, 1–9. [CrossRef]
34. Holly, E.E.; Venkataraman, S.K.; Chambon, F.; Henning Winter, H. Fourier transform mechanical spectroscopy of viscoelastic materials with transient structure. *J. Nonnewton. Fluid Mech.* **1988**, *27*, 17–26. [CrossRef]
35. Hu, X.; Fan, J.; Yue, C.Y. Rheological study of crosslinking and gelation in bismaleimide/cyanate ester interpenetrating polymer network. *J. Appl. Polym. Sci.* **2001**, *80*, 2437–2445. [CrossRef]
36. Nam, G.J.; Yoo, J.H.; Lee, J.W. Effect of long-chain branches of polypropylene on rheological properties and foam-extrusion performances. *J. Appl. Polym. Sci.* **2005**, *96*, 1793–1800. [CrossRef]
37. Liu, G.; Sun, H.; Rangou, S.; Ntetsikas, K.; Avgeropoulos, A.; Wang, S.-Q. Studying the origin of "strain hardening": Basic difference between extension and shear. *J. Rheol.* **2013**, *57*, 89. [CrossRef]
38. Matuana, L.M.; Diaz, C.A. Study of Cell Nucleation in Microcellular Poly(lactic acid) Foamed with Supercritical CO_2 through a Continuous-Extrusion Process. *Ind. Eng. Chem. Res.* **2010**, *49*, 2186–2193. [CrossRef]
39. Okolieocha, C.; Raps, D.; Subramaniam, K.; Altstädt, V. Microcellular to nanocellular polymer foams: Progress (2004–2015) and future directions—A review. *Eur. Polym. J.* **2015**, *73*, 500–519. [CrossRef]
40. Taki, K.; Kitano, D.; Ohshima, M. Effect of growing crystalline phase on bubble nucleation in poly(L-lactide)/CO_2 batch foaming. *Ind. Eng. Chem. Res.* **2011**, *50*, 3247–3252. [CrossRef]

© 2020 by the authors. Licensee MDPI, Basel, Switzerland. This article is an open access article distributed under the terms and conditions of the Creative Commons Attribution (CC BY) license (http://creativecommons.org/licenses/by/4.0/).

Article

Design and Synthesis of Polysiloxane Based Side Chain Liquid Crystal Polymer for Improving the Processability and Toughness of Magnesium Hydrate/Linear Low-Density Polyethylene Composites

Xiaoxiao Guan [1], Bo Cao [1], Jianan Cai [1], Zhenxing Ye [1], Xiang Lu [2,*], Haohao Huang [1,2], Shumei Liu [1,2] and Jianqing Zhao [1,2,*]

[1] School of Materials Science and Engineering, South China University of Technology, Guangzhou 510640, China; guanxiaoxiao.ok@163.com (X.G.); shawncb@126.com (B.C.); mscaijn@mail.scut.edu.cn (J.C.); mrzhxye@163.com (Z.Y.); hhhuang@scut.edu.cn (H.H.); liusm@scut.edu.cn (S.L.)
[2] Key Laboratory of Polymer Processing Engineering, Ministry of Education, Guangzhou 510640, China
* Correspondence: luxiang_1028@163.com (X.L.); psjqzhao@scut.edu.cn (J.Z.)

Received: 10 March 2020; Accepted: 13 April 2020; Published: 14 April 2020

Abstract: In this study, a polysiloxane grafted by thermotropic liquid crystal polymer (PSCTLCP) is designed and synthesized to effectively improve the processability and toughness of magnesium hydroxide (MH)/linear low-density polyethylene (LLDPE) composites. The obtained PSCTLCP is a nematic liquid crystal polymer; the liquid crystal phase exists in a temperature range of 170 to 275 °C, and its initial thermal decomposition temperature is as high as 279.6 °C, which matches the processing temperature of MH/LLDPE composites. With the increase of PSCTLCP loading, the balance melt torque of MH/LLDPE/PSCTLCP composites is gradually decreased by 42% at 5 wt % PSCTLCP loading. Moreover, the power law index of MH/LLDPE/PSCTLCP composite melt is smaller than 1, but gradually increased with PSCTLCP, the flowing activation energy of PSCTLCP-1.0 is lower than that of MH/LLDPE at the same shear rate, indicating that the sensitivity of apparent melt viscosity of the composites to shear rate and to temperature is decreased with the increase of PSCTLCP, and the processing window is broadened by the addition of PSCTLCP. Besides, the elongation at break of MH/LLDPE/PSCTLCP composites increases from 6.85% of the baseline MH/LLDPE to 17.66% at 3 wt % PSCTLCP loading. All the results indicate that PSCTLCP can significantly improve the processability and toughness of MH/LLDPE composites.

Keywords: side chain liquid crystal polymer; magnesium hydroxide; low density polyethylene; toughness; processability

1. Introduction

Plastics modified by various inorganic fillers have been widely used, such as halogen-free flame-retarded polyolefins modified by Al (OH)$_3$, Mg (OH)$_2$ and others, polypropylene reinforced by talc, etc. [1–3]. However, when the inorganic filler loading reaches a certain level, the friction between the inorganic particles and the plastic matrix is exacerbated, the melt viscosity of thermoplastic composites is increased, and the processability is deteriorated [4–6]. Therefore, how to improve the processability of thermoplastic composites with high loading of inorganic fillers has attracted more and more attentions from academia, and especially from industry [7,8].

Adding processing aids is a usual method to improve the processability of thermoplastic composites [9,10]. As known, the linear polydimethylsiloxane (PDMS) is a good flexible polymer, and low molecular weight PDMS (silicone oil) is a good lubricant for plastics [11,12]. A small amount of polysiloxane with long chain alkyl side group obviously increased the melt flow rate and reduced the processing difficulty of thermoplastic composites [13]. Moreover, it was found that the toughness of plastics composites was enhanced with the increase of PDMS loading [14]. However, PDMS is a typical liquid polymer and cannot be conveniently used in industrial production, compared with powdered processing aids.

Thermotropic liquid crystal polymer (TLCP) possesses the excellent mechanical properties and lubrication for thermoplastic polymer melts [15,16]. During the melting process, the rigid TLCP macromolecules are easily oriented and arranged along the flow direction under the action of shear or tensile force, and ultimately improve the processability of thermoplastic composites [17–20].

However, for TLCP, the prerequisite for improving the processability of thermoplastic composites is that the liquid crystal phase transition temperature range is consistent with the processing temperature of thermoplastics [21]. Unfortunately, at present, the melting temperature of most main chain thermotropic liquid crystal polymers is quite high and is just comparable to that of the engineering plastics with high processing temperature [22], but is not well matched with the polyolefins with lower processing temperature [23]. Therefore, it has attracted more and more attention to reduce the melting point of the thermotropic liquid crystal polymer and further to expand its application in polyolefin-based thermoplastic composites with lower processing temperature [24]. Song et al. [25] prepared two series of wholly aromatic thermotropic copolyesters containing 2-(a-phenylisopropyl) hydroquinone (PIHQ) moiety and found that the presence of bulky substituent of PIHQ unit reduced the melting temperature of aromatic thermotropic copolyesters. Zhao et al. [26] prepared a new type of phosphorus-containing thermotropic liquid crystal copolyester with a flexible segment, and its liquid crystal phase existed in the range of 185 to 330 °C, which was matched with the processing temperature of most plastics. Xia et al. [23] synthesized a main chain liquid crystal copolyester with low melting point to match the processing temperature of polypropylene (PP), and it was found that the apparent melt viscosity of PP/TLCP composites was significantly reduced by the addition of the TLCP, compared with original PP. However, the main chain liquid crystal is mainly prepared by polycondensation reaction, and the preparation of large molecular weight polymers requires the high temperature, a complex purification process, and other special conditions. Accordingly, there are few main chain thermotropic liquid crystal polymers that are matched with polyolefins with the lower processing temperature. Moreover, the toughness of plastics composites is usually weakened by the addition of TLCP [27].

Polysiloxane grafted by thermotropic liquid crystal polymer (PSCTLCP) not only has the flexible PDMS segment as main chain, but also owns the liquid crystal performance of thermotropic liquid crystal polymer—although as the side chain [28–30]. The liquid crystal melting temperature of PSCTLCP was adjusted between 15 and 200 °C [31], and the isotropic temperature of PSCTLCP was changed from 84 [32] to 300 °C [33]. It indicates that the liquid crystal phase transition of PSCTLCP is well matched with the processing temperature of MH/LLDPE composites if a reasonably structured PSCTLCP is synthesized by the molecular design [34]. Besides, compared with PDMS, the powdery PSCTLCP is convenient to be used in industrial production. Therefore, PSCTLCP with reasonable structure it is likely to improve the processability and toughness of MH/LLDPE composites. However, to the authors' knowledge, there has hardly been any report on simultaneously improving the processability and toughness of MH/LLDPE composites by PSCTLCP.

In this paper, a PSCTLCP with lower melting temperature matched with the processing temperature of MH/LLDPE/PSCTLCP composites is synthesized and characterized, and the effect of PSCTLCP loading on the processability and tensile properties of MH/LLDPE/PSCTLCP composites is studied. Besides, the modified mechanism of PSCTLCP on MH/LLDPE/PSCTLCP composites is investigated at different processing temperatures, rotation speeds, and shear rates to guide the actual processing in industry. Therefore, the work provides the feasible scheme to design and synthesize the well-structured liquid crystal polymer to meet the special needs in the plastic processing.

2. Experiment

2.1. Materials

Methylparaben, allyl chloride, 4-ethylbenzoic acid, and hydroquinone were purchased from Sain Chemical Technology (Shanghai, China) Co., Ltd. Thionyl chloride, 4-dimethylaminopyridine (DMAP), triethylamine ((Et)$_3$N), and poly(methylhydrogeno)siloxane (PMHS, number-average molecular weight = 1900) were purchased from Aladdin Chemical Reagent Co., Ltd (Shanghai, China). Linear low-density polyethylene (LLDPE, DFDA 7042) was produced by Sinopec Guangzhou Co., Ltd (Guangzhou, China). Magnesium hydroxide (MH, HFR-30A) was produced by Shenzhen Hualin Chemical Co., Ltd. Pyridine, toluene, ethanol, acetone, N, N-dimethylformamide (DMF), tetrahydrofuran (THF) and methanol were purchased from Shenyang Chemical Co., Ltd (China).

2.2. Synthesis of 4-allyloxybenzoic Acid (AOBA)

The mixture of methylparaben (45.60 g, 0.3 mol), allyl chloride (30 mL, 0.37 mol), acetone (120 mL), and ethanol (30 mL) was heated to reflux temperature in a three-necked flask equipped with a magnetic stir bar and a condenser. Then, sodium hydroxide solution (50 wt %, 24 g) was slowly added into the above flask and the reaction lasted for 24 h at 50 °C. The reaction mixture was filtered by the vacuum suction and the solvent was removed by rotary evaporation. The residue is poured into sodium hydroxide solution (10 wt %, 180 mL) under stirring. After obtainment of a clear solution, dilute hydrochloric acid solution was slowly added until pH value of 4. The crude product was obtained by filtration, then washed with water several times, dried in a vacuum oven at 50 °C, further recrystallized from ethanol and vacuum dried again at 50 °C to obtain 4-allyloxybenzoic acid (AOBA) as white crystals. The synthesis route of AOBA is shown in Scheme 1.

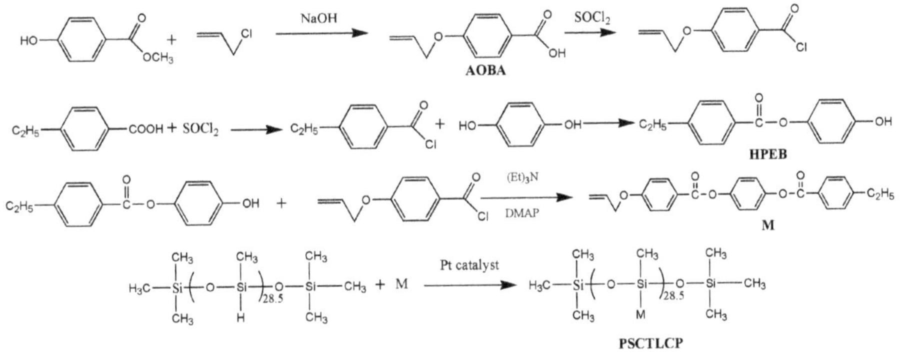

Scheme 1. Synthesis routes of AOBA, 4-hydroxyphenyl-4-ethylbenzoate (HPEB), 4-ethylbenzoic Acid-4-allyloxybenzoic Acid Hydroquinone Diester (M), PSCTLCP.

2.3. Synthesis of 4-allyloxybenzoyl Chloride

AOBA (4.59 g, 0.02 mol), thionyl chloride (20 mL, 0.27 mol), and 4 drops of DMF were added in a single-mouth flask with an absorption instrument of hydrogen chloride. The reaction was stirred for 2 h at room temperature, then for 4 h at 80 °C to ensure the complete reaction. Afterwards, excess SOCl$_2$ was distilled off under reduced pressure to obtain 4-allyloxybenzoyl chloride as pale yellow transparent liquid.

2.4. Synthesis of 4-hydroxyphenyl-4-ethylbenzoate (HPEB)

4-ethylbenzoic acid (50.0 g, 0.33 mol), thionyl chloride (50 mL, 0.68 mol), 4 drops of DMF were added into a one-necked flask with an absorption instrument of hydrogen chloride. The mixture was stirred for 5 h at room temperature, then the reaction was heated to reflux temperature and lasted for 2 h. The mixture was distilled under reduced pressure to obtain ethyl benzoyl chloride. Afterwards, hydroquinone (183.0 g, 1.66 mol) and pyridine (25 mL) were dissolved in THF (220 mL) to form a solution. The obtained ethyl benzoyl chloride was slowly added into the above solution in an ice bath under N_2 atmosphere. The reaction was carried out for 2 h in ice bath, then was heated to reflux temperature and lasted for 8 h. Some solvents were distilled under reduced pressure, and the concentrate was poured into a beaker charged of a large amount of water, precipitated, filtered, washed with hot water several times, recrystallized from ethanol and vacuum dried overnight at 50 °C to obtain 4-hydroxyphenyl-4'-ethylbenzoate (HPEB) as white powder, and the synthetic process as shown in Scheme 1.

2.5. Synthesis of 4-ethylbenzoic Acid-4-allyloxybenzoic Acid Hydroquinone Diester (M)

HPEB (5.20 g, 0.02 mol), triethylamine (8 mL, 0.06 mol), DMAP (0.13g, 0.11 mmol) and dry THF (50 mL) were placed into a 100 mL three-necked flask with a magnetic stir bar under N_2 atmosphere. The flask was equipped with a condenser in an ice bath. The obtained 4-allyloxybenzoyl chloride was slowly added into the above flask inside 30 min. Then, the reaction lasted for 24 h at room temperature. The reaction mixture was filtered, and the filtrate was poured into the acidified aqueous solution to obtain the precipitate. The precipitate was filtered and washed with several times with 5% potassium carbonate solution, deionized water and ethanol, respectively. The obtained solid was dried under vacuum at 50 °C for 24 h to obtain 4-ethylbenzoic acid-4'-allyloxybenzoic acid hydroquinone diester (M) as a white solid and the synthesis route is shown in Scheme 1.

2.6. Synthesis of PSCTLCP

M (5.04 g, 12.54 mmol) and PMHS (0.75 g, 0.04 mmol) were dissolved in toluene (50 mL), then 2.5 mL of $H_2PtCl_6·6H_2O$/isopropyl alcohol (0.5 g of hexachloroplatinic acid hydrate dissolved in 100 ml of isopropyl alcohol) was slowly added into the above solution. The reaction lasted for 72 h at 65 °C under nitrogen and was monitored by FT-IR until the disappearance of the sharp vibrational band at 2165 cm^{-1} assigned as the Si-H stretching. The coarse product was obtained by precipitation of the reaction mixture from methanol, and further purified by dissolving in chloroform, precipitating from methanol several times in order to remove the excessive unreacted monomers. Finally, white powdery polymer (PSCTLCP) was obtained and the synthesis route as shown in Scheme 1.

2.7. Preparation of MH/LLDPE/PSCTLCP Composites

MH and LLDPE were dried in vacuum oven at 80 °C for 12 h prior to processing. The pre-mixed blend of LLDPE, MH and PSCTLCP was conducted using a mixer (HAAKE, RS600, Germany) at a temperature of 200 °C and screw speed of 50 rpm for 5 min. After being pulverized and dried, the well-mixed composite was injection-molded into tensile testing bars using a mini-injection system (Thermo Scientific, USA) at melt temperature of 210 °C and mold temperature of 30 °C. The compositions and code names of the samples were shown in Table 1.

Table 1. Compositions and code names of the samples.

Samples	LLDPE (g)	MH (g)	PSCTLCP (g)
MH/LLDPE	40	60	-
PSCTLCP-0.6	40	60	0.6
PSCTLCP-1.0	40	60	1.0
PSCTLCP-3.0	40	60	3.0
PSCTLCP-5.0	40	60	5.0

2.8. Characterization

2.8.1. Structural Characterization

The structure of AOBA, HPEB, M, PSCTLCP was characterized by Fourier transform infrared spectra (FT-IR) and ^{1}H-NMR spectra. The FT-IR measurements were performed from 400 to 4000 cm^{-1} using a VERTEX70 spectrometer (Bruker, Karlsruhe, Germany) with KBr pellets. ^{1}H-NMR measurements were recorded on a DMX-400 spectrometer (Bruker, Germany) with CDCl$_3$ as the solvent and tetramethylsilane (TMS) as an internal standard.

2.8.2. Differential Scanning Calorimetry (DSC)

Differential scanning calorimetry (DSC) measurements were performed using a Pekin-Elmer Diamond DSC (USA) at 10 °C/min from 30 to 300 °C in a nitrogen atmosphere. About 5 mg of each sample was sealed in an aluminum pan.

2.8.3. Polarized Light Optical Microscopy (POM)

The optical texture of PSCTLCP was examined by a polarized light optical microscopy (POM) (Orthoglan, Leitz, Germany) equipped with a (STC200C, INTEC, USA) hot-stage.

2.8.4. Thermogravimetric (TG)

Thermogravimetric analysis (TGA) was performed using a TG 209 F1 instrument (NETZSCH, Germany) at 10 °C/min from 40 to 800 °C in a nitrogen atmosphere. The samples (10 ± 1 mg) were placed in open platinum pans.

2.8.5. Torque Rheometer

The balance torque of MH/LLDPE/PSCTLCP composite melt was determined by Haake torque XSS-300 rheometer (Thermo, Germany) at a rotation speed from 10 to 50 rpm. The test temperature was set at 170, 180, 195, 210, and 220 °C, respectively.

2.8.6. High Pressure Capillary Rheometer

The rheological properties were measured by a high-pressure capillary rheometer (Rheologic 5000, Ceast, Italy) with a length-to-diameter ratio of 30/1. The tests were performed in a shear rate range from 10 to 5000 s^{-1} at 185, 195, 205, 215, and 225 °C, respectively.

2.8.7. Scanning Electron Microscope (SEM)

Morphology and structure of MH/LLDPE/PSCTLCP composites were observed by scanning electron microscope (SEM, JEOL JSM-5900LV, Japan) equipped with an energy dispersive X-ray spectrometer (EDS, Oxford Isis, UK). The samples were fractured in liquid nitrogen, and the fracture surfaces were coated with gold to prevent charging on the surface.

2.8.8. Tensile Testing

Tensile properties of MH/LLDPE/PSCTLCP composites were measured by an Instron 5967 model materials testing system (USA) according to ASTM D-638 standard. Samples of tension test were dumbbell-shaped and the direction of the tensile force was parallel to the length of samples.

3. Results and Discussion

3.1. Characterization of PSCTLCP

Figure 1a shows the FT-IR spectra of AOBA, HPEB, M, PSCTLCP. Compared with those of AOBA and HPEB, the broad and strong peak of –COOH disappears at 2550 to 3300 cm^{-1}, the strong peak of -OH also disappears at 3500 cm^{-1} in FT-IR spectrum of M, meanwhile there is still the stretching vibration peak of CH$_2$=CH– at 1643 cm^{-1} and the absorption peak of ester group at 1745 cm^{-1}. The results indicate that M is successfully synthesized. For FT-IR spectrum of PSCTLCP, the stretching vibration peak of Si-H disappears at 2174 cm^{-1}, the attribute peak of PMHS, and the stretching vibration peak of CH$_2$=CH– is not found at 1643 cm^{-1} existed in FTIR spectrum of M. Besides, the main characteristic peak of Si-O-Si appears at 1000 to 1120 cm^{-1} with strong and wide bands. All the results indicate that the hydrosilylation reaction has taken place between PMHS and M [35].

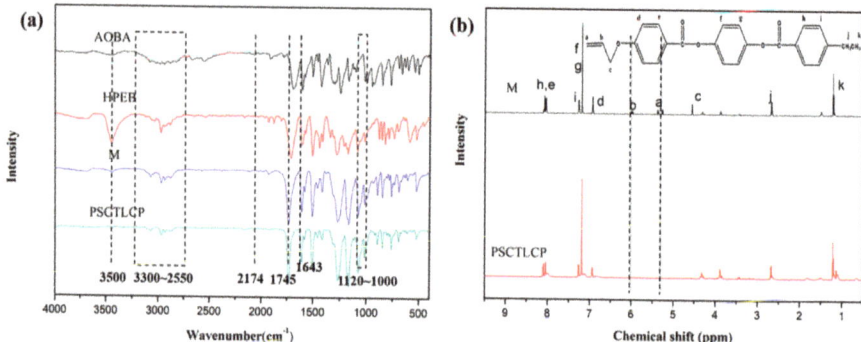

Figure 1. (a) FT-IR spectra of AOBA, HPEB, M, PSCTLCP, (b) ^{1}H NMR spectra of M and PSCTLCP in CDCl$_3$.

Figure 1b shows ^{1}H NMR spectra of M and PSCTLCP. The representative signals of the vinyl group in M at 5.28 and 6.0 ppm are not found in ^{1}H NMR spectrum of PSCTLCP, which further indicates that the excessive M is completely removed and PSCTLCP is successfully obtained [36].

DSC experiment is used to examine the phase transition of PSCTLCP. The heating curve of PSCTLCP and the related data are showed in Figure 2 and Table 2, respectively. It is seen that the glass transition temperature (T_g) of PSCTLCP is 34.5 °C. Furthermore, two heat absorption peaks appear in Figure 2. The temperature corresponding to the top value of the larger endothermic peak is 185.2 °C, that is the melt temperature (T_m) of PSCTLCP. Another one presents smaller and appears at 277.7 °C, related to the liquid crystal phase change from anisotropic to isotropic state, and that is the isotropic temperature (T_i) of PSCTLCP [37]. Besides, the initial melting temperature ($T_{initial}$) PSCTLCP is 173.6 °C.

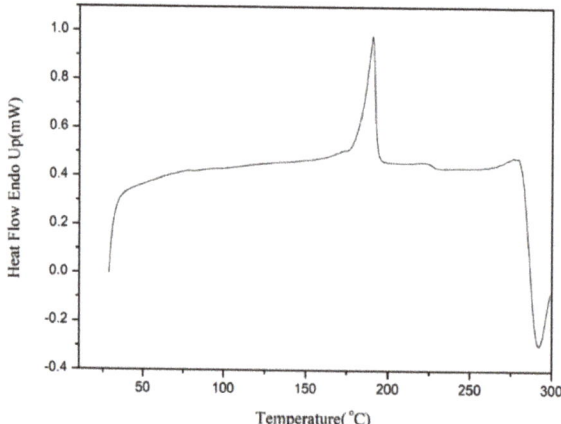

Figure 2. DSC curve of PSCTLCP.

Table 2. DSC data of PSCTLCP in nitrogen atmosphere.

T_g (°C)	$T_{initial}$ (°C)	T_m (°C)	T_i (°C)	ΔT (°C)
34.5	173.6	185.2	277.7	92.5

The polarizing microscope (POM) with a hot stage is used to visually observe the liquid-crystalline transition and optical textures, and Figure 3 shows the POM images at different temperatures. It is found that PSCTLCP begins to transfer from the solid state to liquid state at 170 °C, very close to $T_{initial}$ from DSC test, and the nematic texture of polymer is observed in Figure 3a. The higher temperature, the more obvious texture morphology and color are observed [38]. In Figure 3b, the texture of marbled nematic liquid crystal is seen in bright field at 210 °C. When the temperature is continuously raised to 275 °C, the field of vision is gradually darkened and the liquid crystal phase gradually disappears, indicating PSCTLCP transfers from the anisotropic liquid to isotropic transparent liquid and T_i is about 275 °C, in line with the result from DSC experiment.

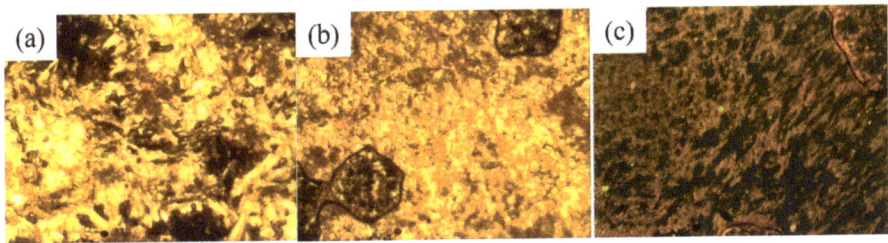

Figure 3. The texture images of PSCTLCP observed by POM at (a) 170, (b) 210, and (c) 275 °C.

Figure 4 shows the thermal decomposition process of PSCTLCP under nitrogen atmosphere. It is found that T_5 (decomposition temperature at 5 wt % mass loss) and T_{50} (at 50 wt % mass loss) are 279.6 and 429.8 °C, respectively, indicating the quite good thermal stability of obtained PSCTLCP.

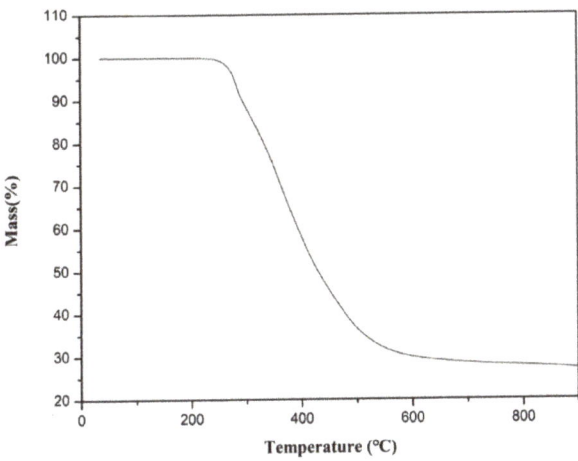

Figure 4. TGA curve of PSCTLCP.

As we know, the processing temperature of MH/LLDPE composites is usually between 160~250 °C, which is well matched with the melt temperature (T_m of 185.2 °C), the isotropic temperature (T_i of 277.7 °C) and thermal decomposition temperature (T_5 of 279.6 °C). Therefore, the obtained PSCTLCP is suitably used as a processing aid to improve the processability of MH/LLDPE composites.

3.2. Processability of MH/LLDPE/PSCTLCP Composites

Torque rheometer is an important instrument to characterize the processability of polymer materials. Figure 5 shows the evolution curves of melt torque values of MH/LLDPE/PSCTLCP composites at 170 °C and the rotation speed of 50 rpm. The balance torque of the MH/LLDPE/PSCTLCP composites is gradually decreased with the increase of PSCTLCP content. Compared with that of MH/LLDPE composite without PSCTLCP (16.59 N·m), the balance torque of MH/LLDPE/PSCTLCP composites is decreased by 15% (14.13 N·m), 18% (13.69 N·m), 29% (11.86 N·m), and 42% (9.67 N·m), respectively, for the samples containing PSCTLCP of 0.6, 1, 3, 5 wt %, respectively. The results indicate that PSCTLCP significantly reduces the melt viscosity and enhances the processability of the composites. After melting, PSCTLCP enters a liquid-like "liquid crystal state" and is easily oriented into microfibers in the shear direction during processing [39], and is regarded as a lubricant, which can reduce the entanglement between LLDPE molecular chains and the aggregation between MH particles, thus reducing the melt viscosity and improving the processability of MH/LLDPE/PSCTLCP composites [40].

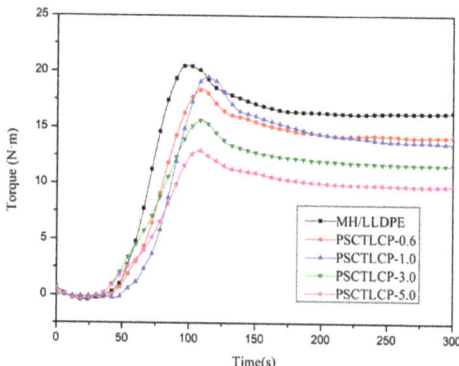

Figure 5. Torque versus time of MH/LLDPE/PSCTLCP composites with various PSCTLCP loading.

In order to investigate the effect of the external conditions, such as stress and temperature, on the usability of PSCTLCP, samples PSCTLCP-1.0 and MH/LLDPE are selected as a partner to be compared the melt torque at different temperatures and rotation speeds. The relationship between balance torque of composites and rotation speed at 195 °C is showed in Figure 6a. It is seen that the balance torque of PSCTLCP-1.0 is higher than that of MH/LLDPE at lower rotation speed, while the former is significantly smaller at higher rotation speed, and their torque difference gradually becomes bigger as the increasement of rotation speed. It is deduced that the molecules of PSCTLCP are oriented in the shear direction during the melt processing when the enough shear force exists. The larger the shear force, the easier the orientation of PSCTLCP and the better the processability [21–23]. Figure 6b shows the balance torque vs. temperature curves for MH/LLDPE and PSCTLCP-1.0 at 50 rpm. With the increasing temperature, the balance torque of MH/LLDPE and PSCTLCP-1.0 is significantly decreased. And the balance torque of PSCTLCP-0.1 is significantly lower than that of MH/LLDPE at the same temperature, indicating that the processability of PSCTLCP-1.0 is superior to that of MH/LLDPE in the whole processing temperature range. When the temperature rises from 170 to 220 °C, the balance torque of MH/LLDPE and PSCTLCP-1.0 is decreased by 32.7% and 18.7%, respectively. It indicates that the processing sensitivity of MH/LLDPE composites to the temperature is reduced by the addition of PSCTLCP.

The above results show that the processability of MH/LLDPE/PSCTLCP composites is not only affected by PSCTLCP content, but also by the rotation speed (shear force) and processing temperature, which is attributed to the orientation of PSCTLCP during the processing [41].

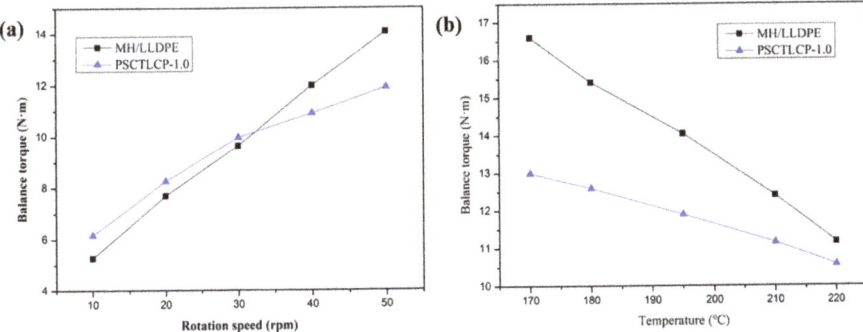

Figure 6. (a) Balance torque versus rotation speed and (b) balance torque versus processing temperature.

In addition, the capillary rheometer is also used to characterize the processability of MH/LLDPE/PSCTLCP composites. Figure 7 shows the relationship between apparent viscosity of the MH/LLDPE/PSCTLCP composites and shear rate, and between shear stress and shear rate at 195 °C. It is seen that the apparent viscosity and shear stress of composites are all gradually decreased with the increase of PSCTLCP content in the experimental range of shear rate, which further proves that the processability of MH/LLDPE/PSCTLCP composites is significantly improved by the addition of PSCTLCP [42].

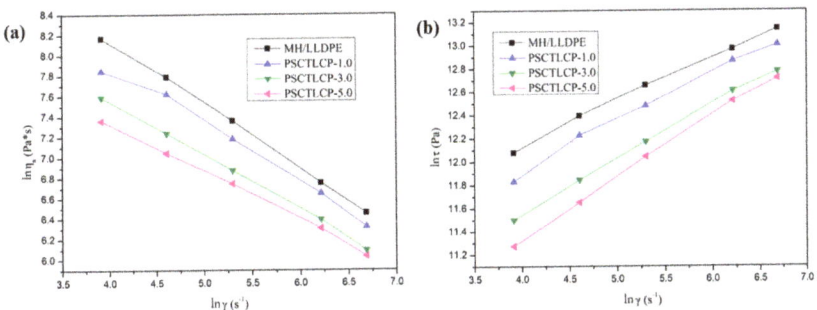

Figure 7. (a) Apparent viscosity, η_a, versus shear rate, γ, temperature: 195 °C; (b) shear stress, τ, versus shear rate, γ, temperature: 195 °C.

Sample PSCTLCP-1.0 is selected to further exhibit the processing temperature on the rheological property. The logarithmic plots of shear stress (τ) versus shear rate (γ) for PSCTLCP-1.0 at 185, 205 and 225 °C are shown in Figure 8a. The logarithm of shear stress is linearly related to that of shear rate, suggesting that the melt flow of PSCTLCP-1.0 follows the power-law equation under experimental conditions [42]. According to the power-law equation ($\tau = K \cdot \gamma^n$), the power law index (n) of MH/LLDPE/PSCTLCP composites with different PSCTLCP loadings is calculated. As shown in Figure 8b, all the n values of MH/LLDPE/PSCTLCP composites are less than 1, it means that the melt of MH/LLDPE/PSCTLCP composites behaves a pseudoplastic fluid. Furthermore, the n value of MH/LLDPE/PSCTLCP composites is increased from 0.38 to 0.53 when the content of PSCTLCP is increased from 0 to 5 wt %. It indicates that the non-Newtonian property of MH/LLDPE/PSCTLCP composites is weakened with the increasing of PSCTLCP loading. The results show that the sensitivity of apparent melt viscosity for composites to shear rate is decreased and the processing window is broadened with the increase of PSCTLCP.

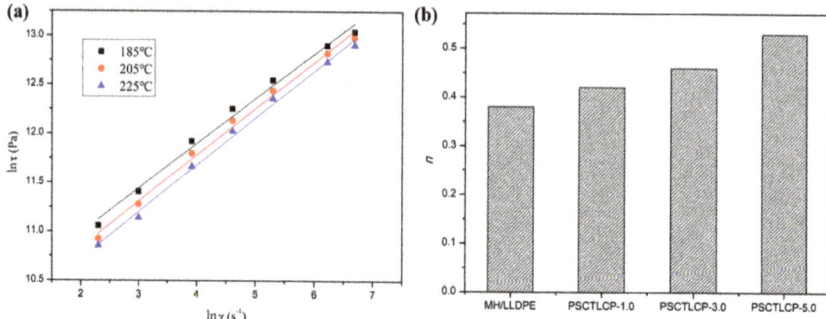

Figure 8. (a) Shear stress versus shear rate of PSCTLCP-1.0, (b) n values of MH/LLDPE/PSCTLCP composites.

According to the Arrhenius equation ($\eta_a = A\exp(E/RT)$), the activation energy (E) of the composite melt is calculated by means of linear regression [6,7,43,44]. Figure 9a,b show the relationship between the logarithm of apparent viscosity ($\ln\eta_a$) and the reciprocal of absolute temperature (1/T) for MH/LLDPE and PSCTLCP-1.0, respectively. $\ln\eta_a$ is increased with 1/T, indicating that the viscosity of the composite melt is gradually decreased with the increase of the processing temperature. With the increase of temperature, the mobility of the molecular chains is increased, and the interaction force between the molecular chains is reduced, thereby the viscosity of the composite is reduced and the fluidity of the composite melt is improved [43]. Furthermore, the values of E and the linear correlation coefficient (R_0) are obtained in Table 3. It is found that E values of both MH/LLDPE and PSCTLCP-1.0 are continuously decreased with the increase of shear rate, which is attributed to the disentanglement between molecular chains at high shear rate [44,45]. On the other hand, the E value of PSCTLCP-1.0 is lower than that of MH/LLDPE at the same shear rate, indicating that PSCTLCP-1.0 melt is easier to flow compared to MH/LLDPE melt, and only 1 wt % PSCTLCP is obvious to improving the processability of MH/LLDPE composites.

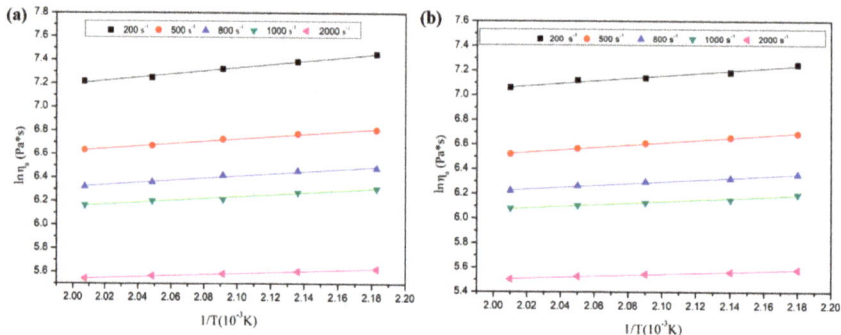

Figure 9. The logarithm of apparent viscosity, η_a, versus 1/T, of (a) MH/LLDPE and (b) PSCTLCP-1.0 at different shear rates.

Table 3. Values of E and R_0 of MH/LLDPE and PSCTLCP-1.0 at different shear rates.

γ (s^{-1})	200	500	800	1000	2000
$E_{\text{MH/LLDPE}}$ (kJ/mol)	11.50	8.50	7.81	6.88	3.95
$E_{\text{PSCTLCP-1.0}}$ (kJ/mol)	8.34	8.03	6.20	5.28	3.54
$R_{0\text{MH/LLDPE}}$	0.9919	0.9924	0.9758	0.9759	0.9977
$R_{0\text{PSCTLCP-1.0}}$	0.9762	0.9941	0.9922	0.9819	0.9899

3.3. Facture Morphology of MH/LLDPE/PSCTLCP Composites

SEM test is used to study the effect of PSCTLCP on the facture morphology of MH/LLDPE/PSCTLCP composites, and the results are shown in Figure 10. According to the basic structure of composite materials [4,19], LLDPE is the continuous phase, MH is the dispersed phase and PSCTLCP should be located at the surface between MH and LLDPE. Some labels are used to clearly show MH and PSCTLCP in Figure 10. Moreover, the mapping scattering images of MH particles and PSCTLCP fibers in PSCTLCP-1.0 are shown in Figure 10d–f. Due to the poor compatibility between PSCTLCP and MH/LLDPE composites, liquid crystal fibers are pulled out from LLDPE matrix and thereby some holes are formed in fracture surface, as shown in Figure 10b,c. It is seen that the number of holes is increased with increasing PSCTLCP loading. The more liquid crystal fibers are formed, the better processability of composites [18–20]. The results further explain that the processability of MH/LLDPE/PSCTLCP composites is significantly improved by the addition of PSCTLCP.

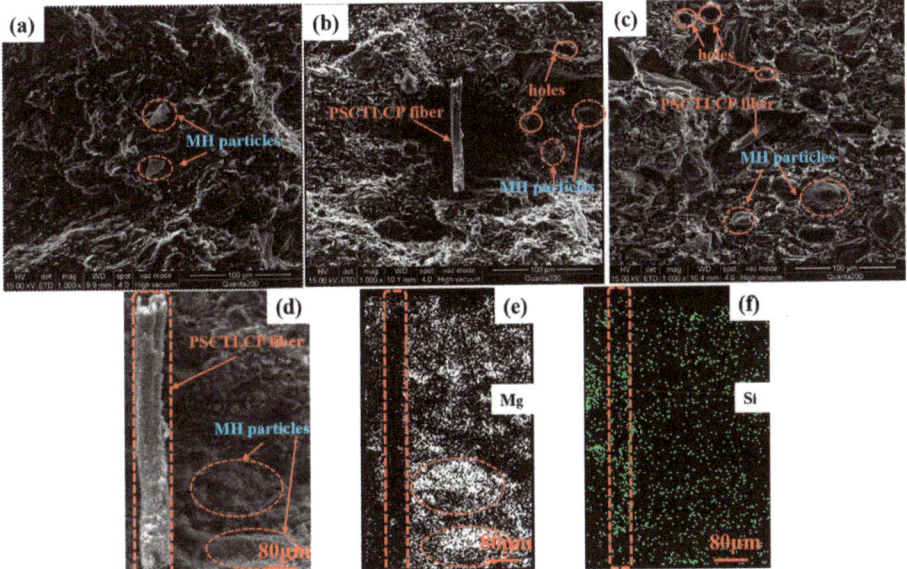

Figure 10. SEM images of (**a**) MH/LLDEP, (**b**) PSCTLCP-1.0, and (**c**) PSCTLCP-3.0. (**d**) Is a magnified image from (**b**), and elemental mapping images of (**e**) Mg, (**f**) Si.

3.4. Mechanical Properties of MH/LLDPE/PSCTLCP Composites

The tensile properties of MH/LLDPE/PSCTLCP composites are shown in Figure 11. Due to the flexibility of the main chains of PSCTLCP molecules, the tensile strength and tensile modulus of composites are slightly decreased with PSCTLCP loading. Meanwhile, the elongation at break of composites is significantly increased, as shown in Figure 11c. Compared with 6.85% of MH/LLDPE composite without PSCTLCP, the elongation at break of MH/LLDPE/PSCTLCP composites is increased to 10.68%, 15.8%, and 17.66%, respectively, and the corresponding loading of PSCTLCP is 0.6, 1, 3 wt %, respectively. The results mean the toughness of MH/LLDPE/PSCTLCP composites is enhanced by the addition of PSCTLCP, most likely due to the polysiloxane structure in the macromolecular chains of PSCTLCP, similar to the toughness enhancement of composites by the addition of polydimethylsiloxane [14].

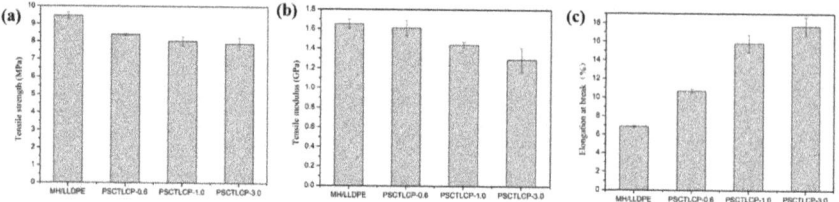

Figure 11. Variation of (**a**) tensile strength, (**b**) tensile modulus, and (**c**) elongation at break with the PSCTLCP content.

4. Conclusions

In order to match the processing temperature and improve the processability of MH/LLDPE composites, polysiloxane grafted by thermotropic liquid crystal polymer (PSCTLCP) was successfully synthesized. The balance melt torque of MH/LLDPE/PSCTLCP composites was obviously decreased with the loading of PSCTLCP and that of the composite (5.0 wt % loading) was decreased from 16.59 N·m of the baseline MH/LLDPE to 9.67 N·m. Besides, PSCTLCP weakened the non-Newtonian property and decreased the flowing activation energy of MH/LLDPE/PSCTLCP composites, and thus broadened the processing window and improved the processability. Considering the flexible polysiloxane as main chains of PSCTLCP, the elongation at break of MH/LLDPE composites was significantly increased, and the tensile strength and modulus were slightly decreased with PSCTLCP loading.

Author Contributions: Conceptualization, X.G., X.L. and J.Z.; data curation, X.G.; formal analysis, X.G.; funding acquisition, J.Z.; investigation, X.G.; methodology, B.C. and J.C.; project administration, S.L.; resources, S.L.; software, Z.Y. and X.L.; supervision, H.H. and J.Z.; writing—original draft, X.G.; writing—review and editing, J.Z. All authors have read and agreed to the published version of the manuscript.

Funding: We are grateful to the National Natural Science Foundation of China (No. 51573054) and the Project for Science and Technology of Guangdong (No. 2019B010940001) for financial support.

Conflicts of Interest: The authors declare no conflict of interest.

Abbreviations

PSCTLCP	polysiloxane grafted by thermotropic liquid crystal polymer;
MH	magnesium hydroxide;
LLDPE	linear low-density polyethylene;
PDMS	polydimethylsiloxane;
TLCP	Thermotropic liquid crystal polymer;
PIHQ	2-(a-phenylisopropyl) hydroquinone;
AOBA	4-allyloxybenzoic acid;
HPEB	4-hydroxyphenyl-4'-ethylbenzoate;
M	4-ethylbenzoic acid-4-allyloxybenzoic acid hydroquinone diester;
DSC	Differential scanning calorimetry;
POM	Polarized light optical microscopy;
TG	Thermogravimetric;
SEM	Scanning electron microscope;
PMHS	poly(methylhydrogeno)siloxane
DMAP	4-dimethylaminopyridine;
$(Et)_3N$	triethylamine;
DMF	N, N-dimethylformamide;
THF	tetrahydrofuran.

References

1. Kartik, B.; Mithilgsh, Y.; Fang, C.C.; Kyong, Y.R. Graphene Nanoplatelet-Reinforced Poly(vinylidene fluoride)/High Density Polyethylene Blend-Based Nanocomposites with Enhanced Thermal and Electrical Properties. *Nanomaterials* **2019**, *9*, 361.
2. Liu, C.; Chan, K.; Shen, J. Polyetheretherketone Hybrid Composites with Bioactive Nanohydroxyapatite and Multiwalled Carbon Nanotube Fillers. *Polymers* **2016**, *8*, 425. [CrossRef] [PubMed]
3. Kartik, B.; Yen, H.C.; Mithilesh, Y.; Fang, C.C. Enhanced thermal stability, toughness, and electrical conductivity of carbon nanotube-reinforced biodegradable poly(lactic acid)/poly(ethylene oxide) blend-based nanocomposites. *Polymer* **2020**, *186*, 122002.
4. Shen, L.; Li, J.; Li, R. A new strategy to produce low-density polyethylene (LDPE)-based composites simultaneously with high flame retardancy and high mechanical properties. *Appl. Surf. Sci.* **2017**, *437*, 75–81. [CrossRef]
5. Shen, L.; Shao, C.; Li, R. Preparation and characterization of ethylene-vinyl acetate copolymer (EVA)-magnesium hydroxide (MH)-hexaphenoxycyclotriphosphazene (HPCTP) composite flame-retardant materials. *Polym. Bull.* **2018**, *76*, 2399–2410. [CrossRef]
6. Liang, J.Z.; Yang, J.; Tang, C.Y. Melt shear viscosity of PP/Al(OH)$_3$/Mg(OH)$_2$ flame retardant composites at high extrusion rates. *J. Appl. Polym. Sci.* **2011**, *119*, 1835–1841. [CrossRef]
7. Yang, J.; Liang, J.Z.; Tang, C.Y. Studies on melt flow properties during capillary extrusion of PP/Al(OH)$_3$/Mg(OH)$_2$ flame retardant composites. *Polym. Test.* **2009**, *28*, 907–911. [CrossRef]
8. Liang, J.Z.; Tang, C.Y.; Zhang, Y.J. Melt density and volume flow rate of polypropylene/Al(OH)$_3$/Mg(OH)$_2$ flame retardant composites. *J. Appl. Polym. Sci.* **2010**, *118*, 331–337. [CrossRef]
9. Cardelli, A.; Ruggeri, G.; Calderisi, M. Effects of poly(dimethylsiloxane) and inorganic fillers in halogen free flame retardant poly(ethylene-co-vinyl acetate) compound: A chemometric approach. *Polym. Degrad. Stabil.* **2012**, *97*, 2536–2544. [CrossRef]
10. Zhou, W.; Osby, J. Siloxane modification of polycarbonate for superior flow and impact toughness. *Polymer* **2010**, *51*, 1990–1999. [CrossRef]
11. Zolper, T.J.; Seyam, A.; Li, Z. Friction and Wear Protection Performance of Synthetic Siloxane Lubricants. *Tribol. Lett.* **2013**, *51*, 365–376. [CrossRef]
12. Kokuti, Z.; Gruijthuijsen, K.V.; Jenei, M. High-frequency rheology of a high viscosity silicone oil using diffusing wave spectroscopy. *Appl. Rheol.* **2014**, *24*, 63984.
13. He, H.M.; Gao, L.; Yang, X.J. Studies on the superhydrophobic properties of polypropylene/polydimethylsiloxane/graphite fluoride composites. *J. Fluorine Chem.* **2013**, *156*, 158–163. [CrossRef]
14. Chen, X.; Yu, J.; Guo, S. Structure and properties of polypropylene composites filled with magnesium hydroxide. *J. Appl. Polym. Sci.* **2006**, *102*, 4943–4951. [CrossRef]
15. Cheng, H.K.F.; Basu, T.; Sahoo, N.G.; Li, L.; Chan, S.H. Current Advances in the Carbon Nanotube/Thermotropic Main-Chain Liquid Crystalline Polymer Nanocomposites and Their Blends. *Polymers* **2012**, *4*, 889–912. [CrossRef]
16. Zeng, L.; Li, R.; Chen, P. Synthesis and characterization of thermotropic liquid crystalline polyarylate with ether ether ketone segments in the main chain. *J. Appl. Polym. Sci.* **2016**, *133*, 43800. [CrossRef]
17. Ren, C.; Gao, P. Synthesis and characterization of thermotropic liquid crystalline copolyester/multi-walled carbon nanotubes composites via in situ polymerization. *Polymer* **2012**, *53*, 3958–3967. [CrossRef]
18. Garcia, M.; Gonzalez, N.; Eguiazabal, J.I. Structure and mechanical properties of new hybrid composites based on polyamide 6,6 reinforced with both glass fibers and a semiaromatic liquid crystalline polyester. *Polym. Compos.* **2004**, *25*, 601–608. [CrossRef]
19. Pisharath, S.; Wong, S.C. Processability of LCP-nylon-glass hybrid composites. *Polym. Compos.* **2003**, *24*, 109–118. [CrossRef]
20. Wu, L.; Chen, P.; Chen, J. Noticeable viscosity reduction of polycarbonate melts caused jointly by nano-silica filling and TLCP fibrillation. *Polym. Eng. Sci.* **2007**, *47*, 757–764. [CrossRef]

21. Kalkar, A.K.; Deshpande, V.D.; Kulkarni, M.J. Nonisothermal crystallization kinetics of poly (phenylene sulphide) in composites with a liquid crystalline polymer. *J. Polym. Sci. Pol. Phys.* **2010**, *48*, 1070–1100. [CrossRef]
22. Yang, R.; Chen, L.; Zhang, W.Q. In situ reinforced and flame-retarded polycarbonate by a novel phosphorus-containing thermotropic liquid crystalline copolyester. *Polymer* **2011**, *52*, 4150–4157. [CrossRef]
23. Xia, Y.; Zhang, H.; Wang, Q. Study on in situ reinforced composites of thermoplastic resins: A novel TLCP with low melting temperature. *J. Thermoplast. Compos.* **2016**, *29*, 37–47. [CrossRef]
24. Antoun, S.; Lenz, R.W.; Jin, J.I. Liquid crystal polymers. IV. Thermotropic polyesters with flexible spacers in the main chain. *J. Polym. Sci. Pol. Chem.* **1981**, *19*, 1901–1920. [CrossRef]
25. Song, J.Y. Synthesis and characterization of a series of wholly aromatic copolyesters containing 2-(α-phenylisopropyl) hydroquinone moiety. *J. Polym. Sci. Pol. Chem.* **1999**, *37*, 881–889. [CrossRef]
26. Zhao, C.S.; Chen, L.; Wang, Y.Z. A phosphorus-containing thermotropic liquid crystalline copolyester with low mesophase temperature and high flame retardance. *J. Polym. Sci. Pol. Chem.* **2008**, *46*, 5752–5759. [CrossRef]
27. Tjong, S.C.; Meng, Y.Z. Morphology and mechanical characteristics of compatibilized polyamide 6-liquid crystalline polymer composites. *Polymer* **1997**, *38*, 4609–4615. [CrossRef]
28. Laurence, N.; Hakima, M.J. Richness of Side-Chain Liquid-Crystal Polymers: From Isotropic Phase towards the Identification of Neglected Solid-Like Properties in Liquids. *Polymers* **2012**, *4*, 1109–1124.
29. Zhang, L.; Yao, W.; Gao, Y. Polysiloxane-Based Side Chain Liquid Crystal Polymers: From Synthesis to Structure-Phase Transition Behavior Relationships. *Polymers* **2018**, *10*, 794. [CrossRef]
30. Chen, X.F.; Shen, Z.; Wan, X.H. Mesogen-jacketed liquid crystalline polymers. *Chem. Soc. Rev.* **2010**, *39*, 3072. [CrossRef]
31. Apfel, M.A.; Finkelmann, H. Synthesis and properties of high-temperature mesomorphic polysiloxane (MEPSIL) solvents: Biphenyl- and terphenyl-based nematic systems. *Anal. Chem.* **1985**, *57*, 651–658. [CrossRef]
32. Jin, Y.; Fu, R.; Guan, Z. Low-temperature side-chain liquid crystalline polysiloxanes used as stationary phases for capillary gas chromatography. *J Chromatogr. A* **1989**, *483*, 394–400. [CrossRef]
33. Finkelmann, H.; Kock, H.J. Investigations on liquid crystalline polysiloxanes 3. Liquid crystalline elastomers-a new type of liquid crystalline material. *Makromol. Rapid Commun.* **1981**, *2*, 317–322. [CrossRef]
34. Dubois, J.C.; Barny, P.L.; Mauzac, M. Behavior and properties of side chain thermotropic liquid crystal polymers. *Acta Polym. Sin.* **1997**, *48*, 47–87. [CrossRef]
35. Hu, J.S.; Zhang, B.Y.; Zhou, A.J. Side-chain cholesteric liquid crystalline elastomers derived from a mesogenic crosslinking agent: I. Synthesis and mesomorphic properties. *Eur. Polym. J.* **2006**, *42*, 2849–2858. [CrossRef]
36. Yao, W.; Gao, Y.; Zhang, C. A series of novel side chain liquid crystalline polysiloxanes containing cyano- and cholesterol-terminated substituents: Where will the structure-dependence of terminal behavior of the side chain reappear? *J. Polym. Sci. Pol. Chem.* **2017**, *55*, 1765–1772. [CrossRef]
37. Wang, G.; Xiong, Y.; Tang, H. Synthesis and characterization of a graft side-chain liquid crystalline polysiloxane. *J. Organomet. Chem.* **2015**, *775*, 50–54. [CrossRef]
38. Zhang, Y.; He, X.Z.; Zheng, J.J. Side-chain cholesteric liquid-crystalline elastomers containing azobenzene derivative as cross-linking agent-synthesis and characterisation. *Liq. Cryst.* **2018**, *45*, 1–12. [CrossRef]
39. Malik, T.M.; Carreau, P.J.; Chapleau, N. Characterization of liquid crystalline polyester polycarbonate blends. *Polym. Eng. Sci.* **1989**, *29*, 600–608. [CrossRef]
40. Chang, J.H.; Choi, B.K.; Kim, J.H. The Effect of Composition on Thermal, Mechanical, and Morphological Properties of Thermotropic Liquid Crystalline Polyester with Alkyl Side-Group and Polycarbonate Blends. *Polym. Eng. Sci.* **2010**, *37*, 1564–1571. [CrossRef]
41. Chan, H.S.; Leng, Y.; Gao, F. Processing of PC/LCP in situ composites by closed-loop injection molding. *Compos. Sci. Technol.* **2002**, *62*, 757–765. [CrossRef]
42. Nanda, M.; Tripathy, D.K. Rheological behavior of chlorosulfonated polyethylene composites: Effect of filler and plasticizer. *J. Appl. Polym. Sci.* **2012**, *126*, 46–55. [CrossRef]
43. Thalib, S.; Huzni, S. The effect of particle compositions on the activation energy of the pa6/bagasse composite. *IOP Conf. Ser. Mater. Sci. Eng.* **2019**, *602*, 012086. [CrossRef]

44. Muksing, N.; Nithitanakul, M.; Grady, B.P. Melt rheology and extrudate swell of organobentonite-filled polypropylene nanocomposites. *Polym. Test.* **2008**, *27*, 470–479. [CrossRef]
45. Hemmati, M.; Rahimi, G.H.; Kaganj, A.B. Rheological and Mechanical Characterization of Multi-Walled Carbon Nanotubes/Polypropylene Nanocomposites. *J. Macromol. Sci. B.* **2008**, *47*, 1176–1187. [CrossRef]

© 2020 by the authors. Licensee MDPI, Basel, Switzerland. This article is an open access article distributed under the terms and conditions of the Creative Commons Attribution (CC BY) license (http://creativecommons.org/licenses/by/4.0/).

Article

TPV: A New Insight on the Rubber Morphology and Mechanic/Elastic Properties

Cindy Le Hel [1], Véronique Bounor-Legaré [1], Mathilde Catherin [1], Antoine Lucas [2], Anthony Thèvenon [2] and Philippe Cassagnau [1],*

[1] Univ-Lyon, Université Claude Bernard Lyon 1, Ingénierie des Matériaux Polymères, CNRS, UMR 5223, 15 Bd Latarjet, 69622 Villeurbanne CEDEX, France; cindy.le-hel@univ-lyon1.fr (C.L.H.); veronique.bounor-legare@univ-lyon1.fr (V.B.-L.); mathilde.catherin@univ-lyon1.fr (M.C.)

[2] Hutchinson, Centre de Recherche, Rue Gustave Nourry-B.P. 31, 45120-Chalette-sur-Loing, France; antoine-p.lucas@hutchinson.com (A.L.); anthony.thevenon@hutchinson.com (A.T.)

* Correspondence: philippe.cassagnau@univ-lyon1.fr

Received: 13 September 2020; Accepted: 7 October 2020; Published: 10 October 2020

Abstract: The objective of this work is to study the influence of the ratio between the elastomer (EPDM) phase and the thermoplastic phase (PP) in thermoplastic vulcanizates (TPVs) as well as the associated morphology of the compression set of the material. First, from a study of the literature, it is concluded that the rubber phase must be dispersed with a large distribution of the domain size in the thermoplastic phase in order to achieve a high concentration, i.e., a maximal packing fraction close to ~0.80. From this discussion, it is inferred that a certain degree of progress in the crosslinking reaction must be reached when the thermoplastic phase is melted during mixing in order to achieve dispersion of the elastomeric phase in the thermoplastic matrix under maximum stress. In terms of elasticity recovery which is measured from the compression set experiment, it is observed that the crosslinking agent nature (DCP or phenolic resin) has no influence in the case of a TPV compared with a pure crosslinked EPDM system. Then, the TPV morphology and the rubber phase concentration are the first order parameters in the compression set of TPVs. Finally, the addition of carbon black fillers leads to an improvement of the mechanical properties at break for the low PP concentration (20%). However, the localization of carbon black depends on the crosslinking chemistry nature. With radical chemistry by organic peroxide decomposition, carbon black is located at the interface of EPDM and PP acting as a compatibilizer.

Keywords: thermoplastic vulcanizates; morphology; compression set

1. Introduction

Thermoplastics vulcanizates (TPVs) are a family of polymer blends with a range of original and varied physical, mechanical, and processing properties. A TPV formulation generally consists of an elastomer phase, a thermoplastic phase, a crosslinking agent, inorganic fillers, plasticizers, stabilizers, and compatibilizing agents. The most developed TPVs to date are those based on polypropylene (PP) and a copolymer of ethylene-propylene-diene monomer (EPDM), but there are some examples in which the elastomer phase consists of another material (poly(butadiene-acrylonitrile), polyisoprene, brominated isoprene, silicone, etc.) [1].

The unique and specific character of TPVs is that the elastomer phase is crosslinked under flow during the mixing process. This leads to significant morphological changes, in particular phase inversion, due to the rapid and significant change in the viscoelasticity of the elastomer, from viscoelastic liquid to viscoelastic solid. The formation of an elastic rubber phase dispersed in the thermoplastic matrix allows the material to present its original properties. Finally, the mechanical properties of a TPV are close to those of a crosslinked rubber, while maintaining the processing characteristics of thermoplastics.

The properties of TPVs depend on various factors such as the formulation (polymers, fillers, plasticizers, crosslinking agent), the morphologies (dispersed and/or continuous phases), and the processing methods (batch and/or continuous processes). The final morphology of these reactive blends depends on the concentration and nature of the main constituents, the interface between the polymers, the crystallinity of the thermoplastic phase, the kinetics of crosslinking reaction, the presence of a compatibilizing agent, and so on. In order to control the morphology development throughout the elaboration process, several aspects must be considered: (i) the crosslinking chemistry of the elastomer phase and its consequences on rheology variation, (ii) the compatibility between phases taking into account the aspects of interfacial tension, (iii) the dispersion of low viscosity compounds (crosslinking agents, plasticizers) and the dispersion/localization of fillers in both phases, and (iv) the manufacturing process (batch mixing and/or continuous extrusion) with the associated viscous dissipation phenomena (self-heating).

Consequently, the morphology development in the TPV has attracted a lot of attention in the last decades and the mechanisms of formation of these rubber domains have been extensively studied from experimental [1,2] and theoretical [3] points of view. It is not possible to report and to exhaustively discuss all the papers published in this domain; however, some reviews are of interest [3,4]. The TPVs were developed to replace conventional elastomers, which once crosslinked can no longer be processed, consequently limiting their fields of applications and their recyclability. As pointed out before, TPVs combine both the processing properties of thermoplastics and the elasticity of crosslinked rubbers. In terms of application, the required main property is the elastic recovery, more commonly called compression set (Cs%). The compression set characterizes the ability of a material to recover its initial shape after being submitted to a constant strain. In the standard tests practiced in the industry, a strain of 25% is applied for several hours at a temperature higher than room temperature, in the ASTMD 395 standards (ISO 815-1) the material is compressed during 24 h at a temperature of 100 °C.

In terms of real industrial applications, the objective is to reach the lowest compression set value. Although we cannot obtain the compression set of the pure elastomeric phase due to the plastic deformation of the thermoplastic phase, the tendency is to reach compression set lower than 30%. It is clear that this property is governed by the formulation and the morphology developed during the TPV processing. It is generally believed that a fine dispersed rubber phase is the key to obtain good mechanical properties and high elasticity of TPV products [5]. However, the concentration in the elastomer phase must be as high as possible while keeping a continuous thermoplastic phase for the processing ability.

The question that must be asked then is: what is the highest concentration of the elastomer phase which can be dispersed in a thermoplastic matrix? It is actually common to observe in various publications and industrial TPVs that the elastomer phase concentration is higher than 60% and up to 80% [6,7]. The objective of this paper is to study the evolution of the morphology at this high level of crosslinked elastomer phase and, first, to explain how such a level of concentration can be reached. Next, we study the influence of the concentration of the elastomer phase on the recovery elasticity of these TPVs as well as the influence of the reinforcing fillers such as carbon black.

1.1. TPV: A Highly Filled Thermoplastic Polymer

In molten state, a TPV can be considered as a fluid with a very high solid phase content. Indeed, it can be assumed that once the TPV has been developed, the crosslinked elastomer domains behave as a weakly deformable phase in a shear thinning fluid. In fact, this rubber phase has a shear modulus greater than ~2.10^5 Pa and behaves as a viscoelastic solid in a suspending viscoelastic liquid. Theoretically, for suspensions, the maximal packing fraction Φ_m of randomly dispersed spherical particles is equal to 0.64. That is to say, if the elastomer domains were perfectly spherical, regardless of their size, a TPV could not contain more than 64 % vol of elastomer phase. However, as illustrated in

Figure 1, we can observe that this morphology is not spherical and that it has a very poorly defined geometrical shape which looks more like an asteroid (see Figure 1) than a planet. This aspect ratio, difficult to quantify precisely from electronic microscopy analysis, is favorable to increase the maximal packing fraction.

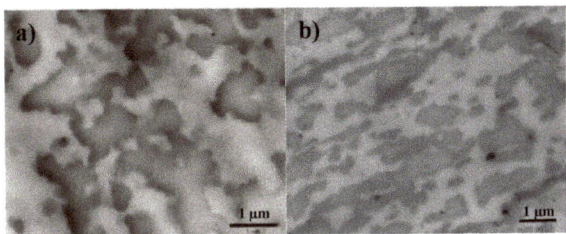

Figure 1. Examples of thermoplastic vulcanizate (TPV) morphology (crosslinked by Resol) from Transmission Electronic Microscopy (TEM) (**a**) PP/EPDM: (30/70) [2] (**b**) PP/EPDM (50/50) [8] with permission from Elsevier.

Actually, as it is well known in a suspension of spherical particles, the main strategy for increasing Φ_m is to control the particle size distribution. The shape of the particles as well as their aspect is of importance [9], but this influence is strongly dependent on the particles' structures. This aspect has been well-studied in the case of suspensions, see, for example, some articles on the rheology of concentrated suspensions of arbitrarily shaped factors [10].

To increase Φ_m for spherical particles, the solution adopted is a bimodal particles population and several models have been developed in this field [11,12]. Bimodal suspensions are characterized in terms of diameter or size ratio λ defined as the ratio between the largest particle diameter (D_L) to the finest particle diameter (D_f), $\lambda = D_L/D_f$ as well as by the volume ratio between each population $\xi = \Phi_L/\Phi_f$. Generally, it is preconized $\lambda \approx 8$ and $\xi = 0.9$.

A bimodal population can be only obtained in the case of theoretical model systems. However, a broad distribution of particle size combined with a more or less ellipsoid shape is obviously favorable to increase Φ_m. Indeed, on the TPV morphology depicted in Figure 1, a broad distribution of domains size from sub-micron scale to the micro-scale (10 µm) can be observed. Indeed, this specific morphology allows the TPV to reach a concentration in the elastomer phase close to 80% whereas the concentration $\Phi_m = 0.64$ cannot be exceeded for spherical particles (Figure 2).

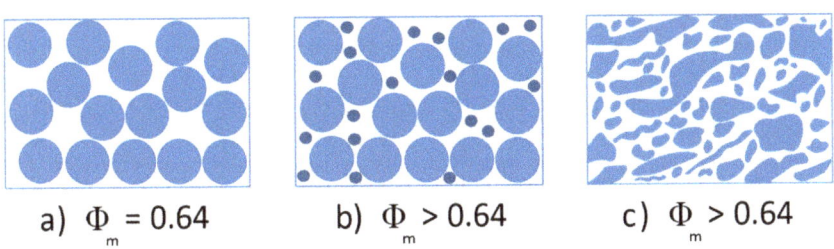

Figure 2. Scheme of different morphologies with the corresponding maximal packing fraction Φ_m that can be expected. (**a**) monodispersed spherical system, (**b**) bimodal spherical system, (**c**) broad distribution of ellipsoid shape system.

Finally, do these morphologies information then give us the conditions to elaborate a TPV with the largest possible Φ_m i.e., with the smallest thermoplastic phase concentration?

1.2. Morphology Development

First, it must be pointed out that most of the time, TPV formulations contain a large amount of oil which can exceed 50 wt %. However, regarding TPVs based on EPDM/PP, this oil is homogeneously distributed between the thermoplastic phase and the EPDM phase. Our latest studies have shown that during the crystallization of PP a part of this oil is expelled and migrates to the elastomer phase which is richer in oil than the thermoplastic phase. However, the global ratio between the elastomer phase (EPDM/oil) and the thermoplastic phase (PP/oil) remains more or less the same. This result has also been demonstrated by the work of Caihong et al. [13]. Thereafter, we will only discuss the solid-elastic elastomer phase and the liquid thermoplastic phase.

Many studies on the development of TPV morphologies have focused on the influence, on the one hand, of the crosslinking reaction on the dispersion of the elastomer phase whose viscoelastic properties change considerably with crosslinking, and, on the other hand, the influence of the shear rate of the mixture on these dispersion mechanisms as well. At the early stage of the processing mixing, there is a blend of two viscoelastic fluids. As the PP is a minor phase with respect to the concentration (<30 wt %), the expected morphology is the dispersion of PP droplets in the EPDM which is governed by the capillary number and Rayleigh instabilities [2]. At the end of the mixing/reaction process, a totally different morphology is observed due to the phase inversion phenomenon induced by the crosslinking reaction. The balance between the crosslinking reaction and mixing intensity is the dominant parameter for the morphology development. If the reaction is faster than the mixing time, the whole material turns into powder due to the macroscopic fragmentation of the EPDM phase. If the reaction is much slower, then the mixing is a priori favorable to control this phase inversion and it is expected to be the right morphology (which does not necessarily mean that the right properties have been achieved).

However, it is difficult to dissociate the mixing from the chemical reaction due to the fact that in these high viscosity media, the viscous dissipation phenomena are extremely high. Thus, the temperature and therefore the chemical reaction is difficult to control, especially if these reactions are of the radical type and are characterized by high activation energies. In order to dissociate these combining effects, Martin et al. [14] studied the dispersion of pre-crosslinked EPDM samples at different extents of crosslinking in the PP phase. They showed that for an EPDM insoluble fraction lower than 0.70, the PP/EPDM blend behaves, from a dynamic point of view, as a usual blend whose morphology is governed by the mechanisms of break up/coalescence. Even if the elasticity ratio changes drastically, this equilibrium remains governed by the capillary number expressed at the applied shear rate. Beyond this level of elasticity, the morphology is established by the fragmentation of the EPDM phase leading to the phase inversion and finally to the TPV morphology. In order to be able to process TPV with a low concentration (<30 wt %) of the thermoplastic phase, this fragmentation of the EPDM must be as fine as possible with a wide size distribution.

In order to achieve this fragmentation, it is therefore imperative to apply the highest possible stresses to the EPDM phase. It is necessary to use a thermoplastic phase of high viscosity and therefore high molar mass. However, it can be seen from a mixing torque curve in a batch mixer that the highest stress is obtained during the melting phase of the polymer pellets. This phenomenon is rarely explained in the literature, but it can be stated that it is due to the friction of the PP pellets that are melting, as in a sintering process. This thermoplastic phase melting event is therefore the one that generates the highest stress on the EPDM phase. Consequently, and from a practical point of view, the crosslinking reaction must be advanced enough to undergo the fragmentation of the EPDM phase when the PP melts.

To conclude, how can one achieve a concentration of the elastomer phase close to 80% while reaching a material with good mechanical properties (elongation at break, for example). On one hand, it is essential to have perfect stress continuity at the interface. In the case of PP/EPDM systems, this stress continuity is favorable due to the low interfacial tension, $\gamma_{1-2} \approx 0.3$ mN/m [15], and the fact that the oil plays the role of the interfacial phase between PP and EPDM [6]. The presence of fillers such as carbon black can also play this role of interfacial compatibility as reported for blend systems [16]. In the case of the TPV, the influence of fillers on the morphology and final properties has not been investigated in the literature, to the best of our knowledge.

2. Experimental

2.1. Materials

The experiments were carried out with EPDM Vistalon 8600 (ExxonMobil Chemical, Houston, TX, USA), in which the diene nature is ethylidene norbornene (ENB). The following values of average molar masses were measured by Size Exclusion Chromatography with a conventional calibration in trichlorobenzene at 150 °C: $\overline{M}_n$ = 31,000 g/mol and $\overline{M}_w$ = 107,000 g/mol. The molar content of each component in the terpolymer was assessed by ^{1}H NMR: 71.6 mol % ethylene, 26.4 mol % propylene and 2.0 mol % ENB. Vistalon 8600 has a density of 0.86 g/cm^3 and a Mooney viscosity ML$_{(1+8)}$ of 81 at 125 °C. Isotactic PP (PPH 3060, Total Petrochemicals, Courbevoie, France) was also used, with a melt flow index MFI = 1.8 g/10 min at 230 °C/2.16 kg. Its average molar masses are $\overline{M}_n$ = 72,000 g/mol and $\overline{M}_w$ = 384,000 g/mol. Additional paraffinic oil (Nypar 330, Nynas, Stockholm, Sweden) was incorporated to mimic industrial compositions. The density of this plasticizer is 0.875 g/cm^3 at 15 °C. Its proportion in the binary EPDM/plasticizer mixture was set to the industrial standard for most samples, i.e., 122 phr (grams per hundred grams of Vistalon 8600).

EPDM crosslinking reaction was carried out with either an organic peroxide (Dicumyl peroxide (DCP) 99% purity, Sigma Aldrich, Saint-Louis, MO, USA) or with an octylphenol-formaldehyde resin (Nures 2055, Newport Industries, ShanXi, China) called Resol in the following text. In order to modulate the degree of crosslinking, the EPDM samples were prepared with diverse amounts of the crosslinking agent. The carbon black (CB) N550 (Lehvoss, Cherisy, France) has a specific surface area of 40 m^2/g and a density of 1.7–1.9 g/cm^3 at 20 °C.

2.2. Samples Preparation

The sample preparation method is based on the one described by Ning et al. [17]. The blends of the polymers, processing oil, and crosslinking system (radical initiator or Resol crosslinker) were prepared in an internal batch mixer (Haake Rheomix 600, Thermo Fisher Scientific). The following protocol was adopted: first, the EPDM, the curing system (Resol or DCP), the carbon black filler (if needed), and the plasticizer were introduced (at time $t = 0$) into the chamber at 60 °C and mixed for 5 min at 50 rpm, and then the PP pellets were introduced at $t = 5$ min and mixed for 5 min. The second phase of the process consisted of an increase of the control temperature up to 180 °C, still at 50 rpm. The dynamic vulcanization occurred during this phase. The time for this mixing process depends on the formulation. The mixing is stopped once the temperature and the torque are stabilized. All this mixing process takes around 20–25 min.

The samples were compression molded into 2 mm-thick sheets for 10 min at 180 °C for samples cured by peroxide and for 20 min at 200 °C for samples cured by Resol. The formulations, i.e., concentration of the PP, EPDM, oil, curing agent, and of all samples prepared for this study are reported in Table 1.

Table 1. Formulations of samples prepared in the batch mixer according to the mixing process described in Figure 3 (in phr).

	EPDM	Oil	PP	DCP	Resol	CB
Variation of the crosslinking agent	100	122	75	1	-	-
	100	122	75	2.75	-	-
	100	122	75	4	-	-
	100	122	75	5.5	-	-
	100	122	75	6.5	-	-
	100	122	75	-	1	-
	100	122	75	-	2.75	-
	100	122	75	-	4	-
	100	122	75	-	5.5	-
	100	122	75	-	6.5	-
Variation of the PP content	100	122	75	4	-	-
	100	122	65	4	-	-
	100	122	55	4	-	-
	100	122	45	4	-	-
	100	122	35	4	-	-
	100	122	25	4	-	-
	100	122	75	-	7	-
	100	122	65	-	7	-
	100	122	55	-	7	-
	100	122	45	-	7	-
	100	122	35	-	7	-
	100	122	25	-	7	-
Addition of CB filler	100	122	75	4	-	41
	100	122	65	4	-	39
	100	122	55	4	-	38
	100	122	45	4	-	37
	100	122	35	4	-	35
	100	122	25	4	-	34
	100	122	75	-	7	41
	100	122	65	-	7	39
	100	122	55	-	7	38
	100	122	45	-	7	37
	100	122	35	-	7	35
	100	122	25	-	7	34

2.3. Compression Set

The elastic capacity of a crosslinked TPV samples to recover its initial size after being subjected to constant deformation for a specified time at a given temperature is evaluated by the compression set test. This test is standardized by ASTMD 395 standards (ISO 815-1).

The cylindrical sample, 13 mm in diameter and l_0 = 6 mm thick, is 25% compressed (ε_0 = 0.25) in an oven for 24 h at 100 °C and then removed and left to cool for 30 min. The compression set is then expressed by:

$$Cs\ (\%) = \frac{1}{\varepsilon_0}\left(1 - \frac{l_1}{l_0}\right).100 \tag{1}$$

l_1 the final thickness of the sample and ε_0 the nominal deformation imposed to 0.25. The residual deformation or elastic recovery of the material can then be determined. The following limits of the compression set data under these experimental conditions (temperature and loading time) are then: Cs = 0%, as $l_1 = l_0$, where the sample shows perfect elastic recovery and Cs = 100%, as $l_1 = l_0 (1 - \varepsilon_{0,n})$, where the sample reflects no elastic recovery.

2.4. Electron Microscopy

2.4.1. Scanning Electron Microscopy (SEM)

Each sample was first cryogenically surfaced by ultramicrotomy, then the EPDM phase was selectively extracted by immersion in THF for 4 days. After drying, the samples were covered with a homogeneous 10 nm copper deposit by plasma metallization. The microscopic observation was performed using a QUANTA 250 SEM at a voltage of 10 kV.

2.4.2. Transmission Electron Microscopy (TEM)

The samples were first ultra-microtomed into thin pieces of about 70 nm in thickness with a Leica UC 7 under liquid nitrogen atmosphere at −160 °C. The lamellae were then deposited on copper grids (Mesh 300). The observation was carried out using a Philips CM 120 TEM transmission electron microscope. The EPDM phase appears dark on the images, while the polypropylene is very light in appearance.

2.4.3. Mechanical Properties

Standard tensile tests were conducted on dumbbell-shaped H2 specimens using a tensile test machine (Shimadzu AGS-X, force sensor: SSM-DAK-5000N) at room temperature. The test speed was kept at 500 mm/min according to ISO 37-2017. The tensile strength, the elongation at break, and the Young modulus were calculated from at least five specimens for each sample and the results were averaged.

3. Results and Discussion

Figure 3 shows the variation of the torque and temperature versus the mixing time under a rotor speed of 50 rpm. In the first mixing steps, the increase in torque is mainly due to the diffusion and homogeneous mixing of the oil with the EPDM. This increase in torque is also due to the introduction of solid PP pellets which increases the filling rate of the mixer chamber. The blend is finally homogeneous when the torque curve passes through this maximum. Due to the increase of the temperature, the torque decreases as expected after this maximum. However, a second maximum corresponding to the melting temperature of the PP (T ≈ 162 °C) emerges. It is, however, surprising that the melting of the PP leads to an increase in torque as the pellets change from a solid state to a molten liquid state. In fact, this phenomenon is due to interaction/friction (plastic deformation) between the PP pellets and is comparable to a sintering phenomenon. The stress intensity obtained in this way is higher than the stress level in the liquid state. This transition zone during mixing can be then expected to be the most favorable for optimal EPDM phase dispersion during the phase inversion phenomenon. In a quantitative way, the effect of the melt temperature cannot be dissociated from the effect of the shear rate as the temperature of the system results from the viscous dissipation phenomenon. In the case of EPDM crosslinking with a radical initiator, the quantity of free radical generated can be calculated from the thermal kinetic decomposition of DCP according to the following equation:

$$\frac{d[DCP]}{dt} = -k_d [DCP] \qquad (2)$$

with k_d the dissociation constant: $k_d = A.e^{-E_a/RT}$, A the frequency factor: $A = 7.47 \times 10^{15}$ s^{-1}; and E_a the activation energy of the reaction: $E_a = 153.5$ kJ/mol.

The crosslinking is initiated by the thermal decomposition of the peroxide. Next, the active radicals formed generally react with the elastomer chains by removing hydrogen atoms from the carbon backbone of the polymer, creating highly active radicals on the chain. Finally, crosslinking results by the combination of two macroradicals or by the addition of a macroradical to the unsaturated portion of another major elastomer chain. This produces a carbon–carbon crosslink.

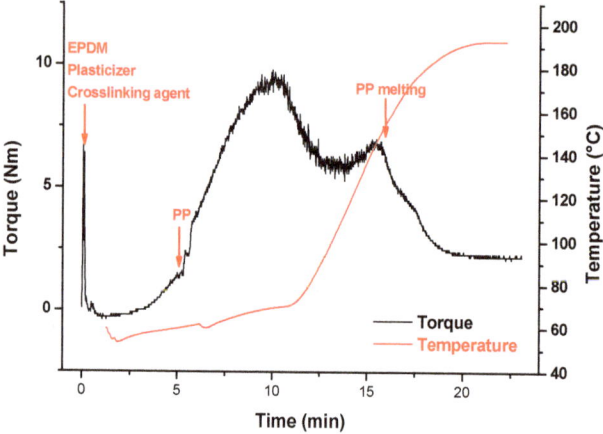

Figure 3. Mixing conditions of TPV in batch mixer. Variation of the torque and temperature versus mixing time. Formulation (phr): EPDM = 100, Oil Plasticizer = 122; PP = 75, DCP = 1. N = 50 rpm.

As a result of this prediction of free radicals concentration, it can be then observed that the phase inversion, near 160 °C and thus the melting point of the PP, takes place at a concentration of ~10 mol/m^3 of free radicals created (Figure 4). This result proves that the phenomenon of phase inversion occurs for the early steps of the crosslinking reaction while the total creation of radicals at the end of the mixing process is 83 mol/m^3. However, in terms of crosslinking density, 10 mol/m^3 corresponds to a level of crosslinking for an EPDM with a certain level of elasticity which, according to our previous work [18], results in an elastic modulus of 6×10^4 Pa, a swelling rate of 20, and a compression set of pure crosslinked EPDM of 60%. Most importantly, from the point of view of viscoelasticity and deformation of the elastomer phase, this level of crosslinking corresponds to the regime for which the morphology is developed by the erosion and fragmentation of weakly crosslinked elastomers domains as already proven by Martin et al. [2]. The crosslinking kinetics of the type of phenolic resin used in this study is not known in the literature and we are therefore unable to extend this study to TPVs crosslinked with phenolic resin.

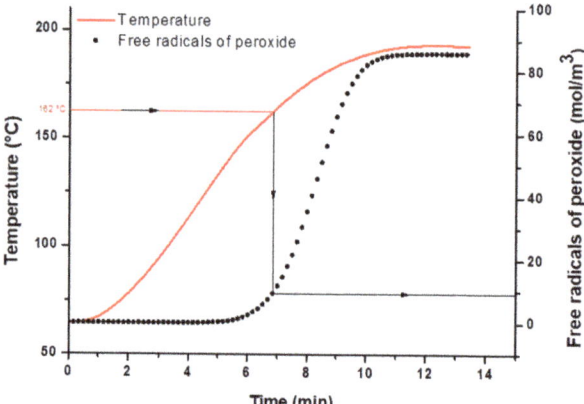

Figure 4. Prediction of free radical concentration generated during mixing in the EPDM phase from thermal dicumyl peroxide (DCP) composition. The temperature curve is from melt temperature recorded during the mixing experiment (Formulation depicted in Figure 3).

As we have seen in our previous work [18], the compression set of the elastomeric phase is mainly influenced by the crosslinking chemistry. In fact, better elasticity recovery properties as shown in Figure 5 are obtained with the peroxide radical initiator in regard to the Resol-based crosslinking system because DCP results in a more statistical crosslinking reaction. If we consider that only the elastomeric phase is a major influence on the elasticity recovery of the TPV, we can easily first imagine that the tendency will be similar for a TPV. However, in Figure 5, it can be observed for the TPVs that both crosslinking ways lead to the same trend and that there is not a significant difference between the two in term of compression set. Moreover, it can be clearly observed that the elasticity recovery of TPV is lower (higher Cs) than for neatly crosslinked EPDM.

For example, we can notice that with around 40 mol·m^{-3} of the crosslinker, in the EPDM matrix, our previous work showed a compression set of ~20% for the system crosslinked by DCP and of ~30% for the system crosslinked by Resol, whereas, in the TPV with this same amount of crosslinking agent, the compression set is around ~60%. It can be then concluded that there are other factors, such as PP concentration and morphology that influence the elasticity recovery of the TPV.

These observations mean that the addition of the PP has an important influence on the elasticity recovery property of the TPV material. The role of the PP in the TPV is to make it processable like a thermoplastic while maintaining the mechanical properties, for example, elongation at break. Accordingly, the concentration of the PP was decreased from 75 (25 wt %) to 25 phr (10 wt %) in order to study the influence of the thermoplastic phase on the compression set experiments.

Figure 6 clearly shows that the concentration of PP has a strong influence on the elasticity recovery of TPVs: the less the PP concentration is, the better compression set is. The values of the compression are reported in Table 2. For example, a compression set of 30% is reached for peroxide-based TPV with 25 phr of PP. Overall, from the set of values, it seems that the use of DCP as a radical initiator, compared to the use of the Resol crosslinker, leads to better results in terms of elastic recovery.

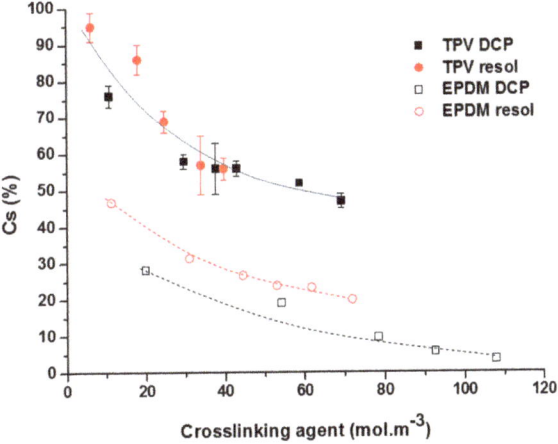

Figure 5. Variation of the compression set versus the molar concentration of the crosslinking agent for TPV with 75 phr of PP. Comparison between the TPV (full symbols) and the neat EPDM systems (open symbols from our previous work [18]).

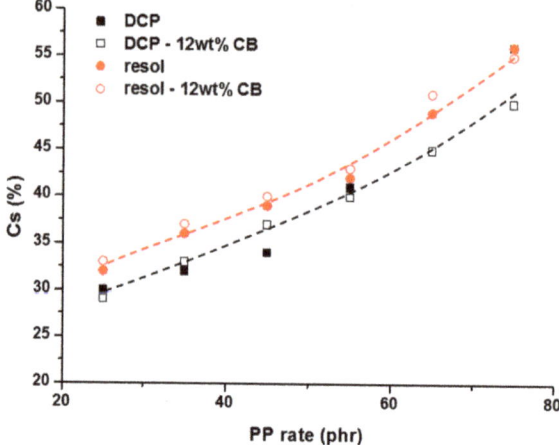

Figure 6. Variation of the compression set (Table 2) versus the PP concentration in the TPV formulation. Amount of chemical reagent in the formulations: 4 phr (DCP) and 7 phr (Resol).

Table 2. Values of compression set for TPV based on a decreasing concentration of the PP phase. Amount of crosslinker in the formulations: 4 phr (DCP) and 7 phr (Resol). Amount of CB: 12 wt %.

DCP (4 phr)			Resol (7 phr)		
PP (phr)	Cs (%) Unfilled Samples	Cs (%) Filled Samples	PP (phr)	Cs (%) Unfilled Samples	Cs (%) Filled Samples
75	56 ± 2	50 ± 2	75	56 ± 1	55 ± 1
65	45 ± 4	45 ± 5	65	49 ± 2	51 ± 2
55	41 ± 2	40 ± 2	55	42 ± 1	43 ± 3
45	34 ± 1	37 ± 3	45	39 ± 0	40 ± 1
35	32 ± 5	33 ± 2	35	36 ± 3	37 ± 2
25	30 ± 1	29 ± 2	25	32 ± 2	33 ± 1

The morphologies of the unfilled peroxide-based TPV were observed by SEM, and the images are shown in Figure 7. As expected, TPVs have a heterogeneous morphology in terms of the shape of both phases. In fact, it can be observed both large and small domains of crosslinked EPDM dispersed in the thermoplastic phase. As the PP concentration decreases, and thus the EPDM concentration increases, the larger EPDM domains predominate. In fact, as discussed in the introduction part, the concentration of EPDM is much higher (up to 80 wt %) than the maximum packing fraction we would have with spheres. These results show that only this type of morphology allows dispersing so much EPDM (60–80%) in PP from the phase inversion phenomenon.

However, when the PP concentration is decreased, a significant loss of the mechanical properties at break (stress and elongation) can be observed (See Figure 8). This result shows that the two phases (PP and EPDM) are not compatible as one might have expected although the interfacial tension between PP and EPDM is low (~0.3 mN/m) and the oil plasticizer also acts as a compatibilizer agent by favoring physical entanglements between PP and EPDM chains at the interface of both phases [6]. The decrease in Young modulus with the increase of the EPDM phase is expected because the TPV tends towards the Young modulus of pure crosslinked EPDM for the lowest amount of PP.

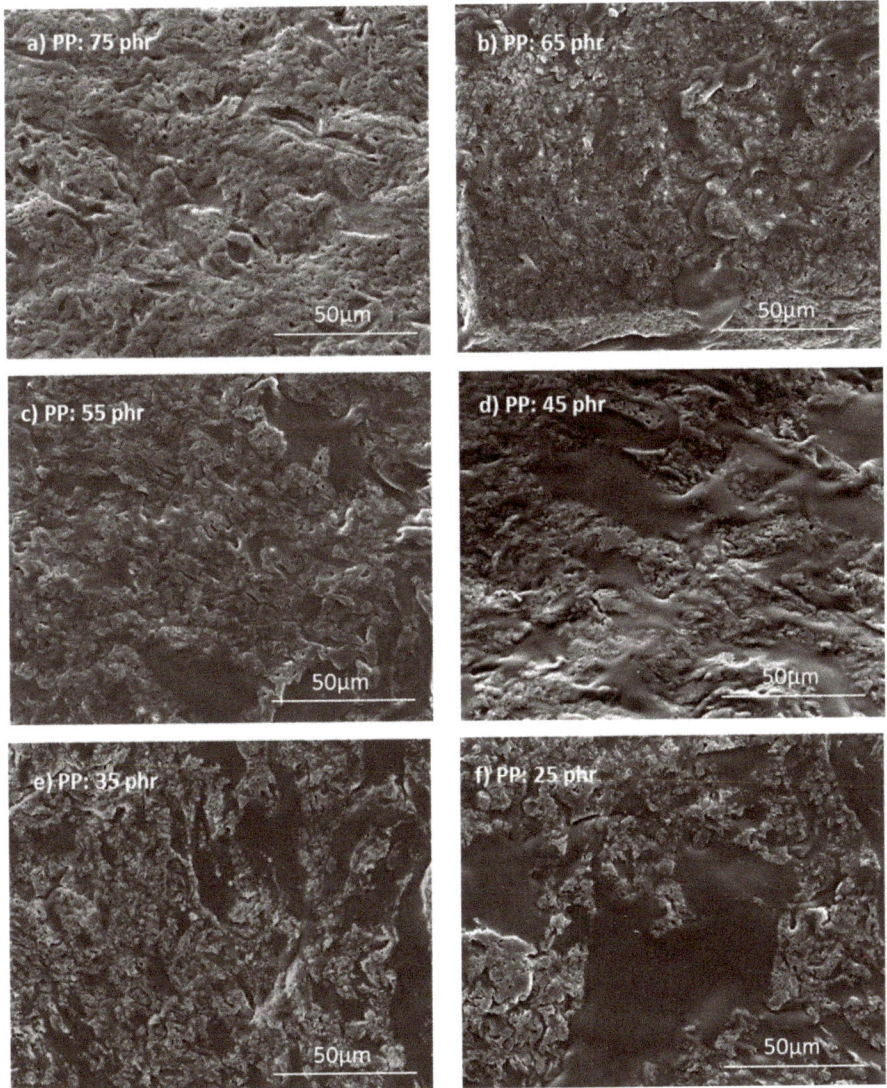

Figure 7. Evolution of the morphologies observed by SEM for TPV with different amounts of PP and crosslinked by DCP (4 phr).

One strategy to compatibilize polymer blends that has been developed these last years is the addition of nanofillers [16]. Sumita et al. [19] proposed that the localization of CB particles can be determined by the wetting parameter: the nanofiller can be located in one phase or at the interface. Other studies [6,14] reported that the viscosity ratio of the two phases of the blend can also play a major role in the dispersion of the CB particles. Regarding PP/EPDM blends, few works have been devoted to their compatibilization from the addition on nanofillers. Yang et al. [20] performed a study on PP/EPDM/SiO$_2$ nanocomposites by varying the amount of EPDM phase (10–30 wt %) and of the silica (1–5 wt %). Hydrophobic silica (particle size: 10–30 nm; S = 200 m^2/g) and hydrophilic silica (particle size: 15–20 nm; S = 150 m^2/g) were used. They showed that PP/EPDM/SiO$_2$ systems exhibit the formation of a filler network structure in the PP matrix that lead to a super toughened ternary composite with Izod impact strength 2–3 times higher than PP/EPDM binary blends. The morphology of PP/EPDM blends at different proportions, especially at the co-continuity of both phases was studied by Martin et al. [21]. More particularly, they studied the influence of silica particles (hydrophilic or hydrophobic) on the co-continuous morphology. They observed that hydrophilic silica particles tend to migrate within the EPDM phase and form huge aggregates (0.5–1 µm). On the other hand, hydrophobic particles are dispersed homogeneously and can be found both at the interface and within the EPDM phase. The most interesting results have been obtained with carbon black [22–24] dispersed in TPV based on PP/EPDM. For example, Ma et al. [24] observed that with the addition of CB, the uncrosslinked EPDM/PP blend (Thermoplastic elastomer, TPE) and dynamically vulcanized blend (TPV) showed a notable difference in the conductive electrical properties, which is mainly caused by the different localization of CB particles resulting from the dynamic vulcanization process. Particularly, they found that the CB particles in the TPE were preferably dispersed in the EPDM phase, whereas the CB particles in the TPV were almost located in the PP matrix due to the high viscosity ratio of cured EPDM compared with PP.

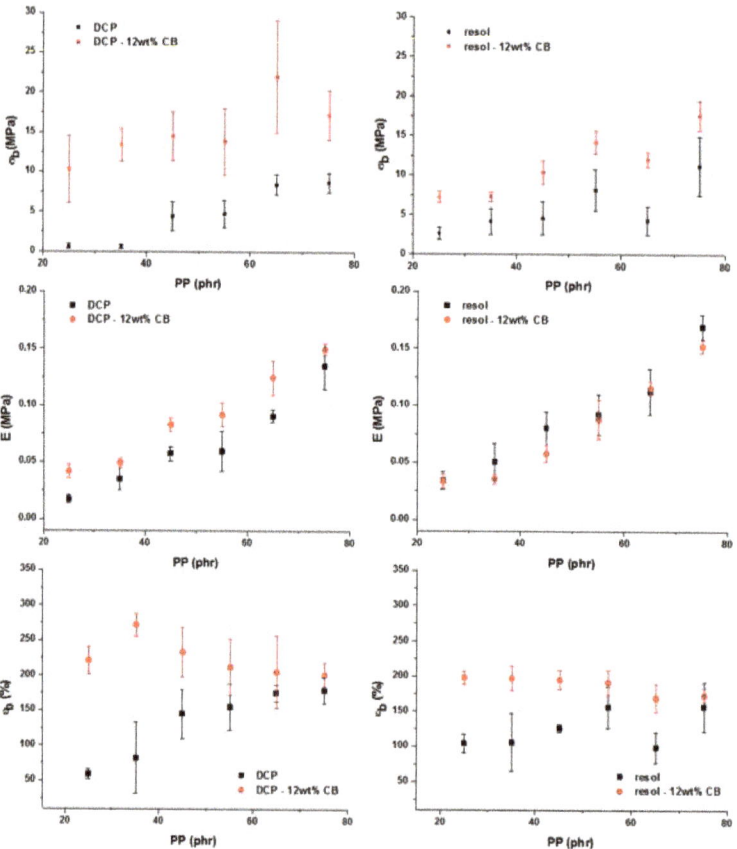

Figure 8. Evolution of the mechanical properties: tensile strength, Young modulus and elongation at break in function of the PP amount for unfilled and carbon black filled TPV.

In the present study, 12 wt % of carbon black was added in the TPV with the aim to improve the mechanical properties. According to the experimental protocol, CB was first mixed with

EPDM/plasticizer before the addition of PP and the phase inversion phenomenon. From TEM pictures in Figure 9, it can be observed that the CB particles are located in the case of peroxide-based TPV at the interface between the PP and the EPDM, while the filler is located mainly in the EPDM phase in the case of Resol crosslinker. This result means that the carbon black localization is not governed by thermodynamic aspect such as wettability criteria or by viscosity aspects. Indeed, we have demonstrated in our previous work [18] that carbon black interacts with the chemical reaction of crosslinking due to the radical initiator and the chemical grafting of EPDM chains on the surface of CB can be expected.

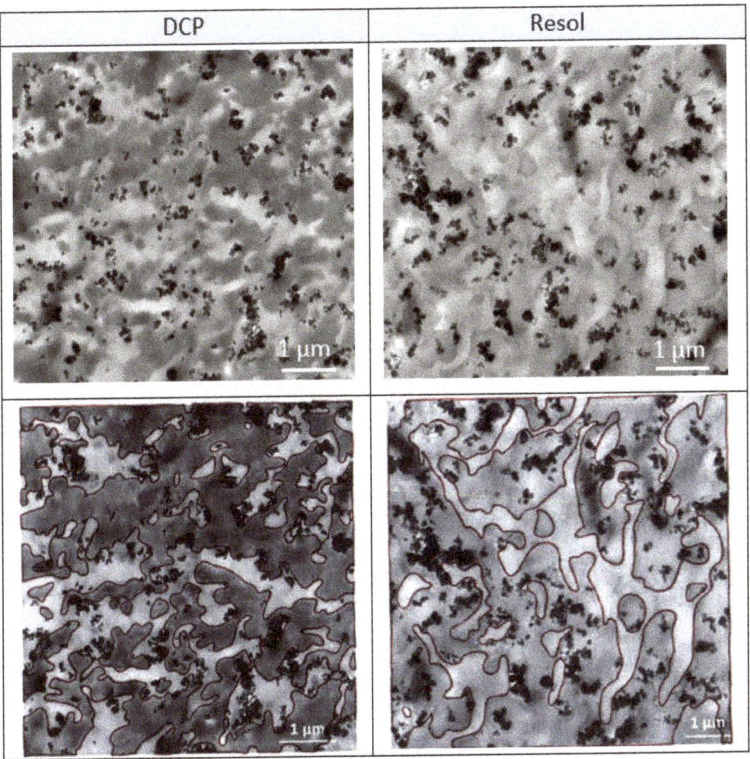

Figure 9. TEM pictures of filled TPV (PP: 75 phr) with 12 wt % of carbon black (**top**). Highlighting of phase limits (**bottom**) to help visualize the location of carbon black fillers.

Furthermore, a significant enhancement of tensile stress at 300% has been also reported with the addition of CB in TPV [23]. Actually, the mechanical properties can be then improved significantly when the fillers are located at the interface, because such a filler localization, like a compatibilizer, is helpful to refine the phase morphology and enhance interfacial adhesion.

The mechanical properties of filled TPV in Figure 8 show a significant improvement of the mechanical properties at break, more particularly for the lowest concentrations of PP. This improvement of the mechanical properties is even more important for peroxide-based TPV which can be explained with the localization of the CB at the interface between the two phases which enhance the interfacial adhesion. This phenomenon leads to a change in the morphology development and compatibilization. However, the phenomenon of carbon black localization is not well understood in such complex formulations.

4. Conclusions

The interesting feature of TPV is their ability to recover the elasticity measured by the compression set. The development of the morphology requires then the highest possible concentration in the crosslinked phase well above the maximum packing fraction of 0.64 for spheres. In order to achieve a high concentration of dispersed rubber phase, the EPDM phase must be dispersed with a large distribution of the domain size.

The main conclusions of the present work are the following:

The phase inversion between both phases takes place when the EPDM phase has reached a certain level of crosslinking (swelling ratio ~20 corresponding to a crosslinking density of 10 mol/m^{-3}) which allows the stress applied by the PP matrix to disperse this phase by fragmentation. The maximal stress is observed at the melting temperature of the PP pellets.

The crosslinking chemistry nature (radical initiator or phenolic resin crosslinker) have less influence on the elasticity recovery in the case of the TPV compared with a pure EPDM system. This result means that the morphology plays an important role in the compression set of TPVs. Morphology is the first order parameter whereas crosslinking density of the EPDM phase is a second order parameter.

The morphologies of the peroxide-based TPV, observed by SEM, show heterogeneous systems with large EPDM domains. However, this large distribution of EPDM particles is the only way to increase the maximal packing fraction of crosslinked EPDM phase which behave as solid particles.

The addition of carbon black fillers allows the mechanical properties at break to be significantly improved at the lowest PP concentration (20%). This improvement of mechanical properties is still not well understood.

Finally, to reach the best compression set while keeping good mechanical properties, a peroxide-based TPV containing carbon black and the less PP as possible should be formulated. Therefore, the low quantity of PP allows to have a good compression set; the peroxide permits the carbon black to be located at the interface between the two phases and thus the filler can reinforce the mechanical properties of the material.

Author Contributions: Data curation, M.C.; Methodology, C.L.H.; Supervision, V.B.-L.; Validation, A.L. and A.T.; Writing—original draft, P.C.; Writing—review & editing, P.C. All authors have read and agreed to the published version of the manuscript.

Funding: This research received no external funding.

Acknowledgments: This work was carried out in the joint laboratory with the Hutchinson company. The authors would like to thank the Hutchinson company for their support in this research.

Conflicts of Interest: The authors declare no conflict of interest.

References

1. Coran, A.Y.; Patel, R. Rubber-Thermoplastic Compositions. Part VIII. Nitrile Rubber Polyolefin Blends with Technological Compatibilization. *Rubber Chem. Technol.* **1983**, *56*, 1045–1060. [CrossRef]
2. Martin, G.; Barres, C.; Sonntag, P.; Garois, N.; Cassagnau, P. Morphology development in thermoplastic vulcanizates (TPV): Dispersion mechanisms of a pre-crosslinked EPDM phase. *Eur. Polym. J.* **2009**, *45*, 3257–3268. [CrossRef]
3. Wu, H.; Tian, M.; Zhang, L.; Tian, H.; Wu, Y.; Ning, N. New understanding of microstructure formation of the rubber phase in thermoplastic vulcanizates (TPV). *Soft Matter* **2014**, *10*, 1816–1822. [CrossRef] [PubMed]
4. Babu, R.R.; Naskar, K. Recent Developments on Thermoplastic Elastomers by Dynamic Vulcanization. *Adv. Polym. Sci.* **2010**, *239*, 219–247.
5. Li, S.; Tian, H.; Wu, H.; Ning, N.; Tian, M.; Zhang, L. Coupling effect of molecular weight and crosslinking kinetics on the formation of rubber nanoparticles and their agglomerates in EPDM/PP TPVs during dynamic vulcanization. *Soft Matter* **2020**, *16*, 2185–2198. [CrossRef]
6. Litvinov, V.M. EPDM/PP Thermoplastic Vulcanizates As Studied by Proton NMR Relaxation: Phase Composition, Molecular Mobility, Network Structure in the Rubbery Phase, and Network Heterogeneity. *Macromolecules* **2006**, *39*, 8727–8741. [CrossRef]

7. Wang, Z.; Li, S.; Wei, D.; Zhao, J. Mechanical properties, Payne effect and Mullins effect of thermoplastic vulcanizates based on high-impact polystyrene and styrene—Butadiene rubber compatibilized by styrene—Butadiene—Styrene block copolymer. *J. Thermoplast. Compos. Mater.* **2015**, *28*, 1154–1172. [CrossRef]
8. Antunes, C.; Machado, A.; Van Duin, M. Morphology development and phase inversion during dynamic vulcanisation of EPDM/PP blends. *Eur. Polym. J.* **2011**, *47*, 1447–1459. [CrossRef]
9. Baule, A.; Mari, R.; Bo, L.; Portal, L.; Makse, H.A. Mean-field theory of random close packings of axisymmetric particles. *Nat. Commun.* **2013**, *4*, 2194. [CrossRef]
10. Santamaria-Holek, I.; Mendoza, C.I. The rheology of concentrated suspensions of arbitrarily-shaped particles. *J. Colloid Interface Sci.* **2010**, *346*, 118–126. [CrossRef]
11. Farris, R.J. Prediction of the Viscosity of Multimodal Suspensions from Unimodal Viscosity Data. *Trans. Soc. Rheol.* **1968**, *12*, 281–301. [CrossRef]
12. Qi, F.; Tanner, R.I. Random close packing and relative viscosity of multimodal suspensions. *Rheol. Acta* **2011**, *51*, 289–302. [CrossRef]
13. Caihong, L.; Xianbo, H.; Fangzhong, M. The distribution coefficient of oil and curing agent in PP/EPDM TPV. *Polym. Adv. Technol.* **2007**, *18*, 999–1003. [CrossRef]
14. Martin, G.; Barres, C.; Cassagnau, P.; Sonntag, P.; Garois, N. Viscoelasticity of randomly crosslinked EPDM networks. *Polymer* **2008**, *49*, 1892–1901. [CrossRef]
15. Bhadane, P.A.; Champagne, M.F.; Huneault, M.A.; Tofan, F.; Favis, B.D. Erosion-dependant continuity development in high viscosity ratio blends of very low interfacial tension. *J. Polym. Sci. Part B Polym. Phys.* **2006**, *44*, 1919–1929. [CrossRef]
16. Taguet, A.; Cassagnau, P.; Lopez-Cuesta, J. Structuration, selective dispersion and compatibilizinf effect of (nano)fillers in polymer blends. *Prog. Polym. Sci.* **2014**, *39*, 1526–1563. [CrossRef]
17. Ning, N.; Li, S.; Wu, H.; Tian, H.; Yao, P.; Hu, G.-H.; Tian, M.; Zhang, L. Preparation, microstructure, and microstructure-properties relationship of thermoplastic vulcanizates (TPVs): A review. *Prog. Polym. Sci.* **2018**, *79*, 61–97. [CrossRef]
18. Le Hel, C.; Bounor-Legaré, V.; Lucas, A.; Cassagnau, P.; Thèvenon, A. Elasticity Recovery of Crosslinked EPDM: Influence of the Chemistry and Nanofillers. *Rheol. Acta* **2020**, in press. [CrossRef]
19. Sumita, M.; Sakata, K.; Asai, S.; Miyasaka, K.; Nakagawa, H. Dispersion of fillers and the electrical conductivity of polymer blends filled with carbon black. *Polym. Bull.* **1991**, *25*, 265–271. [CrossRef]
20. Yang, H.; Zhang, X.; Qu, C.; Li, B.; Zhang, L.; Fu, Q. Largely improved toughness of PP/EPDM blends by adding nano-SiO2 particles. *Polymer (Guildf)* **2007**, *48*, 860–869. [CrossRef]
21. Martin, G.; Barres, C.; Sonntag, P.; Garois, N.; Cassagnau, P. Co-continuous morphology and stress relaxation behaviour of unfilled and silica filled PP/EPDM blends. *Mater. Chem. Phys.* **2009**, *113*, 889–898. [CrossRef]
22. Le, H.H.; Heidenreich, D.; Kolesov, I.S.; Ilisch, S.; Radusch, H.-J. Effect of carbon black addition and its phase selective distribution on the stress relaxation behavior of filled thermoplastic vulcanizates. *J. Appl. Polym. Sci.* **2010**, *117*, 2622–2634. [CrossRef]
23. Ma, L.-F.; Bao, R.-Y.; Dou, R.; Liu, Z.-Y.; Yang, W.; Xie, B.-H.; Yang, M.-B.; Fu, Q. A high-performance temperature sensitive TPV/CB elastomeric composite with balanced electrical and mechanical properties via PF-induced dynamic vulcanization. *J. Mater. Chem. A* **2014**, *2*, 16989–16996. [CrossRef]
24. Ma, L.-F.; Bao, R.; Huang, S.; Liu, Z. Electrical properties and morphology of carbon black filled PP/EPDM blends: Effect of selective distribution of fillers induced by dynamic vulcanization. *J. Mater. Sci.* **2013**, *48*, 4942–4951. [CrossRef]

© 2020 by the authors. Licensee MDPI, Basel, Switzerland. This article is an open access article distributed under the terms and conditions of the Creative Commons Attribution (CC BY) license (http://creativecommons.org/licenses/by/4.0/).

Article

Effects of Hydrothermal Aging of Carbon Fiber Reinforced Polycarbonate Composites on Mechanical Performance and Sand Erosion Resistance

Mei Fang [1], Na Zhang [1,*], Ming Huang [1], Bo Lu [1], Khalid Lamnawar [2], Chuntai Liu [1] and Changyu Shen [1,3]

1. National Engineering Research Center for Advanced Polymer Processing Technology, Zhengzhou University, Zhengzhou 450002, China; 16603869720@163.com (M.F.); huangming@zzu.edu.cn (M.H.); bolu@zzu.edu.cn (B.L.); ctliu@zzu.edu.cn (C.L.); shency@zzu.edu.cn (C.S.)
2. Ingénierie des Matériaux Polymères, UMR 5223, CNRS, INSA Lyon, Université de Lyon, 69622 Villeurbanne, France; khalid.lamnawar@insa-lyon.fr
3. State Key Laboratory of Structural Analysis for Industrial Equipment, Dalian University of Technology, Dalian 116023, China
* Correspondence: nazhang@zzu.edu.cn

Received: 8 September 2020; Accepted: 14 October 2020; Published: 23 October 2020

Abstract: Carbon fiber reinforced polycarbonate (CF/PC) composites have attracted attention for their excellent performances. However, their performances are greatly affected by environmental factors. In this work, the composites were exposed to hydrothermal aging to investigate the effects of a hot and humid environment. The mechanical properties of CF/PC composites with different aging times (0, 7, 14, 21, 28, 35, and 42 days) were analyzed. It was demonstrated that the storage modulus of CF/PC composites with hot water aged for seven days has the highest value in this sampling period and frequency. Through the solid particle erosion experiment, it was found that the hydrothermal aging causes the deviation of the maximum erosion angle of composites, indicating the composites underwent ductile–brittle transformation. Furthermore, the crack and cavity resulting from the absorption of water was observed via the scanning electron microscope (SEM). This suggested that the hydrothermal aging leads to the plasticization and degradation of CF/PC composites, resulting in a reduction of corrosion resistance.

Keywords: carbon fiber reinforced polycarbonate composites; hydrothermal aging; solid particle erosion; mechanical property

1. Introduction

Carbon fiber reinforced polymer composites (CFRP) have been widely used in construction, transportation, and sports equipment fields, profiting from their high durability and strength, light weight, and excellent thermodynamic stability [1–5]. Zhang et al. [6] demonstrated that the average tensile and compression strength of glass fiber (GF) composites is almost 50% lower than CF composites. Batuwitage et al. [7] reported that the strength of CFRP composite is ten times as much as steel at the same volume.

Nevertheless, the properties of CFRP composites will change during their operating life because of the complex work environment [8–10]. Many works have been published about the aging resistance of CFRP. Some researches demonstrated that the matrix and interface of composites are greatly influenced by hydrothermal effect [11,12]. Generally, there are three types of composites absorbing water: transporting of water molecules within the matrix, permeating at the matrix–fiber interface, and absorbing into the crack produced by the action of high temperature [13–15]. The entry of water molecules causes the matrix to expand, which induces the residual or hydrothermal stress in CFRP composites [16]. Moreover, moisture accelerates the degradation of composites through the breaking up of molecular chains and deterioration of the interface of matrix/fiber [17,18]. High temperature environments accelerate the diffusion rate of moisture and result in the plasticization of CF/PC composites [8,11,19].

In many industrial applications, sand erosion is of wide concern due to its serious friction drag, structural integrity, and high maintenance costs, e.g., in helicopter rotor blades and high-speed vehicles, whose surfaces are usually exposed to dusty environments (contain erodent flux conditions, erosive particle characteristics). Extreme conditions, such as sand and dust environments, may even accelerate erosion and wear processes [20]. Hydrothermal aging also has an important effect on impact property for composites. Lu et al. [21] investigated the impact behavior of unidirectional CF- reinforced epoxy resin compounds after hydrothermal aging and found that the entry of water molecules decreased the impact resistance of CF/PC compounds. Ahmad et al. [22] concluded that the water molecules have a negative effect on the impact resistance of composite plate. Hanan et al. [23] found the impact damage for CF/PC composite was influenced by hydrothermal aging via ultrasonic technology. Based on the above-mentioned factors, it is extremely urgent to explore the effects of hydrothermal aging on the mechanical properties and solid particle erosion resistance for CFRP composites.

In this study, polycarbonate (PC) was chosen as matrix for its excellent fatigue resistance. The aging process of CF/PC composites was carried out under a humid and hot environment. The moisture absorption rate was measured and calculated during the aging process. The changes of mechanical properties before and after hydrothermal aging were measured through tensile and flexural tests. Underlying mechanisms for hydrothermal aging induced changes in mechanical properties and sand erosion property were surveyed. Meanwhile, the effect of hydrothermal aging on thermodynamic stability of CF/PC was examined by dynamic mechanical analysis (DMA). The surface morphologies of CF/PC were analyzed through SEM and three-dimensional hyper depth of field. It is expected that, in the case of CF/PC material being chosen as a surface material, hydrothermal aging protection will be applied, so as not to affect the product performance.

2. Experimental

2.1. CF/PC Composites

T700SC-3K Carbon fiber (CF) supplied by Covestro Company (Tokyo, Japan) is adopted as reinforcing CF/PC composites. The Makrolon series of Polycarbonate (PC) 2407 was purchased from Covestro Company in Germany. Figure 1 shows the pictures of the lamination diagram and its preparation process. The specific preparation process is as follows: Liquid PC infiltrated CF and solidified to make CF/PC unidirectional single layer belt. According to the design requirements, the single layer belt was cut at different angles in the light of the laying direction of CF. To avoid the warping deformation of products, 8-layer unidirectional tapes were assembled at angles of 0°/90°/+45°/−45°/−45°/+45°/90°/0°. In the end, CF/PC composites were prepared and molded by hot processing at 240 °C for 3 min.

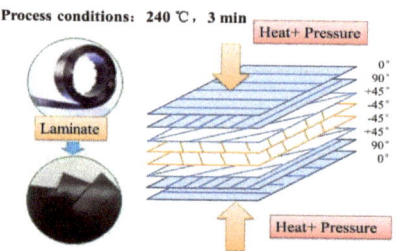

Figure 1. Schematic diagram of forming process for prepreg laminate.

2.2. Hydrothermal Aging

Samples were divided into seven groups and immersed in deionized water at 80 °C. The aging time of the seven groups was 0, 7, 14, 21, 28, 35, and 42 days, respectively. After hydrothermal aging, specimens were taken out and dried for 24 h at 25 °C.

2.3. Moisture Absorption Test

Seven specimens of un-aged were selected and marked, each specimen was weighed by 0.1 mg accuracy electronic scale and labeled as M_0. The sample weight was measured regularly during the aging process at the following time interval: every day in the earlier stage and every seven days in the later period. During testing, specimens were taken out, dried, weighed, and put back. Moisture uptake was measured from the average weight at time t, and moisture absorption, W_t, is defined in the following equation.

$$W_t(\%) = \frac{M_t - M_0}{M_0} \times 100 \tag{1}$$

where M_0 is the initial weight and M_t is the weight at time t.

2.4. Mechanical Performance Testing

Universal tensile tester (model INSTRON 5585, Boston, MA, USA) was applied to study the tensile property of CF/PC of 170 mm × 12 mm × 2 mm dimension according to GB-T 1040, the tests were performed at room temperature (RT). The bending tests were conducted using an INSTRON 5585 machine at a span length of 31.5 mm according to GB-T 9341-2000. The tensile and bending tests were performed at the same rate of 1 mm/min. At least seven specimens were measured for each test and the average value was derived.

2.5. Dynamic Mechanical Analysis (DMA)

The DMA analyzer (Q800) was used to characterize the dynamic thermo-mechanical properties of un-aged and hydrothermal aged CF/PC specimens. The single cantilever mode was used during the test. The specimens with size of 30 mm × 12 mm × 2 mm were heated from 50 °C to 200 °C at a heating rate of 3 °C/min with a frequency of 1 Hz. At least five specimens were scanned for each sample.

2.6. Solid Particle Erosion

STR-9060 model sand-blasting equipment (Zhangjiagang Stell Coating Equipment Co.LTD, Suzhou, China) was employed to explore the influence of hydrothermal aging on the specimen erosion resistance. The erodent was silicon carbide (SiC) with sharp edge, and the average size was 300 μm to 800 μm. The SiC particles impinged on the sample surface under the acceleration of high pressure gas. The mass flow of the particles is 16.7 g/s in 0.345 MPa pressure. The distance between the specimen holder and the nozzle was 30 mm and the impact angle was 30°, the inner diameter of the nozzle is 6 mm, and each specimen was eroded for 1 min. After erosion, in order to clear away the SiC particles, specimens were washed by ethyl alcohol and dried by air blasting. The weight loss was measured and calculated by at least seven specimens. All tests were done at RT.

2.7. Scanning Electron Microscope (SEM)

The surface morphology of specimens with different aging time was analyzed by a JEOL JSM-7500F (Tokyo, Japan) scanning election microscopy (SEM). A thin layer of gold was sprayed on the surfaces of specimens to make them more conductive and visible.

3. Results and Discussion

3.1. Moisture Absorption Analysis

The relationship between water absorption and aging time was displayed in Figure 2. The carbonate base of PC molecular chain with strong polarity could interact with water molecules [24–26], allowing the CF/PC composites to absorb moisture. As clearly seen in Figure 2, the moisture absorption is increasing linearly with the square root of five days of aging. Of note, the specimens approached the saturation point after five days, and the water absorption capacity decreases from the fifth to seventh day, when it reaches water absorption saturation state. After seven days, the hygroscopic equilibrium line reached a flat state.

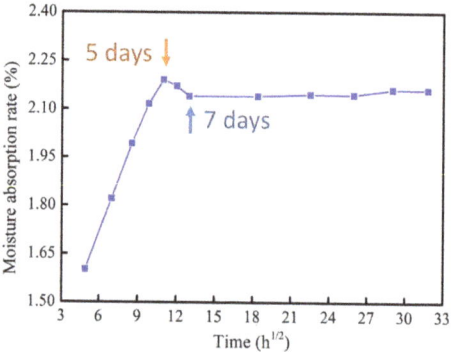

Figure 2. Water absorption of CF/PC composites.

Figure 3a presents a smooth and compact surface for un-aged specimen. Surprisingly, after 42 days, the surfaces of the hydrothermally aging samples (Figure 3b) showed deep cracks and small debris. Clearly, water molecules entered CF/PC composites along the interface of the fiber matrix and occupied additional volume. Once the water molecules evaporate from the composite, cracks and voids appear both inside and on the surface.

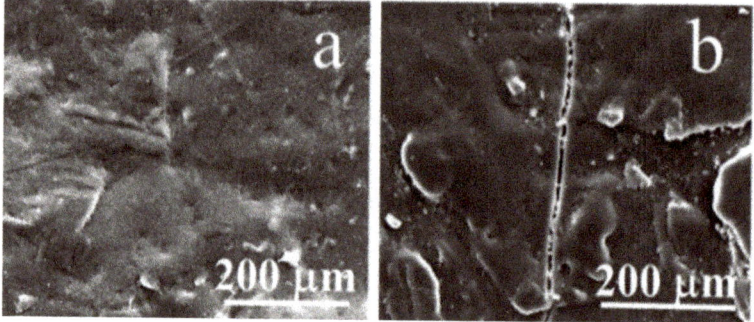

Figure 3. Surface topography of CF/PC composites. (**a**) un-aged composites surface, (**b**) hydrothermal aged for 42 days surface of CF/PC composites. 80 °C.

3.2. Tensile Property Analysis

Representative tensile behavior of CF/PC at different aging times is displayed in Figure 4. The elongation at break (from Figure 4a) of CF/PC decreased with the aging time. On the premise of sampling period and frequency every seven days, the tensile strength (Figure 4b) of specimens reached its peak on the 7th day of aging in hot water at 80 °C. A tighter structure owing to the entanglement of molecular chains in a hot environment is beneficial to the improvement of tensile properties.

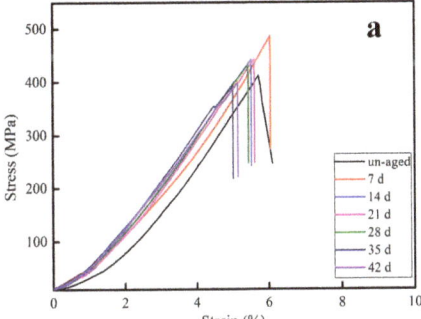
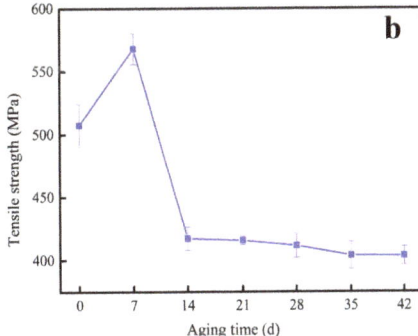

Figure 4. Tensile properties for CF/PC composites. (**a**) the representative stress strain curves of CF/PC composites with different hydrothermal aged, (**b**) tensile strength of CF/PC composites with different hydrothermal aged. 1 mm/min.

However, after seven days aging, the tensile properties of CF/PC decreased rapidly with the increasing of aging time. It was also found that the tensile strength of CF/PC after 14 days aging was inferior to virgin samples. The maximum tensile stress of the CF/PC composites decreased slightly. There are two reasons for the decline in mechanical properties of CF/PC composites: the presence of absorbed water and the degradation of CF/PC composites.

Polycarbonate could react with water molecules due to their strong polarity. The schematic diagram of hydrolysis mechanism for PC was seen as Figure 5. Hydroxyl groups in water molecules bind to PC chains, leading to the breaking of molecular chains and hydrolysis of the polymer [27]. Macroscopically, the polymer performance decreases.

Figure 5. Schematic diagram of hydrolysis mechanism for PC.

3.3. Three-Point Bending Analysis

Figure 6 shows the flexural property of CF/PC composites as a function of aging time. It can be seen from Figure 6a–c that the flexural performance of specimens increased first and then decreased with the aging time. On the premise of sampling period and frequency every seven days, the maximum bending performance was at 14 days for hydrothermal aging. Figure 6d displayed that the flexural strain declines as the aging time increases. According to the result of three-point bending, the tensile strength and flexural strength values are very close. This is because, for the test of flexural strength, the upper part of the specimen is compressed and the lower part is stretched. The lower part subjected to stretch damaged first, indicating that the material is brittle. Combined with the result of sand erosion in Figure 8, it is shown that the ductile–brittle transition occurs in CF/PC composites.

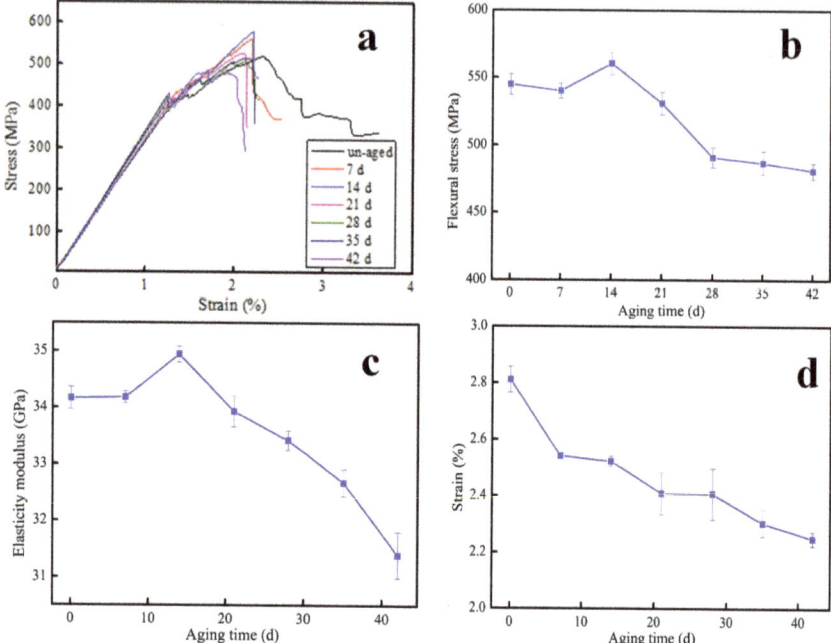

Figure 6. Flexural properties for CF/PC composites as a function of aging time. (**a**) the representative stress-strain curves of CF/PC composites with different hydrothermal aged, (**b**) flexural stress of CF/PC composites with different hydrothermal aged, (**c**) elasticity modulus with different hydrothermal aged, (**d**) strain with different hydrothermal aged. 1 mm/min.

On one hand, the matrix of CF/PC would shrink when aged at 80 °C, resulting in a closer attachment with CF and matrix. In addition, the residual stress of CF/PC composites was eliminated, and the regularity of CF/PC composites structure was improved when aged in high temperature environment. Thus, the flexural property increased in the early aging. On the other hand, the absorption of water molecule caused voids and defects within CF/PC composites. The matrix would separate from carbon fiber owing to the swelling effects. On the whole, the flexural strength and modulus decreased with further aging.

3.4. Dynamic Thermal Mechanical Analysis

Figure 7 presents the storage modulus variation with the aging time. Results reveal an increase of composites modulus for seven days aging. However, the moduli of CF/PC composites aging for 14 days to 42 days are lower than those un-aged. This results further confirmed the results in Figure 6. In the early aging, composites structure became tight, facilitating the increase of modulus. During the later aging time, the water molecules caused cavities and cracks in the matrix of CF/PC composites.

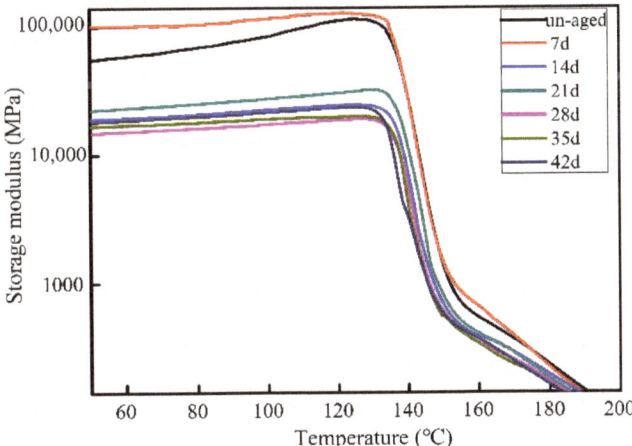

Figure 7. The storage modulus of CF/PC composites at different aging time.

3.5. Solid Particle Erosion Performance

Solid particle erosion test was performed on CF/PC for the purpose of studying the sand erosion behavior. Under the same impact conditions, the weight loss of CF/PC composites increased and then decreased (Figure 8). Compared with un-aged composites, the samples aging for seven days exhibited the worst sand erosion resistance performance.

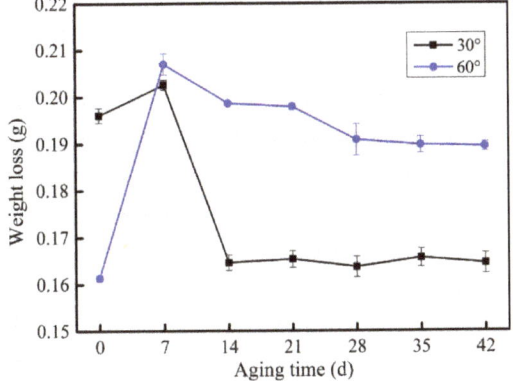

Figure 8. Sand erosion test of CF/PC composites, 1 min, 50 psi.

The un-aged CF/PC composites were severely eroded at 30°, and the weight loss decreased at an erosion angle of 60°. Nevertheless, the maximum erosion behavior of plastic material occurs at a low angle [28]. Therefore, it could be concluded that the PC matrix indicated a ductile characteristic after hydrothermal aging (7–42 days).

With the increase of hydrothermal aging time, water molecules occupied more additional volume inside the CF/PC composite, resulting in the structure become compact and shown the characteristic of brittleness. Hence, the maximum erosion angle of CF/PC composites has changed. Therefore, the weight loss of CF/PC composites aging for 14 to 42 days decrease compared with un-aged CF/PC when eroded at 30°. Conversely, the weight loss of CF/PC composites after hydrothermal aging (7–42 days) increased obviously when eroded at 60°. This result further proves the ductile and brittle transition of the CF/PC composites during hydrothermal aging.

3.6. Surface Morphology Analysis

The eroded surfaces of un-aged CF/PC composites were presented in Figure 9a,b. A mass of matrix fragments was attached to the fiber, indicating a perfect combination of fibers and matrix. Large amounts of cracks and matrix fragments were on CF/PC surface, and a few fibers are exposed and broken. Under a 30° erosion angle, CF/PC composite is subjected to significant shear stress. Repeated cutting of high-speed particles will cause deformation, scratches, and pits on CF/PC surfaces, causing a weight loss of CF/PC composite. In addition, some tiny particles embedded in the cracks would accelerate the crack propagation and mass loss of CF/PC composites.

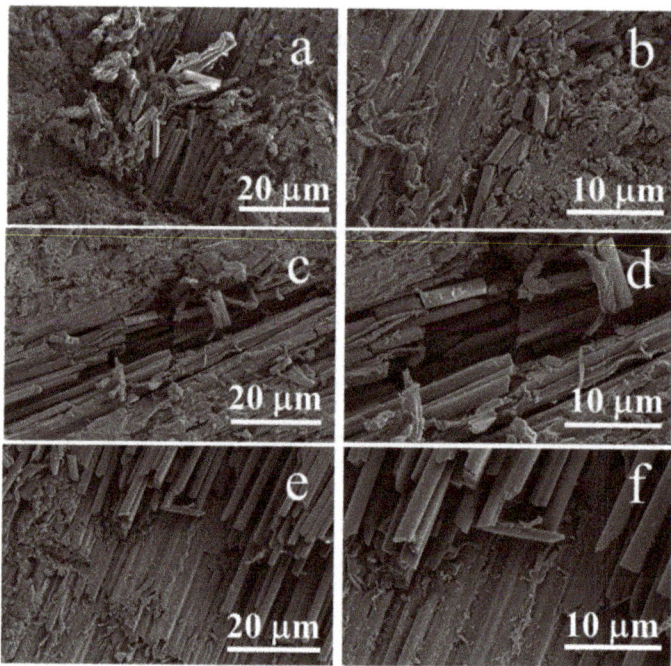

Figure 9. Surface topography of CF/PC composites after erosion. (**a,b**) un-aged composites surface after erosion at an angle of 30°, (**c,d**) hydrothermal aged for 7 days composites surface after erosion at an angle of 30°, (**e,f**) hydrothermal aged for 42 days after erosion at an angle of 30° [29].

Figure 9c–f displayed that the matrix stripping fiber fracture and crack propagation occurred on aged specimens under the erosion of sand. Compared to un-aged composites, a small amount of matrix was attached to the fiber surface (see Figure 9c,d) and the fiber (see Figure 9e,f) has a smooth surface with no matrix attached. On one hand, the infiltration of water molecules increases the distance between the molecules of the CF/PC composites. On the other hand, the combination between water molecules and PC matrix shielded the interaction of CF and matrix. The eroded surface of CF/PC composites aging for seven days are shown in Figure 9c,d. When CF/PC was eroded at the angle of 30°, deep pits and broken fiber can be clearly observed. Under the action of hot water aging, the terminal segments of the PC chemical groups are easily degraded, resulting in plasticization of the matrix, which is manifested as the decrease of the sand erosion resistance of CF/PC composites at the macro level.

Combined with Figures 8 and 9e,f, it was found that CF/PC samples aged 42 days eroded at of 30° was the lowest. Exposed to hot and humid environment for a long time, the CF/PC would undergo ductile-brittle transition, and its maximum erosion angle transfers to 90°, the vertical impact of solid particles is the most serious damage to the material [30]. The particles generated less vertical force when impinging on the sample at 30°, and therefore less erosion was caused to the sample.

The three-dimensional morphology on CF/PC surface after solid particle erosion was exhibited as Figure 10. Lots of scratches formed on the surface of un-aged samples after sand erosion (see Figure 10a). The height from crest to trough was 0.46 mm. Compared with un-aged samples, extensive pit was formed on the surface of CF/PC composites aged for seven days under the solid particle erosion. And the depth of the pit reached 0.62 mm (Figure 10b). The CF/PC composites show a poor solid particle erosion resistance after 42 days hydrothermal aging. As shown in Figure 10c, there are a lot of 'lips' on the CF/PC composites surface. However, the weight loss of the CF/PC composites is less, indicating that the shear stress has little influence on the erosion of the CF/PC composites. It was also indicated that the CF/PC composites undergo a ductile–brittle transition and the maximum erosion angle has changed.

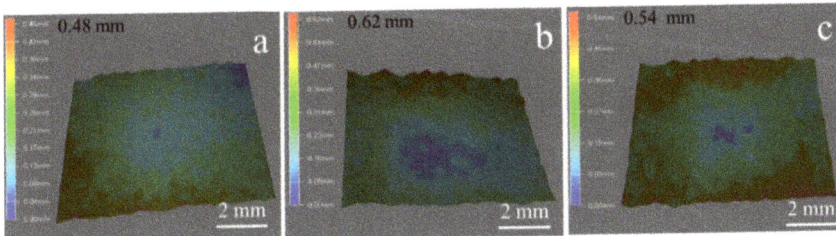

Figure 10. Three-dimensional picture of CF/PC composites at same experimental conditions after sand erosion. (**a**) un-aged composites surface after erosion at an angle of 60°, (**b**) hydrothermal aged for 7 days composites surface after erosion at an angle of 60°, (**c**) hydrothermal aged for 42 days after erosion at an angle of 60°.

4. Conclusions

This work presents the effects of hydrothermal aging on the mechanical properties and erosion resistance for CF/PC composites. The moisture absorption rate of CF/PC composites grows linearly with the square root of aging time during the first five days, and it stays flat when reaching saturation. On the premise of a sampling period and frequency every seven days, the tensile properties reached their maximum at the seventh day and the peak value of bending flexural performance was reached on the 14th day. The effects of aging time on storage modulus and solid particle erosion resistance are consistent with the stretching results. Hydrothermal aging causes holes to form on the surface of CF/PC composites and reduces the specimen's sand erosion resistance. Meanwhile, the physical property of CF/PC composites was changed.

Author Contributions: Writing original draft & data curation, M.F.; supervision & editing, N.Z.; methodology, M.H.; review & editing, B.L., validation, K.L.; supervision & editing, C.L. and C.S. All authors have read and agreed to the published version of the manuscript.

Funding: This work was supported by the National Natural Science Foundation of China (11432003), 111 project (D18023), the Key R&D Program of Jiangsu (BE2019096) and the Open Fund of State Key Laboratory of Structural Analysis for Industrial Equipment of DUT (GZ18203).

Conflicts of Interest: The authors declare no conflict of interest.

References

1. Xian, G.; Karbhari, V.M. DMTA based investigation of hygrothermal ageing of an epoxy system used in rehabilitation. *J. Appl. Polym. Sci.* **2007**, *104*, 1084–1094. [CrossRef]
2. Karbhari, V.M.; Xian, G. Hygrothermal effects on high VF pultruded unidirectional carbon/epoxy composites: Moisture uptake. *Compos. Part B Eng.* **2009**, *40*, 41–49. [CrossRef]

3. He, J.; Xian, G. Bond-slip behavior of fiber reinforced polymer strips-steel interface. *Constr. Build. Mater.* **2017**, *155*, 250–258. [CrossRef]
4. Jawali, N.D.; Siddaramaiah; Siddeshwarappa, B.; Lee, J.H. Polycarbonate/Short Glass Fiber Reinforced Composites—Physico-mechanical, Morphological and FEM Analysis. *J. Reinf. Plast. Compos.* **2007**, *27*, 313–319. [CrossRef]
5. Song, J.H.; Lim, J.K. Fatigue crack growth behavior and fiber orientation of glass fiber reinforced polycarbonate polymer composites. *Met. Mater. Int.* **2007**, *13*, 371–377. [CrossRef]
6. Zhang, J.; Chaisombat, K.; He, S.; Wang, C.H. Hybrid composite laminates reinforced with glass/carbon woven fabrics for lightweight load bearing structures. *Mater. Des.* **2012**, *36*, 75–80. [CrossRef]
7. Batuwitage, C.; Fawzia, S.; Thambiratnam, D.; Liu, X.; Al-Mahaidi, R.; Elchalakani, M. Impact behaviour of carbon fibre reinforced polymer (CFRP) strengthened square hollow steel tubes: A numerical simulation. *Thin-Walled Struct.* **2018**, *131*, 245–257. [CrossRef]
8. Alessi, S.; Pitarresi, G.; Spadaro, G. Effect of hydrothermal ageing on the thermal and delamination fracture behaviour of CFRP composites. *Compos. Part B Eng.* **2014**, *67*, 145–153. [CrossRef]
9. Marom, G. Environmental effects on fracture mechanical properties of polymer composites. In *Composite Materials Series*; Elsevier: Amsterdam, The Netherlands, 1988; Volume 6, pp. 397–424.
10. Shen, C.H.; Springer, G.S. Environmental effects on the elastic moduli of composite CF/PC composites. *J. Compos. Mater.* **1977**, *11*, 250–264. [CrossRef]
11. Tsai, Y.; Bosze, E.; Barjasteh, E.; Nutt, S. Influence of hygrothermal environment on thermal and mechanical properties of carbon fiber/fiberglass hybrid composites. *Compos. Sci. Technol.* **2009**, *69*, 432–437. [CrossRef]
12. Jefferson, G.D.; Farah, B.; Hempowicz, M.L.; Hsiao, K.T. Influence of hygrothermal aging on carbon nanofiber enhanced polyester CF/PC composites systems. *Compos. B Eng.* **2015**, *78*, 319–323. [CrossRef]
13. Bian, L.; Xiao, J.; Zeng, J.; Xing, S. Effects of seawater immersion on water absorption and mechanical properties of GFRP composites. *J. Compos. Mater.* **2012**, *46*, 3151–3162. [CrossRef]
14. Alomayri, T.; Assaedi, H.; Shaikh, F.U.A.; Low, I.M. Effect of water absorption on the mechanical properties of cotton fabric-reinforced geo-polymer composites. *J. Asian Ceram. Soc.* **2014**, *2*, 223–230. [CrossRef]
15. Li, Y.; Li, R.; Huang, L.; Wang, K.; Huang, X. Effect of hygrothermal aging on the damage characteristics of carbon woven fabric/epoxy laminates subjected to simulated lightning strike. *Mater. Des.* **2016**, *99*, 477–489. [CrossRef]
16. Meng, M.; Rizvi, J.; Grove, S.; Le, H. Effects of hygrothermal stress on the failure of CFRP composites. *Compos. Struct.* **2015**, *133*, 1024–1035. [CrossRef]
17. Grammatikos, S.; Evernden, M.C.; Mitchels, J.; Zafari, B.; Mottram, J.T.; Papanicolaou, G.C. On the response to hygrothermal aging of pultruded FRPs used in the civil engineering sector. *Mater. Des.* **2016**, *96*, 283–295. [CrossRef]
18. Ellyin, F.; Maser, R. Environmental effects on the mechanical properties of glass-fiber epoxy composite tubular specimens. *Compos. Sci. Technol.* **2004**, *64*, 1863–1874. [CrossRef]
19. Visco, A.; Campo, N.; Cianciafara, P. Comparison of seawater absorption properties of thermoset resins based composites. *Compos. Part A Appl. Sci. Manuf.* **2011**, *42*, 123–130. [CrossRef]
20. Harsha, A.; Jha, S.K. Erosive wear studies of epoxy-based composites at normal incidence. *Wear* **2008**, *265*, 1129–1135. [CrossRef]
21. Lu, X.J.; Zhang, Q. Effect of Water Absorption on the Impact Properties of Carbon Fiber/Epoxy Composites. *Chin. J. Aeronau.* **2016**, *19*, 14–18.
22. Ahmad, F.; Hong, J.W.; Choi, H.S.; Park, M.K. Hygro effects on the low-velocity impact behavior of unidirectional CFRP composite plates for aircraft applications. *Compos. Struct.* **2016**, *135*, 276–285. [CrossRef]
23. Mokhtar, H.; Sicot, O.; Rousseau, J.; Aminanda, Y.; Aivazzadeh, S. The Influence of Ageing on the Impact Damage of Carbon Epoxy Composites. *Procedia Eng.* **2011**, *7*, 2615–2620. [CrossRef]
24. Joseph, E.A.; Paul, D.R.; Barlow, J.W. Boiling water aging of a miscible blend of polycarbonate and a copolyester. *J. Appl. Polym. Sci.* **1982**, *27*, 4807–4819. [CrossRef]
25. Narkis, M.; Nicolais, L.; Apicella, A.; Bell, J.P. Hot water aging of polycarbonate. *Polym. Eng. Sci.* **1984**, *24*, 211–217. [CrossRef]
26. Ito, E.; Kobayashi, Y. Changes in physical properties of polycarbonate by absorbed water. *J. Appl. Polym. Sci.* **1978**, *22*, 1143–1149. [CrossRef]

27. Barkoula, N.M.; Karger, K. Processes and influencing parameters of the solid particle erosion of polymers and their composites. *Chemin. Form.* **2003**, *37*, 3807–3820. [CrossRef]
28. Schilling, F.C.; Ringo, W.M.; Sloane, N.J.A.; Bovey, F.A. Carbon-13 nuclear magnetic resonance study of the hydrolysis of bisphenol a polycarbonate. *Macromolecules* **1981**, *14*, 532–537. [CrossRef]
29. Fang, M.; Ma, Y.; Zhang, N.; Huang, M.; Lu, B.; Tan, K.; Liu, C.; Shen, C. Solid particle erosion resistance and electromagnetic shielding performance of carbon fiber reinforced polycarbonate composites. *Mater. Res. Express* **2020**, *7*, 045305. [CrossRef]
30. Chen, J.; Hutchings, I.M.; Deng, T.; Bradley, M.S.; Koziol, K. The effect of carbon nanotube orientation on erosive wear resistance of CNT-epoxy based composites. *Carbon* **2014**, *73*, 421–431. [CrossRef]

Publisher's Note: MDPI stays neutral with regard to jurisdictional claims in published maps and institutional affiliations.

© 2020 by the authors. Licensee MDPI, Basel, Switzerland. This article is an open access article distributed under the terms and conditions of the Creative Commons Attribution (CC BY) license (http://creativecommons.org/licenses/by/4.0/).

Article

Multiscale Structural Evolution and Its Relationship to Dielectric Properties of Micro-/Nano-Layer Coextruded PVDF-HFP/PC Films

Jie Wang [1], Daniel Adami [2], Bo Lu [1,*], Chuntai Liu [1], Abderrahim Maazouz [2,3] and Khalid Lamnawar [2]

1. Key Laboratory of Materials Processing and Mold (Ministry of Education), National Engineering Research Center for Advanced Polymer Processing Technology, Zhengzhou University, Zhengzhou 450002, China; jjwangnet@163.com (J.W.); ctliu@zzu.edu.cn (C.L.)
2. CNRS, UMR 5223, Ingénierie des Matériaux Polymères, INSA Lyon, Université de Lyon, F-69621 Villeurbanne, France; daniadami_@hotmail.com (D.A.); abderrahim.maazouz@insa-lyon.fr (A.M.); khalid.lamnawar@insa-lyon.fr (K.L.)
3. Hassan II Academy of Science and Technology, 10100 Rabat, Morocco
* Correspondence: bolu@zzu.edu.cn

Received: 19 October 2020; Accepted: 31 October 2020; Published: 5 November 2020

Abstract: An understanding of the structural evolution in micro-/nano-layer co-extrusion process is essential to fabricate high-performance multilayered products. Therefore, in this work, we reveal systematically the multiscale structural development, involving both the layer architecture and microstructure within layers of micro-/nano-layer coextruded polymer films, as well as its relationship to dielectric properties, based on poly(vinylidene fluoride-*co*-hexafluoropropylene) (PVDF-HFP)/polycarbonate (PC) system. Interestingly, layer architecture and morphology show strong dependences on the nominal layer thicknesses. Particularly, with layer thickness reduced to nanometer scale, interfacial instabilities triggered by viscoelastic differences between components emerge with the creation of micro-droplets and micro-sheets. Films show an enhanced crystallization with the formation of two-dimensional (2D) spherulites in microlayer coextruded systems and the oriented in-plane lamellae in nanolayer coextruded counterparts, where layer breakup in the thinner layers further changes the crystallization behaviors. These macro- and microscopic structures, developed from the co-extrusion process, substantially influence the dielectric properties of coextruded films. Mechanism responsible for dielectric performance is further proposed by considering these effects of multiscale structure on the dipole switching and charge hopping in the multilayered structures. This work clearly demonstrates how the multiscale structural evolution during the micro-/nano-layer coextrusion process can control the dielectric properties of multilayered products.

Keywords: micro-/nano-layer coextrusion; multilayer films; multiscale structure; dielectric properties

1. Introduction

Multilayer polymer films have been increasingly used in electrics, energy, display devices, packaging, construction and other applications. Methods of fabricating polymer multilayer films generally include layer-by-layer assembly (LbL) [1], and multilayer coextrusion technology [2,3]. Among them, the multilayer coextrusion technology has aroused widespread interest over the past 20 years, because it has become a reliable technology for continuous production of micro- and nanolayers. Different from layer-by-layer assembly, multilayer coextrusion is a top-down method that can manufacture thousands of layers of films with layer thicknesses controllable down to nanometer scale, thus endowing the final films with high degree of flexibility, tunable gas/liquid barrier [4], mechanical [5,6], optical [7], and electrical properties [8,9].

Generally, a stable and continuous multilayer morphology in films is necessary to achieve the tailor-made end-use properties that are superior to those of conventional polymer blend films. Many efforts have been, thus, made to prepare microlayer and nanolayer films with a great number of layers alternating of polymer pairs, such as polystyrene (PS)/poly(methyl methacrylate) (PMMA) [10], polycarbonate (PC)/PMMA [11], PC/polyvinylidene fluoride (PVDF) [12], etc. Although, many polymer pairs could be combined to produce multilayer films via layer multiplying coextrusion technology, layer instabilities can occur under the laminar flow conditions, which deteriorates the layer continuity and uniformity [13,14]. This concern holds particularly true for polymer pairs with a large difference in rheological properties between components [2]. Also, it has been observed that below a certain layer thickness normally of several nanometers, layers tend to lose their integrity with severe interfacial distortions [15]. Some polymer layers can even break into nanosheets and nanodroplets. The presence layer instabilities and layer breakup can greatly influence the properties of films [15]. Apart from the macroscopic layer structure, the microstructure inside the layers are also important in defining the macroscopic properties of multilayer products. With a number of layers multiplied in a limited space, various microstructure and dynamics from molecular aggregation within layers, including the macromolecular alignment, crystallization, and glass transition, can be developed with the decrease in layer thicknesses [16–18]. For example, a confined crystallization with highly oriented crystals stacked between layers was recently reported by reducing the layer thicknesses, which resulted an enhanced dielectric and other physicochemical properties [19–21]. Despite of the huge importance of multilayer films in the daily life, the evolutions in both the macroscopic layer structure and microstructure inside layers are far from being completely understood, as well as their effects on the resulting macroscopic properties. The question of understanding how the multiscale structure develops all along the multilayer coextrusion processing line then emerges as a crucial issue, in order to optimize the process parameters and manufacture multilayer films with excellent properties.

Therefore, the objective of this present work is to demonstrate the development of multiscale structure along the multilayer coextrusion process and its influence on the resulting dielectric properties. To this end, films alternating components of poly(vinylidene fluoride-*co*-hexafluoropropylene) (PVDF-HFP) and polycarbonate (PC) were fabricated by micro-/nano-layer coextrusion process. The evolution in the layer structure and microstructure of layers with varying the number of layers or layer thicknesses was investigated. By combining a high dielectric constant polymer of PVDF-HFP with super electrochemical properties [22–24], and a linear dielectric polymer of PC with low dielectric loss and high breakdown strength, dielectric properties of the obtained films were further investigated. The relationship between the multiscale structure and the final dielectric properties was thus established, and the corresponding mechanisms was elucidated. This work will give some new guidelines for controlling the multiscale structure development and macroscopic properties of multilayer products from multilayered assembly processing.

2. Experimental Section

2.1. Materials

Two kinds of polymers were used in this work. Poly(vinylidene fluoride-*co*-hexafluoropropylene) (PVDF-HFP, Kynar Flex 2500-20, ARKEMA, Colombes, France) with weight-average molecular weight M_w of 500,000 g/mol and melt flow index of 5.8 cm^3/10 min at 230 °C/3.8 kg was supplied by Arkema (Colombes, France). PC (Calibre 303EP) with M_w of 23,000 g/mol and melt flow index of 5.8 cm^3/10 min at 300 °C/1.2 kg, were kindly provided by Trinseo (Stade, Germany). Both virgin polymers were in the granular pellet form. The glass transition temperatures of PVDF-HFP and PC are −34, and 150 °C, respectively, as determined using a differential scanning calorimetry. All the pristine products were dried in a vacuum oven at 80 °C for 48 h before usage.

2.2. Sample Preparation

PVDF-HFP/PC multilayer films were prepared using a homemademultilayer coextrusion system that is capable of preparing microlayer and nanolayer films. As schematically illustrated by Figure 1, this setup is composed of two single-screw extruders (A and B), a feedblock, a set of layer multiplying elements, an exit die and a thermally regulated chill roll. The layer multiplying element used here has a constant cross-sectional area. More details about this multilayer coextrusion system has been given in our recent report [25]. During processing, two component polymer melts of PVDF-HFP and PC are extruded from two extruders, and brought together at the feedblock, followed by going through a series of layer multiplying elements. Each element multiplies the melt while keeping the total melt thickness constant, thus doubling the number of layers and reducing the individual layer thickness by a factor of 2. During coextrusion process, the extruders, multipliers and die temperatures were prescribed to 250 °C. Films containing the nominal number layers of 2-16384 layers were fabricated by a given number of multipliers (Table 1). The as-prepared multilayer films were abbreviated as "NL" where N represents the nominal number of layers. Films were collected by a chill roll at a temperature of 80 °C with a negligible speed to take them without stretching. The nominal layer thicknesses were calculated by the total thickness divided by the nominal number of layers and listed in Table 1. Layer thicknesses of all films were further determined by the morphological investigations.

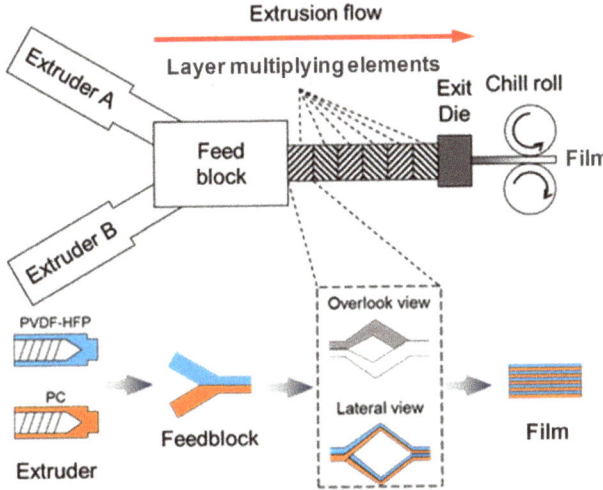

Figure 1. Schematic of the layer multiplying process in the multilayer coextrusion system.

Table 1. Characteristics of coextruded PVDF-HFP/PC films.

Samples	No. of Multipliers (n)	No. of Layers (N)	Total Film Thickness (μm)	Nominal Layer Thickness (nm)
2 L	0	2	243	121,250
8 L	2	8	245	30,630
32 L	4	32	248	7730
256 L	7	256	313	1220
2048 L	10	2048	325	158
16,384 L	13	16,384	415	25

2.3. Characterizations

2.3.1. Rheological Measurements

Linear viscoelasticity of PVDF-HFP and PC was evaluated using a shear rheometer (DHR-2, TA Instruments, New Castle, DE, USA) with a 25 mm parallel-plate geometry at different temperatures from 220 to 260 °C under a nitrogen atmosphere. Dynamic frequency sweep tests were performed from the angular frequency of 628–0.05 rad/s and under a strain of 5%.

2.3.2. Morphological Observations

Cross-sectional morphology of as-coextruded films were observed by an atomic force microscope (AFM, Bruker Multimode 8, Santa Barbara, CA, USA) at room temperature. Before our observation, the films were embedded in the standard epoxy, and cured at room temperature for overnight. A flat and smooth cross section of the cured film was obtained by cryo-ultramicrotoming at −80 °C with a diamond knife blade. Phase and height images of the cross sections were taken during observations.

2.3.3. Differential Scanning Calorimetry (DSC)

Thermal properties of the as-coextruded films were examined using a differential scanning calorimeter (DSC, Q20, TA Instruments, New Castle, DE, USA) under a nitrogen atmosphere. Specimens around ~5 mg taken from films were loaded into the DSC aluminum pans. Samples were first heated from −80 to 240 °C and then equilibrated at 240 °C for 3 min, followed by being cooled to −80 °C. Both the heating and cooling rates were set as 10 °C/min. The crystallinity (X_c) of PVDF-HFP in films was determined from the enthalpy of fusion (ΔH_m) in the heating scan according to,

$$X_c = \frac{\Delta H_m}{w \Delta H_m^0} \times 100\% \tag{1}$$

where w is the weight fraction of PVDF-HFP, and ΔH_m^0 is the theoretical melting enthalpy of 100% crystalline polymer of PVDF-HFP, estimated as 104.6 J/g from the literature data [26].

2.3.4. X-ray Analyses

The crystalline morphology and structure in as-coextruded films were explored using two-dimensional wide-angle X-ray diffraction (WAXD, D8 Discover, Bruker, Karlsruhe, Germany) and small-angle X-ray scattering (SAXS, NanoSTAR-U, Bruker, Karlsruhe, Germany). The X-ray wavelength was 0.154 nm with Cu Kα radiation source. Signals were obtained by aligning the incident X-ray beam parallel to the extrusion direction (ED) of the film.

2.3.5. Fourier Transform Infrared Spectroscopy (FTIR)

Fourier transform infrared spectroscopy was carried out using an infrared spectrometer (Nicolet 6700, Thermo Fisher Scientific, Waltham, MA, USA). Infrared spectra were collected in transmission mode at a resolution of 4 cm^{-1} with 64 scans. A thin slice about 4 μm in thickness was cut from the as-extruded films using a Rotary microtome (Leica RM2235, Leica Microsystems GmbH, Wetzlar, Germany) at room temperature. Samples, same to those for morphological measurements, were embedded in the standard epoxy and cured at room temperature prior to cutting.

2.3.6. Dielectric Measurements

Dielectric properties of as-coextruded films were measured using a precision LCR meter (Agilent E4980A, Agilent Technologies Inc., Penang, Malaysia) equipped with an Environmental Test Chamber (ETC, TA Instruments, New Castle, DE, USA) for temperature control. The silver electrode is applied to both sides of the film samples with a diameter of 25 mm. Frequency scans from 20 Hz to 2 MHz were carried out with a constant voltage of 1 V. For temperature scans, temperature was ramped from −60 to 150 °C at a rate of 5 °C/min. All tests were repeated at least three times with fresh samples to ensure the reproducibility.

3. Results and Discussion

3.1. Viscoelastic Ratios between PVDF-HFP and PC

It is known that the melt viscosity and/or elasticity ratios among components are highly important in determining the layer architecture and uniformity for multilayered polymer structures. To determine the viscoelastic ratios, the linear viscoelasticity of PVDF-HFP and PC were investigated using dynamic shear rheology at the processing temperature of 250 °C (Figure 2a), and the frequency dependent viscosity and elasticity ratios between two polymers were computed accordingly (Figure 2b). PVDF-HFP shows the higher elasticity with the larger storage modulus relative to PC within the measured frequency range. Also, PVDF-HFP is more viscous with the larger complex viscosity and loss modulus than PC, and it shows the more pronounced shear thinning (see inset of Figure 2a). Due to the different frequency dependencies of PVDF-HFP and PC a decrease in viscosity and elasticity ratios can be noticed by increasing the frequency. Typically, the shear rate range of the coextrusion process lies between 1 and 10 s^{-1} [27]. Therefore, assuming the validity of Cox-Merz rule [28], the viscosity ratio of PVDF-HFP to PC within 1–10 s^{-1} is estimated to be ranging from 2.8 to 6.8, while the elasticity ratio is from 15 to 200 depending on the shear rates of coextrusion processing window (Figure 2b).

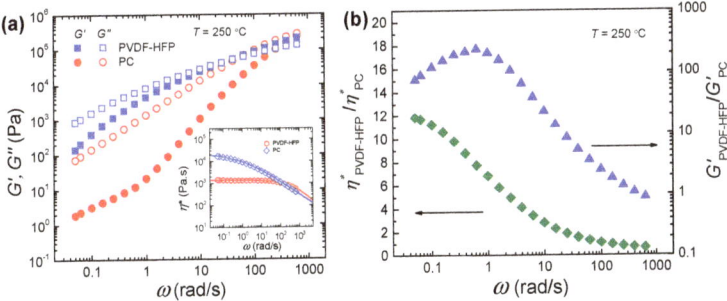

Figure 2. (a) Comparison of storage modulus (G') and loss modulus (G'') against angular frequency for PVDF-HFP and PC at 250 °C. The inset in (a) shows the complex viscosity versus frequency. (b) Viscosity and elasticity ratios of PVDF-HFP to PC against angular frequency at 250 °C.

3.2. Layer Morphology of As-Coextruded Films

The layer structure and uniformity in as-coextruded PVDF-HFP/PC films were investigated using AFM. Figure 3a–d display the morphological features of cross-sections of films with various nominal number of layers. In the phase images, the dark phase represents PVDF-HFP and the bright phase represents PC, due to the higher modulus of PC than PVDF-HFP at room temperature. As visualized by AFM, continuous PVDF-HFP and PC layers are noticeable for films with nominal number of layers ranging from 2 to 256. Using image analysis, the average layer thickness for both components were determined (Figure 3e). In this instance, for the 32 L film PVDF-HFP and PC layer thickness were identified to be 7.79 ± 2.12 µm, and 5.05 ± 1.22 µm, respectively. The thicker PVDF-HFP layers than PC layers are ascribed to the higher viscosity and modulus during layer multiplication process. Constituent layer thicknesses were further decreased by increasing the number of layers, e.g., 703 ± 121 nm for PVDF-HFP and 365 ± 192 nm for PC in the 256 L film. However, at the nominal layer thicknesses below 160 nm, about half of the layers remains continuous and some layers even broke up into micro-sheets. It is evident that these elongated micro-sheets are still parallel to the survived continuous layers, even when some of the micro-sheets coalesced into droplets (see dashed circles in Figure 3c–e). Those micro-sheets and droplets result from interfacial instabilities in the unstable continuous layers that are triggered by the melt viscosity and elasticity differences between two constituent polymers during layer multiplying coextrusion process [2,15]. With the further decrease in the layer thicknesses, the layers are prone to having a gradual breakup process. At 25 nm nominal layer thickness (i.e., 16384 L film, Figure 3d), almost no continuous layers can be discerned across the film. Such a nominal layer thickness is close to the critical layer thickness (~10 nm) for the occurrence of layer breakup as recently reported for immiscible PS/PMMA multilayer systems [29]. It is also worthwhile that at the locations where layer breakup occurred (Figure 3d), some droplets display a diameter up to ~1.0 µm, and some sheets have a thickness of several microns, of which the sizes are unprecedently larger than the expected nominal layer thickness. Similar morphological feature has been also observed in PMMA/PS nanolayer films when subjected to a post-annealing to the molten state [10]. The larger sizes of droplets and sheets could be explained by the physical properties of polymers. PVDF-HFP is a semicrystalline polymer with a melting temperature of 124 °C and crystallization temperature of 92 °C, while PC is an amorphous polymer with a glass transition of 150 °C. It could be expected that phase coalescence of thin layers took place as the polymer flow exited the extrusion die (250 °C), thus leading to the larger sized micro-droplets and micro-sheets.

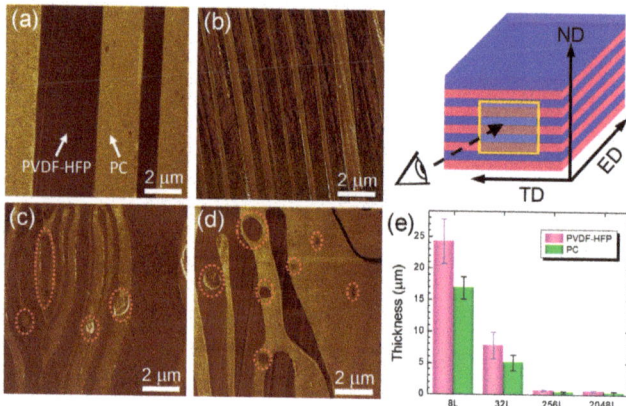

Figure 3. AFM phase images for PVDF-HFP/PC films: (**a**) 32 L, (**b**) 256 L, (**c**) 2048 L and (**d**) 16384 L. (**e**) is the statistical layer thicknesses determined by AFM analysis. Regions inside the dashed circles in (c) and (d) indicate the broken layers. The scheme appended in the upper right corner shows the configuration of observed cross-sections relative to the extrusion direction (ED), transverse direction (TD), and normal direction (ND) of the film.

3.3. Microstructure Evolutions in As-Coextruded Films

The microstructure inside the as-coextruded PVDF-HFP/PC films was further investigated. Figure 4a shows the X-ray diffraction profiles captured by WAXD for all films. In all as-extruded films, PVDF-HFP crystallized typically in the dominated α-phase as noticed from the diffraction peaks indexed as (020), (110) and (021) planes [30]. Further analysis of crystalline polymorphism using FTIR reveals that strong α-phase with some traces of β-phase coexists in these films as evidenced by peaks at 766 and 1041 cm^{-1} for the α-phase, and 1279 and 840 cm^{-1} for the β-phase (Figure 4b) [31]. Such polar β-phase crystals that are electrically active phases are beneficial for application in dielectric and piezoelectric devices. The formation of β-phase should be related to the strong polarized interatomic C-F bonds from the extreme electronegativity of the fluorine atoms compared to that of carbon atoms in the copolymer of PVDF-HFP [32]. It is also interesting that an increase in the diffraction intensity is noticeable with increasing the nominal number of layers, indicating an enhanced crystallization of PVDF-HFP at the higher number of layers. This observation was further validated by the crystallinity (X_c) determined by DSC analysis (Figure 4c, Table 2). Quantitatively, the crystallinity of coextruded films is gradually increased to be 20.4% in the 2048 L film and then further reduced to be 13.3%, which is clearly larger than that of pure PVDF-HFP. The observed increase in crystallinity is explained by the microstructure evolution of crystallization in the layered systems. As illustrated by the scheme in Figure 5, for microlayer films the crystals have relatively larger space to crystallize into three-dimensional (3D) spherulites. However, with the decrease in layer thicknesses, these spaces allowing for crystallization are reduced, which would result in the formation of 2D disc-like oriented spherulites (see the bottom panel of Figure 5). As the layer thickness is further reduced to nanoscale at the higher number of layers, the highly oriented in-plane lamellae will be formed between two neighboring rigid PC layers [16,18]. These stacked lamellae are believed to have the more crystallinity than spherulites composed of crystallization and interlamellar amorphous regions. This could explain the increased crystallinity at the higher number of layers.

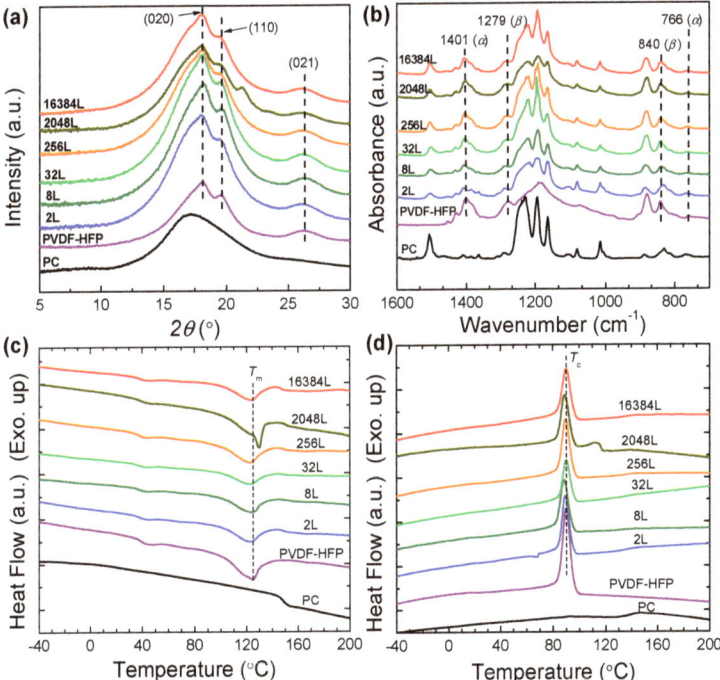

Figure 4. (a) WAXD profiles for coextruded PVDF-HFP/PC films varying the nominal number of layers. The intensity was normalized with film thicknesses. (b) FTIR spectra of PVDF-HFP/PC films. DSC thermographs of; (c) heating scan; and (d) cooling scan for coextruded PVDF-HFP/PC films.

Table 2. Melting temperature (T_m), enthalpy of fusion (ΔH_m), crystallinity (X_c) and crystallization temperature (T_c) obtained from the first heating scan of DSC.

Samples	T_m (°C)	ΔH_m (J/g)	X_c (%)	T_c (°C)
Pure PVDF-HFP	124.81	13.57	12.9	92.08
Pure PC	-	-	-	-
2 L	122.54	8.77	13.9	91.94
8 L	123.14	9.098	14.4	91.52
32 L	122.31	10.69	16.9	91.78
256 L	122.24	10.78	17.1	91.26
2048 L	128.69	12.85	20.4	91.13
16,384 L	123.02	8.35	13.3	91.72

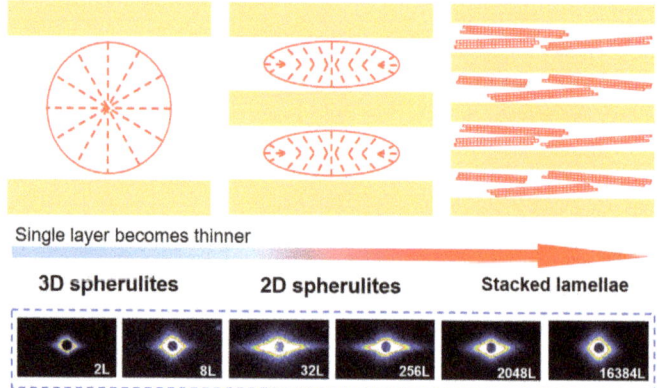

Figure 5. Schematic of the morphology evolution of PVDF-HFP crystallization with the decrease in layer thicknesses from micro- to nanoscale. The panel appended at the bottom shows the SAXS patterns recorded with X-ray beam parallel to the extrusion direction for coextruded PVDF-HFP/PC films.

It is also surprising to view that the melting behavior for the 16,384 L film resembles those for 2 L to 256 L films (Figure 4c). For nanolayer coextruded films with a larger amount of layer breakup, the broken thin layers and sheets would relax and coalesce into a larger phase structure like droplets before melt crystallization as discussed earlier. The confinement of rigid PC layers on the crystallization of PVDF-HFP would be therefore alleviated, which is confirmed by the reduced structural orientations shown by the SAXS pattern of 16384 L film (see the bottom panel of Figure 5). This led to the formation of crystallization structure close to that in microlayer films, as evidenced by the similar crystallinity as the latter cases (Table 2). This argument was further supported by checking the crystallization behaviors of films during the DSC cooling scan after eliminating all layer structures through an annealing at an elevated temperature of 240 °C for 3.0 min to (Figure 4d). Notably, the crystallization temperatures (T_c) for all annealed films almost coincides with these of microlayers and bulk PVDF-HFP component (Table 2). This, in turn, suggests that different crystallization behaviors of PVDF-HFP in a layered system from that in a blend-like system.

3.4. Dielectric Properties of As-Coextruded Films

Coextruded multilayer films are known as the promising materials in dielectric capacitor applications. Therefore, dielectric properties of coextruded PVDF-HFP/PC films were evaluated using dielectric spectroscopy. Figure 6a–c compares the dielectric spectra as a function of frequency for as-coextruded PVDF-HFP/PC films measured at the room temperature (30 °C). The corresponding dielectric spectra versus temperature at the tested frequency of 1 kHz are plotted in Figure 6d–f. Clearly, the dielectric properties of these films show a strong dependency on the nominal number of layers. As shown in Figure 6a, the values of ε' (dielectric constant $k = \varepsilon'/\varepsilon_0$ with ε_0 being the vacuum permittivity) at 30 °C for PVDF-HFP/PC films fall in between neat polymers. The ε' is gradually improved by increasing the temperature due to the fact that dipoles become more activated upon heating. To clearly demonstrate the dependency of dielectric properties on the nominal number of layers, the theoretical spectra of the multilayer system are also calculated (dashed lines in Figure 5) according to the formula as detailed in our recent study [25,33]. It is obvious that the value of ε' is continuously increased with nominal number of layers increasing up to 8. Herein, the increased ε' is related to the dipole polarizations contributed by the effects from crystallizations and molecular orientations. As aforementioned, increasing the nominal number of layers enhanced the crystallizations that would lead to less mobile dipoles, which are detrimental for improving dipole polarization. Instead, the improved structural orientations with the reduction in the layer thicknesses result in the alignment of the *c*-axis of crystals along the film in-plane direction, which facilitates the preferential switching of dipole moments parallel to the electric field. Therefore, such improvement in ε' is ascribed primarily to the enhanced molecular orientations. However, we can also note that ε' is further suppressed with increasing the nominal number of layers to 16,384. This should be caused by the reduced orientations by the presence of layer breakup, as demonstrated above. It is also worth mentioning that the dielectric performance obtained in this study is higher than that reported in the literature. Taking the 32 L film as an example, the dielectric constant k and dielectric loss tanδ are 6.7 and 0.01, at 1000 Hz, respectively, which is superior to literature values for 32 L film with K being 4.5 and tanδ being 0.02 [34,35]. This should be attributed to the higher structural orientation in the multilayer system of this study.

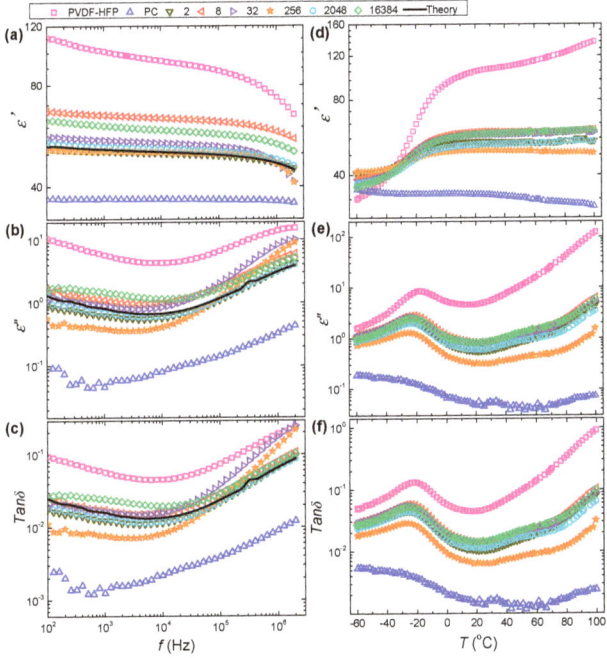

Figure 6. (a) Storage permittivity, (b) loss permittivity, and (c) loss tangent spectra versus frequency at 30 °C for PVDF-HFP/PC films. (d–f) Temperature dependent dielectric spectra measured at 1 kHz. Solid lines in (a–c) are theoretical predictions for layered dielectrics without interphase layers.

It is also worth mentioning that the dielectric loss of the microlayer films (i.e., 2 L to 256 L) are lower than that of 2048 L and 16,384 L films. This could be interpreted by invoking the structural evolutions in PVDF-HFP/PC films. As known, dielectric loss for polymer dielectrics are mainly related to the hopping of charge carriers and the energy dissipation from the switching of dipoles [36]. As schematically shown in Figure 7, under the action of an external electric field, charge ions, free electrons and polarized impurity ions in the more conducting component of PVDF-HFP will migrate and accumulate at the PVDF-HFP/PC interfaces. Namely, the more insulating PC layers act as charge-blocking layers to form interfacial polarization. The interfacial charges thus alter the local electric field, hinder the hopping of electrons, and eventually minimize the leakage current going through the whole films. Therefore, the thicker the PC layer, the more interface charges will be effectively blocked. However, when a large number of layers are broken in nanolayer extruded films, the charger carriers can move easily across physical thickness of film, resulting in a large increase in dielectric loss (e.g., 16,384 L). Also, for nanolayer coextruded films, with the occurrence of layer breakup the energy dissipation from the switching of dipoles is higher due to the decreased molecular orientations along the in-plane direction as aforementioned. These effects thus accounts for the higher dielectric loss in 2048 and 16,384 L films. Therefore, it is concluded that the dielectric performance of multilayer films is strongly dependent on the layer architecture and microstructure. A better control of multilayer structure via processing offers a strategy for tailoring their dielectric properties.

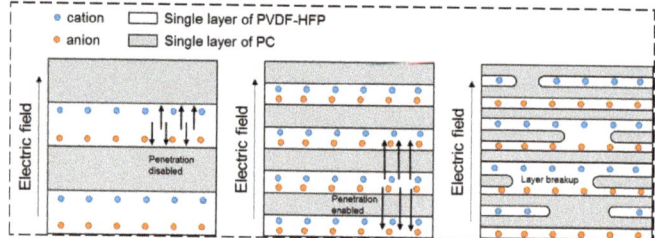

Figure 7. Schematic illustration of the interfacial polarization of space charges in coextruded PVDF-HFP/PC films varying layer thickness and structure under an external electric field.

4. Conclusions

In summary, we investigated the multiscale structural development and its relationship with the dielectric properties of micro-/nano-layer coextruded polymer films based on PVDF-HFP/PC system. Multiscale structural evolutions involving both the layer architecture and microstructure within layers were revealed systematically. The layers were stable and continuous for microlayer coextruded films with nominal layer thickness. Whereas, for nanolayer coextruded films at nominal layer thickness below 160 nm, layer integrity was reduced by interfacial instabilities triggered by viscoelastic differences between component melts. Layers even broke into micro-droplet and micro-sheets due to the coalescence of thin layers. Besides, with the reduction in nominal layer thickness, films displayed an enhanced crystallization, with the formation of 2D oriented spherulite structure in microlayer coextruded systems and highly oriented in-plane lamellae in nanolayer coextruded counterparts. Layer breakup in thinner layers further resulted in the crystallization and structural orientation similar to that in microlayer films, which was attributed to the relaxation and phase coalescence of thin layers during processing. Furthermore, dielectric properties of films were strongly dependent on these multiscale structures. The gradually increased storage permittivity with reducing layer thickness was ascribed to the enhanced molecular orientations that could facilitate the dipole switching. The lower storage permittivity in thin layers with breakup was caused by the reduced orientations. In addition, the dielectric loss of microlayer coextruded films was lower than that of nanolayer coextruded analogues. This was due to the increased hopping of charge carriers and the higher energy dissipation from dipole switching when layers were broken. The results of this study will enable a better understanding of the multiscale structure evolution in micro-/nanolayer coextrusion to optimize the target macroscopic properties.

Author Contributions: Conceptualization, B.L. and K.L.; methodology, B.L.; software, J.W.; validation, B.L., C.L. and K.L.; formal analysis, D.A.; investigation, J.W. and D.A.; resources, C.L., K.L. and A.M.; data curation, J.W.; writing—original draft preparation, J.W.; writing—review and editing, B.L.; visualization, D.A.; supervision, B.L.; project administration, B.L.; funding acquisition, B.L. and K.L. All authors have read and agreed to the published version of the manuscript.

Funding: This research was funded by the financial support from the National Key R&D Program of China (grant no. 2019YFA0706801), China Postdoctoral Science Foundation (grant no.2019M650174), the China 111 project (grant no. D18023), the French National Research Agency (ANR, grant no. ANR-11-RMNP-0002) and Auvergne-Rhône-Alpes Region Council (ARC, AURA 2017-2020).

Acknowledgments: The authors acknowledge Arkema for providing the virgin polymers. This manuscript is in honor of the 50th anniversary of the French Polymer Group (Groupe Français des Polymères-GFP).

Conflicts of Interest: The authors declare no conflict of interest.

References

1. Richardson, J.J.; Cui, J.; Björnmalm, M.; Braunger, J.A.; Ejima, H.; Caruso, F. Innovation in Layer-by-Layer Assembly. *Chem. Rev.* **2016**, *116*, 14828–14867. [CrossRef] [PubMed]
2. Langhe, D.; Ponting, M. *Manufacturing and Novel Applications of Multilayer Polymer Films*; Elsevier: Oxford, UK, 2016.
3. Zhang, X.; Xu, Y.; Zhang, X.; Wu, H.; Shen, J.; Chen, R.; Xiong, Y.; Li, J.; Guo, S. Progress on the layer-by-layer assembly of multilayered polymer composites: Strategy, structural control and applications. *Prog. Polym. Sci.* **2019**, *89*, 76–107. [CrossRef]
4. Feng, J.; Li, Z.; Olah, A.; Baer, E. High oxygen barrier multilayer EVOH/LDPE film/foam. *J. Appl. Polym. Sci.* **2018**, *135*, 46425. [CrossRef]
5. Ji, X.; Chen, D.; Zheng, Y.; Shen, J.; Guo, S.; Harkin-Jones, E. Multilayered assembly of poly(vinylidene fluoride) and poly(methyl methacrylate) for achieving multi-shape memory effects. *Chem. Eng. J.* **2019**, *362*, 190–198. [CrossRef]
6. Zheng, Y.; Dong, R.; Shen, J.; Guo, S. Tunable Shape Memory Performances via Multilayer Assembly of Thermoplastic Polyurethane and Polycaprolactone. *ACS Appl. Mater. Interfaces* **2016**, *8*, 1371–1380. [CrossRef] [PubMed]
7. Li, Z.; Sun, R.; Rahman, M.A.; Feng, J.; Olah, A.; Baer, E. Scaling effects on the optical properties of patterned nano-layered shape memory films. *Polymer* **2019**, *167*, 182–192. [CrossRef]
8. Zhu, J.; Shen, J.; Guo, S.; Sue, H.-J. Confined distribution of conductive particles in polyvinylidene fluoride-based multilayered dielectrics: Toward high permittivity and breakdown strength. *Carbon* **2015**, *84*, 355–364. [CrossRef]
9. Baer, E.; Zhu, L. 50th Anniversary Perspective: Dielectric Phenomena in Polymers and Multilayered Dielectric Films. *Macromolecules* **2017**, *50*, 2239–2256. [CrossRef]
10. Ania, F.; Baltá-Calleja, F.J.; Henning, S.; Khariwala, D.; Hiltner, A.; Baer, E. Study of the multilayered nanostructure and thermal stability of PMMA/PS amorphous films. *Polymer* **2010**, *51*, 1805–1811. [CrossRef]
11. Abdel-Mohti, A.; Garbash, A.N.; Almagahwi, S.; Shen, H. Effect of layer and film thickness and temperature on the mechanical property of micro-and nano-layered PC/PMMA films subjected to thermal aging. *Materials* **2015**, *8*, 2062–2075. [CrossRef]
12. Chen, X.; Tseng, J.-K.; Treufeld, I.; Mackey, M.; Schuele, D.E.; Li, R.; Fukuto, M.; Baer, E.; Zhu, L. Enhanced Dielectric Properties due to Space Charge-induced Interfacial Polarization in Multilayer Polymer Films. *J. Mater. Chem. C* **2017**, *5*, 10417–10426. [CrossRef]
13. Schrenk, W.J.; Bradley, N.L.; Alfrey, T.; Maack, H. Interfacial flow instability in multilayer coextrusion. *Polym. Eng. Sci.* **1978**, *18*, 620–623. [CrossRef]
14. Han, C.D.; Shetty, R. Studies on multilayer film coextrusion II. Interfacial instability in flat film coextrusion. *Polym. Eng. Sci.* **1978**, *18*, 180–186. [CrossRef]
15. Feng, J.; Zhang, Z.; Bironeau, A.; Guinault, A.; Miquelard-Garnier, G.; Sollogoub, C.; Olah, A.; Baer, E. Breakup behavior of nanolayers in polymeric multilayer systems—Creation of nanosheets and nanodroplets. *Polymer* **2018**, *143*, 19–27. [CrossRef]
16. Wang, H.; Keum, J.K.; Hiltner, A.; Baer, E.; Freeman, B.; Rozanski, A.; Galeski, A. Confined Crystallization of Polyethylene Oxide in Nanolayer Assemblies. *Science* **2009**, *323*, 757–760. [CrossRef] [PubMed]
17. Casalini, R.; Zhu, L.; Baer, E.; Roland, C.M. Segmental Dynamics and the Correlation Length in Nanoconfined PMMA. *Polymer* **2016**, *88*, 133–136. [CrossRef]
18. Carr, J.M.; Langhe, D.S.; Ponting, M.T.; Hiltner, A.; Baer, E. Confined Crystallization in Polymer Nanolayered Films: A review. *J. Mater. Res.* **2012**, *27*, 1326–1350. [CrossRef]
19. Huang, H.; Chen, X.; Li, R.; Fukuto, M.; Schuele, D.E.; Ponting, M.; Langhe, D.; Baer, E.; Zhu, L. Flat-On Secondary Crystals as Effective Blocks To Reduce Ionic Conduction Loss in Polysulfone/Poly(vinylidene fluoride) Multilayer Dielectric Films. *Macromolecules* **2018**, *51*, 5019–5026. [CrossRef]

20. Yin, K.; Zhou, Z.; Schuele, D.E.; Wolak, M.; Zhu, L.; Baer, E. Effects of Interphase Modification and Biaxial Orientation on Dielectric Properties of Poly(ethylene terephthalate)/Poly(vinylidene fluoride-co-hexafluoropropylene) Multilayer Films. *ACS Appl. Mater. Interfaces* **2016**, *8*, 13555–13566. [CrossRef] [PubMed]
21. Li, Z.; Olah, A.; Baer, E. Micro- and nano-layered processing of new polymeric systems. *Prog. Polym. Sci.* **2020**, *102*, 101210. [CrossRef]
22. Ranjani, M.; Yoo, D.J.; Gnana kumar, G. Sulfonated Fe_3O_4@SiO_2 nanorods incorporated sPVdF nanocomposite membranes for DMFC applications. *J. Membr. Sci.* **2018**, *555*, 497–506. [CrossRef]
23. Hariprasad, R.; Vinothkannan, M.; Kim, A.R.; Yoo, D.J. SPVdF-HFP/SGO nanohybrid proton exchange membrane for the applications of direct methanol fuel cells. *J. Dispersion Sci. Technol.* **2019**, 1–13. [CrossRef]
24. Kim, A.R.; Vinothkannan, M.; Yoo, D.J. Fabrication of Binary Sulfonated Poly Ether Sulfone and Sulfonated Polyvinylidene Fluoride-Co-Hexafluoro Propylene Blend Membrane as Efficient Electrolyte for Proton Exchange Membrane Fuel Cells. *Bull. Korean Chem. Soc.* **2018**, *39*, 913–919. [CrossRef]
25. Lu, B.; Lamnawar, K.; Maazouz, A.; Sudre, G. Critical Role of Interfacial Diffusion and Diffuse Interphases Formed in Multi-Micro-/Nanolayered Polymer Films Based on Poly(vinylidene fluoride) and Poly(methyl methacrylate). *ACS Appl. Mater. Interfaces* **2018**, *10*, 29019–29037. [CrossRef]
26. Teyssedre, G.; Bernes, A.; Lacabanne, C. Influence of the Crystalline Phase on the Molecular Mobility of PVDF. *J. Polym. Sci. Part B* **1993**, *31*, 2027–2034. [CrossRef]
27. Li, Z.; Zhou, Z.; Armstrong, S.R.; Baer, E.; Paul, D.R.; Ellison, C.J. Multilayer coextrusion of rheologically modified main chain liquid crystalline polymers and resulting orientational order. *Polymer* **2014**, *55*, 4966–4975. [CrossRef]
28. Cox, W.P.; Merz, E.H. Correlation of dynamic and steady flow viscosities. *J. Polym. Sci.* **1958**, *28*, 619–622. [CrossRef]
29. Bironeau, A.; Salez, T.; Miquelard-Garnier, G.; Sollogoub, C. Existence of a Critical Layer Thickness in PS/PMMA Nanolayered Films. *Macromolecules* **2017**, *50*, 4064–4073. [CrossRef]
30. Martins, P.; Lopes, A.C.; Lanceros-Mendez, S. Electroactive phases of poly(vinylidene fluoride): Determination, processing and applications. *Prog. Polym. Sci.* **2014**, *39*, 683–706. [CrossRef]
31. Cai, X.; Lei, T.; Sun, D.; Lin, L. A critical analysis of the α, β and γ phases in poly(vinylidene fluoride) using FTIR. *RSC Adv.* **2017**, *7*, 15382–15389. [CrossRef]
32. Cui, Z.; Hassankiadeh, N.T.; Zhuang, Y.; Drioli, E.; Lee, Y.M. Crystalline polymorphism in poly(vinylidenefluoride) membranes. *Prog. Polym. Sci.* **2015**, *51*, 94–126. [CrossRef]
33. Lu, B.; Alcouffe, P.; Sudre, G.; Pruvost, S.; Serghei, A.; Liu, C.; Maazouz, A.; Lamnawar, K. Unveiling the Effects of In Situ Layer–Layer Interfacial Reaction in Multilayer Polymer Films via Multilayered Assembly: From Microlayers to Nanolayers. *Macromol. Mater. Eng.* **2020**, *305*, 2000076. [CrossRef]
34. Tseng, J.-K.; Yin, K.; Zhang, Z.; Mackey, M.; Baer, E.; Zhu, L. Morphological effects on dielectric properties of poly(vinylidene fluoride-*co*-hexafluoropropylene) blends and multilayer films. *Polymer* **2019**, *172*, 221–230. [CrossRef]
35. Mackey, M.; Schuele, D.E.; Zhu, L.; Flandin, L.; Wolak, M.A.; Shirk, J.S.; Hiltner, A.; Baer, E. Reduction of Dielectric Hysteresis in Multilayered Films via Nanoconfinement. *Macromolecules* **2012**, *45*, 1954–1962. [CrossRef]
36. Kremer, F.; Schönhals, A. *Broadband Dielectric Spectroscopy*; Springer: Berlin/Heidelberg, Germany, 2003.

Publisher's Note: MDPI stays neutral with regard to jurisdictional claims in published maps and institutional affiliations.

© 2020 by the authors. Licensee MDPI, Basel, Switzerland. This article is an open access article distributed under the terms and conditions of the Creative Commons Attribution (CC BY) license (http://creativecommons.org/licenses/by/4.0/).

Article

Reliability of Free Inflation and Dynamic Mechanics Tests on the Prediction of the Behavior of the Polymethylsilsesquioxane–High-Density Polyethylene Nanocomposite for Thermoforming Applications

Fouad Erchiqui [1,*], Khaled Zaafrane [1], Abdessamad Baatti [1], Hamid Kaddami [2] and Abdellatif Imad [3]

1. École de Génie, Université du Québec en Abitibi-Témiscamingue, 445, boul. de l'Université, Rouyn-Noranda, QC J9X 5E4, Canada; Khaled.Zaafrane@uqat.ca (K.Z.); Abdessamad.Baatti@uqat.ca (A.B.)
2. Faculté des Sciences et Techniques, Université Caddi Ayad, Marrakech 40000, Morocco; h.kaddami@uca.ma
3. Unité de Mécanique de Lille, Université de Lille, UML, Joseph Boussinesq, F5900 Lille, France; abdellatif.imad@polytech-lille.fr
* Correspondence: fouad.erchiqui@uqat.ca

Received: 21 October 2020; Accepted: 17 November 2020; Published: 21 November 2020

Abstract: Numerical modeling of the thermoforming process of polymeric sheets requires precise knowledge of the viscoelastic behavior under conjugate effect pressure and temperature. Using two different experiments, bubble inflation and dynamic mechanical testing on a high-density polyethylene (HDPE) nanocomposite reinforced with polymethylsilsesquioxane HDPE (PMSQ–HDPE) nanoparticles, material constants for Christensen's model were determined by the least squares optimization. The viscoelastic identification relative to the inflation test seemed to be the most appropriate for the numerical study of thermoforming of a thin PMSQ–HDPE part. For this purpose, the finite element method was considered.

Keywords: thermoforming; PMSQ–HDPE; viscoelastic; experimental; bubble inflation test; DMA; Christensen's model; FEM

1. Introduction

The forming of thermoplastics in the plastic processing industry generally requires a high number of experimental tests to detect optimal conditions for mass production of products or optimizing of the manufacturing process. These experimental tests are costly and time-consuming. To circumvent the costs associated with these tests, many manufacturers are deploying computer-assisted analysis for product design [1]. However, computer-assisted analysis of the processing of polymers and composites demonstrates the need for an accurate description of the behavior of these materials under the combined effect of applied forces and temperature [2]. The quality of behavioral characterization depends largely on the tools used in experimentation, modeling, and optimization. Regarding the behavior of thermoplastics used in thermoforming, associated with the manufacture of thin parts, it is generally of a viscoelastic nature and the generated strains can be linear or nonlinear [3]. Several behavioral laws are available in the scientific literature to represent thermoplastic polymers. Among them are Maxwell [4], Christensen [5], K-BKZ [6], and Lodge [7]. These laws are generally constructed by combining the elastic and viscous responses of thermoplastics, in terms of spring and damper-based models.

For the numerical characterization of the viscoelastic behavior of materials, experimental data from rheological and mechanical tests are often used [8]. Concerning the experimental tests used for viscoelastic identification, there are two classes in particular: unidirectional tests [9–11] (dynamic mechanical tests in shear and extension, compression, etc.) and multidirectional tests [12–16] (inflation of circular and cylindrical membranes, equibiaxial stretching of membranes, extensions and simultaneous inflations of membranes, etc.).

At the level of the numerical identification of non-linear mechanical parameters, associated with the laws of viscoelastic behavior of thermoplastics, it is often necessary, with the help of mathematical modeling (analytical or numerical) and optimization, to reproduce, as faithfully as possible, the data measured in experimentation. Among the methods used for numerical modeling, the finite element method [2,11] and finite difference method [2] are used to model the experimental tests. Concerning the problem of identifying mechanical parameters by optimization algorithms, two classes are encountered: the class based on least squares algorithms [2,15,16] and the approach using artificial intelligence (neural networks) [12].

The deformations induced in thermoplastics, in the thermoforming process, are significant and, in general, of a biaxial nature. However, several works encountered in the literature on the construction of viscoelastic constitutive laws are based on experimental data from dynamic mechanical test (DMA). Thus, the following question arises: are the rheological data resulting from DMA tests reliable for the construction of a viscoelastic law? It is in this context that the present work is oriented and aims at a study on the reliability of the results obtained from two experimental tests: one was based on the inflation of the membrane and the other on a dynamic mechanical test (DMA). The two experimental tests were carried out at a temperature of 130 °C. For the viscoelastic characterization, we considered the Christensen model [5]. The mechanical parameters were identified using the Levenberg–Marquardt algorithm [17].

For the comparative study of the reliability of the results of the viscoelastic identification, compared to each experimental test, we considered the numerical modeling of the thermoforming of a thin part in PMSQ–HDPE. For this purpose, the finite element method was considered.

2. Material

This work is part of the work carried out on the development of a nanocomposite family based on polymethylsilsesquioxane (PMSQ, synthetized in our previous work [18]) and a high-density polyethylene (HDPE Hival-500354 with a melt flow index of 0.03 g min-1 (ASTM D1505) and a density of 0.954 g cm-3 (ASTM D1238) was supplied by IDES Prospector North America) matrix [19]. The method for the development of HDPE–PMSQ nanocomposites is based on a fusion mixing process. To this end, PMSQ nanoparticles were swollen in an organic solvent using an UltraTurax system (IKA, Wilmington, NC, USA) and sonication, then mixed with molten HDPE using a twin-screw extruder (Coperion corporation, Sewell, NJ, USA). Then, the solvent was removed. Nanocomposites with different PMSQ contents (from 0 to 1%) were manufactured. Then, the nanocomposites were characterized (Fourier transformation in the infrared, Perkin Elmer, Woodbridge, ON, Canada; transmission electron microscopy, JEOL, Tokyo, Japan; differential scanning calorimetry, Mettler Toledo, Greifensee, Suisse; scanning electron microscopy, JEOL, Tokyo, Japan; mechanical tests, TA Instruments, New Castle, DE, USA; thermophysical characterization, TA Instruments, New Castle, DE, USA). The mechanical properties obtained from HDPE–PMSQ nanocomposites were compared with the barrier effect of PMSQ nanoparticles. The elastic modulus, yield stress, and elongation at break of the neat HDPE and its nanocomposites are shown in Table 1. Compared to HDPE, the modulus of elasticity of HDPE–PMSQ was slightly improved.

Table 1. Tensile mechanical properties of high-density polyethylene–polymethylsilsesquioxane (HDPE–PMSQ) nanocomposites [19].

% PMSQ–HDPE	Elastic Modulus (MPa)	Yield Stress (MPa)	Elongation at Break (%)
0.0%	1031 ± 26	26.8 ± 0.2	39.2 ± 2.3
0.5%	1064 ± 60	27.9 ± 0.3	47.2 ± 3.1
1.0%	1115 ± 54	30.1 ± 0.1	41.1 ± 2.3

In this present work, only the rheological properties obtained for a concentration of 1% of PMSQ were considered to detect the viscoelastic behavior of HDPE–PMSQ nanocomposites in the semi-solid state.

3. Experimental Testing

3.1. Bubble Inflation Testing

For the free blowing test, we considered a circular PMSQ–HDPE composite membrane. The diameter and thickness of the membrane were 80 and 1.5 mm, respectively (Figure 1). The description of the set-up and the experimental test procedures are described in [20]. Figure 2, extracted from [20], shows the experimental set-up diagram. Figure 3 shows the experimental results of the evolution, over time, of the internal pressure and the height at the pole, respectively, of the PMSQ–HDPE membrane.

Figure 1. PMSQ–HDPE membrane.

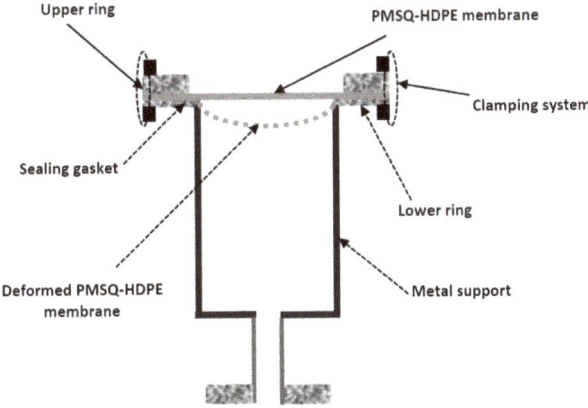

Figure 2. Fixation module.

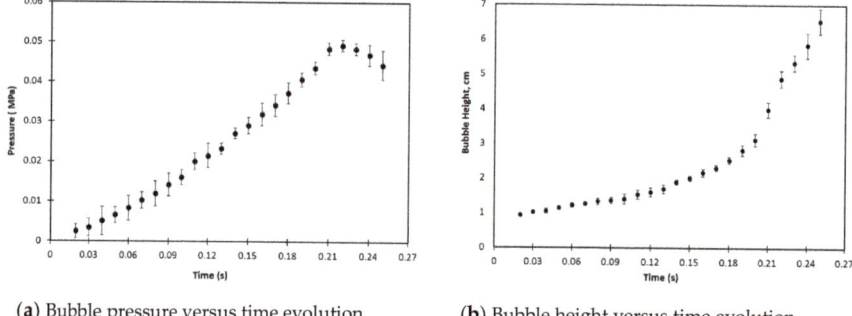

(**a**) Bubble pressure versus time evolution   (**b**) Bubble height versus time evolution

Figure 3. Bubble pressure and bubble height versus time evolution.

3.2. Dynamic Mechanical Testing

In our study, the oscillatory shear experiment was performed to determine the elasticity or storage modulus (G′) and the loss modulus (G″) of the PMSQ–HDPE material. The results obtained with respect to the frequencies are given in Figure 4 at a temperature of 130 °C.

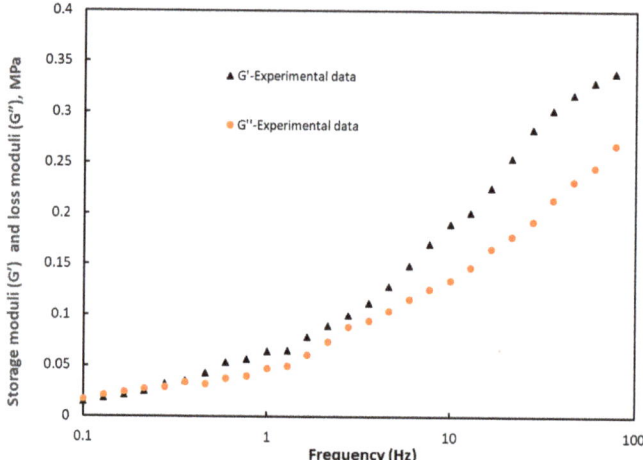

Figure 4. Experimental results of the storage moduli (G′) and loss moduli (G″) as a function of the frequency.

4. Viscoelastic Behavior Model

In this work, Christensen's model [5], suitable for representing the viscoelastic behavior of thermoplastics in the semi-solid state, was used. For this model, in Lagrangian formulation, the second Piola–Kirchhoff stress tensor **S** at time t is given by:

$$S(t) = -p(t)\mathbf{C}^{-1} + g_0\mathbf{I} + \int_{-\infty}^{t} g_1(t-\tau)\frac{\partial \mathbf{E}(t-\tau)}{\partial \tau} d\tau \quad (1)$$

E is the Lagrangian strain tensor **E**. g_0 is the hyperelastic modulus and g_1 is the material relaxation function given by equation:

$$g_1(t-\tau) = \sum_k C_k e^{-\frac{t-\tau}{\tau_k}} \quad (2)$$

where C_k is the stiffness modulus. The Lagrangian strain history **E** is related to Cauchy tensor deformation **C** by **E** = 1/2(**C**−**I**) and **I** is identity tensor.

The tensor **S** is related to the Cauchy stress tensor σ by the following relationship:

$$\mathbf{S}(t) = \mathbf{J}(t)\mathbf{F}^{-1}(t)\sigma(t)\mathbf{F}^{-T}(t) \tag{3}$$

J(t) and F(t) are, respectively, the Jacobian of the transformation and deformation gradient tensor. For incompressible materials, det(J(t)) = 1. For the blowing modeling of the PMSQ–HDPE membrane, we considered the following two assumptions:

1. The state of plane stress;
2. Material is incompressible.

The first hypothesis induces the following forms for the matrices E(t) and S(t):

$$\mathbf{C}(t) = \begin{bmatrix} C_{xx}(t) & C_{xy}(t) & 0 \\ C_{yx}(t) & C_{yy}(t) & 0 \\ 0 & 0 & C_{zz}(t) \end{bmatrix}; \quad \mathbf{S}(t) = \begin{bmatrix} S_{xx}(t) & S_{xy}(t) & 0 \\ S_{yx}(t) & S_{yy}(t) & 0 \\ 0 & 0 & S_{zz}(t) \end{bmatrix} \tag{4}$$

With the assumption of incompressibility of the PMSQ–HDPE composite, the term $C_{zz}(t)$, appearing in Equation (4), can be directly calculated from the other components of the strain tensor **C**:

$$C_{zz}(t) = \lambda_3^2(t) = \frac{1}{C_{xx}(t)C_{yy}(t) - C_{xy}(t)C_{yx}(t)} \tag{5}$$

λ_3 is the principal stretch ratio in thickness direction defined by:

$$\lambda_3(t) = \frac{h(t)}{h_0} \tag{6}$$

where $h(t)$ and h_0 represent the PMSQ–HDPE membrane thicknesses in the deformed and undeformed configurations, respectively.

5. Viscoelastic Model Identification

5.1. PMSQ–HDPE Viscoelastic Behavior Identification Conforms to Bubble Testing

The mathematical formulation of the problem is described in [2,12]. To this end, the deformation of the circular membrane was assumed to remain axisymmetric during inflation. The strategy used for identification was as follows: first, for a given experimental thickness, the theoretical blowing pressure of the membrane, compatible with the measured thickness, was determined. For this purpose, we used the finite difference method with variable pitch. Then, using a modified Levenberg–Marquardt algorithm [17], the difference between the calculated and measured inflation pressure was minimized. Using this procedure, the material constants C_0, g_b, and τ_b were determined. However, it is important to note that the resolution of the equilibrium equations, which govern membrane inflation, can induce instabilities [12] that affect the numerical resolution. The choice of the initial values of the material constants is crucial for the convergence of the problem. As the experiment was based on a single average air flow rate for blowing the membrane, we considered a single relaxation time. In other words, three parameters for Christensen's model were determined: C_0, C_1, and τ_1.

The pressures and heights measured from the bubble to the pole were interpolated by polynomial functions and used in the identification problem. The predictions obtained with Christensen's model gave very satisfactory results and are presented in relation to the experimental data in Figure 5. The mechanical properties obtained by numerical identification are given in Table 2.

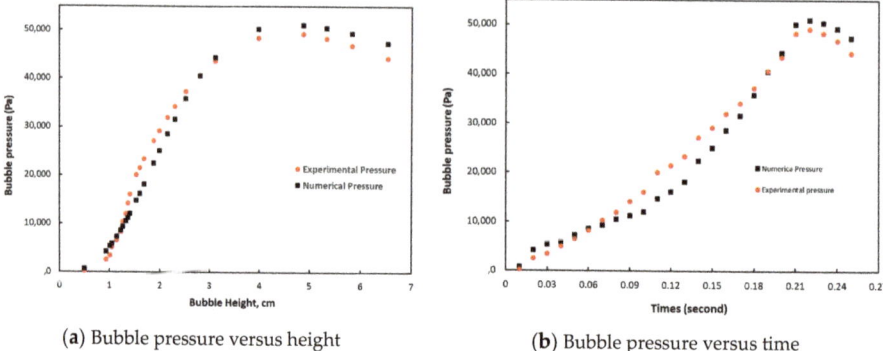

(a) Bubble pressure versus height   (b) Bubble pressure versus time

Figure 5. Results of the optimization with the experimental data: (**a**) bubble pressure vs bubble height and (**b**) bubble pressure vs times.

Table 2. Materials constants for PMSQ–HDPE at 130 °C.

C_0 (MPa)	C_1 (MPa)	τ_1 (s)
0.71694	0.00001	772.00037

Figure 6 illustrates, according to Christensen's viscoelastic model, the main geometrical results relative to the PMSQ–HDPE membrane trace, at 0.05, 0.10, 0.15, 0.20, and 0.25 s: bubble height (Figure 6a), thickness (Figure 6b), meridian extension (Figure 6c), and circumferential extension (Figure 6d).

5.2. PMSQ–HDPE Viscoelastic Behavior Identification Conform to DMA Testing

The least squares method was used to minimize the discrepancies between the experimental and theoretical values during the identification of the relaxation spectrum for PMSQ–HDPE material. This method is described by reducing the objective function defined by Equation (7) where N is the number of experimental data points:

$$Z = \sum_{i=1}^{N}\left[\frac{G'_{i,exp}-G'_{i,th}}{G'_{i,exp}}\right]^2 + \left[\frac{G''_{i,exp}-G''_{i,th}}{G''_{i,exp}}\right]^2 \qquad (7)$$

The parameters $G'_{i,\,exp}$ and $G''_{i,\,exp}$ represent the dynamic moduli from the experimental data while $G'_{i,\,th}$ and $G''_{i,th}$ represent the theoretical values given by Christensen's model (Equation (1)).

$$G_{th}'(\omega) = C_0 + \sum_{i=1}^{N}\frac{C_i\tau_i^2\omega^2}{2(1+\tau_i^2\omega^2)} \text{ and } G_{th}''(\omega) = \sum_{i=1}^{N}\frac{C_i\tau_i\omega}{2(1+\tau_i^2\omega^2)} \qquad (8)$$

The parameter C_i is the stiffness constant and τ_i the relaxation time associated with the mode i, while ω is the frequency. The results obtained are given in Table 3. Figure 7 shows the results of the optimization in comparison with those of the experiment.

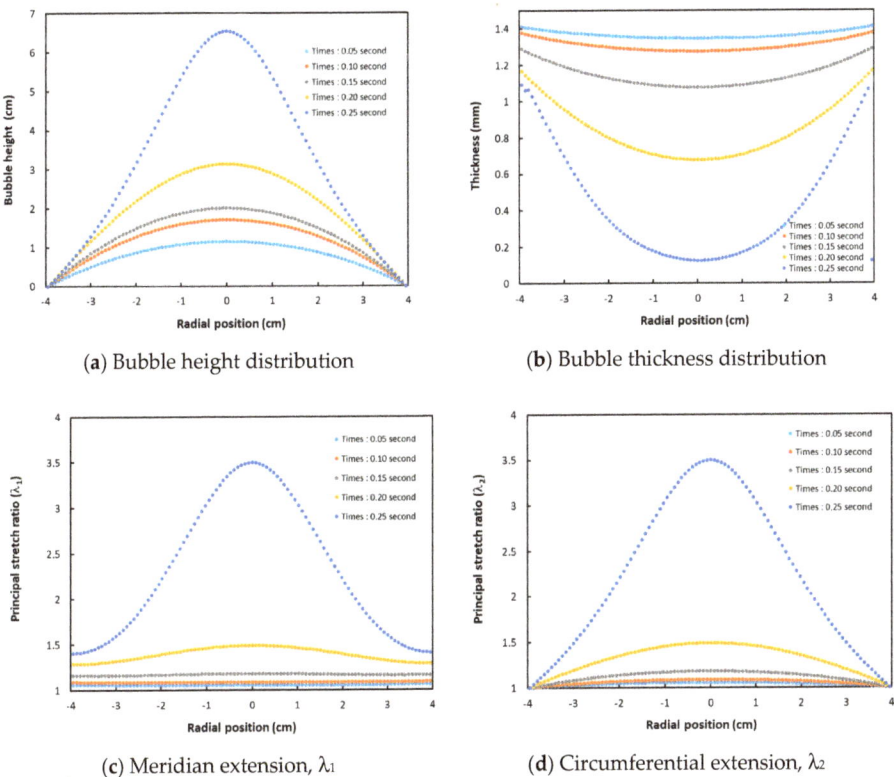

Figure 6. Geometrical PMSQ–HDPE membrane according to Christensen's model at 0.05, 0.10, 0.15, 0.20, and 0.25 s: (**a**) Bubble height, (**b**) Bubble thickness, (**c**) Stretch ration λ_1 and (**d**) Stretch ration λ_2.

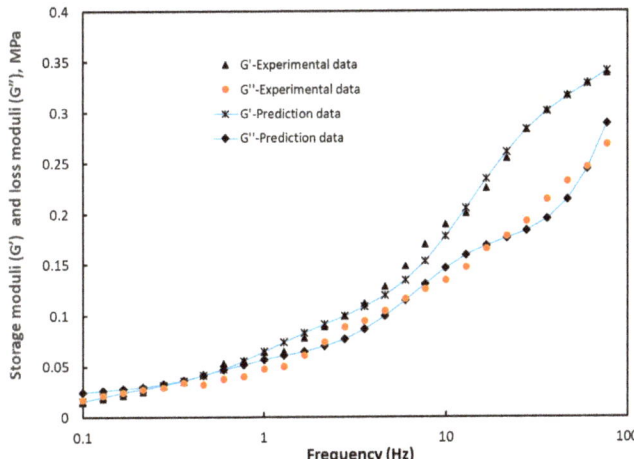

Figure 7. Results of the optimization with the experimental storage moduli G′ and loss moduli G″.

Table 3. Stiffness modulus and relaxation time for the PMSQ–HDPE nanocomposite at T = 130 °C.

PMSQ–HDPE at 130 °C					
C_0 (MPa)	C_1 (MPa)	C_2 (MPa)	C_3 (MPa)	C_4 (MPa)	C_5 (MPa)
−0.000633	6.414907	0.294631	0.170894	0.132937	0.062403
	τ_1(s)	τ_2(s)	τ_3(s)	τ_4(s)	τ_5(s)
	0.01	0.06	0.1	1.0	10.0

6. Reliability of Tests on the Viscoelastic Behavior of the PMSQ–HDPE on Thermoforming

In order to identify the most reliable experimental test to characterize the viscoelastic behavior of PMSQ–HDPE for thermoforming applications, we considered the problem of numerical modeling of the thermoforming of a PMSQ–HDPE membrane. For this, we used a circular membrane, similar to the one used in the experiment (see Section 3.1), and a conical mold (see Figure 8). The nonlinear mechanical properties identified for Christensen's model, with both DMA and biaxial inflation approaches, was used.

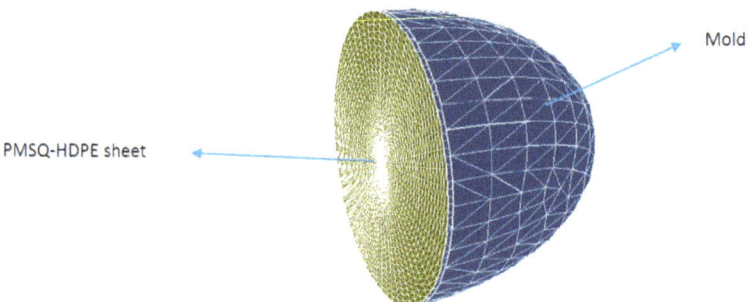

Figure 8. Geometries of the mold and the wood plastic composite (WPC) sheet.

6.1. Finite Element Analysis

For the analysis, the explicit dynamic finite element method with discretization in space and time was used to simulate the thermoforming of the PMSQ–HDPE membrane. The principle of virtual work was expressed on the undeformed configuration for the inertial effects and internal work.

The spatial and temporal discretizations were both necessary for the virtual work due to the presence of the force of inertia. In the case of spatial discretization, the finite element method approach was considered [21]. However, for temporal discretization, the centered finite difference method, which is conditionally stable, was used. Consequently, the system of equations governing the blowing problem is given by [21]:

$$\mathbf{M}\ddot{\mathbf{u}}(t) = \mathbf{F}_{ext} + \mathbf{F}_{grav} - \mathbf{F}_{int} \tag{9}$$

where

$\mathbf{F}_{ext}$: Global nodal external force vectors
$\mathbf{F}_{grav}$: Global nodal body force vectors
$\mathbf{F}_{int}$: Global nodal internal force vectors
$\mathbf{M}$: Global mass matrix

The mass matrix **M** can be reduced to a diagonal matrix, **Md**, by using the diagonalization method. For the temporal scheme, we used the finite difference method centered. In this case, Equation (9) can be rewritten as Equation (10):

$$u_i(t + \Delta t) = \frac{\Delta t^2}{M_{ii}^d}\left(F_i^{ext}(t) + F_i^{grav}(t) - F_i^{int}(t)\right) + 2u_i - u_i(t - \Delta t) \tag{10}$$

For the stability criterion of system (10), we used the Courant–Friedrichs–Lewy criterion [22].

6.2. Plane Stress Assumption and Constitutive Equation

For this, the hypothesis of plane stress and incompressibility of the thermoplastic material was considered. The behavior model used in the simulation was that of Christensen (see Section 3).

6.3. Pressure Loading and Van der Waals Equation of State

For the blower modeling of the PMSQ–HDPE membrane, we considered an air flow load. For this purpose, the Redlich–Kwong gas equation of state was considered [23]:

$$P(t) = \frac{n(t)RT_g}{V(t) - b\,n(t)} - \frac{n^2(t)a}{V(t)[V(t) + b\,n(t)]\sqrt{T_g}} \tag{11}$$

where:

$n(t)$: the number of gas moles introduced to inflate the thermoplastic-based composite membrane
$P(t)$: the internal pressure
$V(t)$: the volume occupied by the membrane at time t,
T_g: the absolute gas temperature
R: the universal gas constant (=8.3145 J mol^{-1} K^{-1})
a and b: constants evaluated from the critical state of the gas [23]:

$$a = 0.42748\,\frac{\overline{R^2}T_c^{2.5}}{P_c} \text{ and } b = 0.08664\,\frac{\overline{R^2}T_c}{P_c} \tag{12}$$

where T_c and P_c are the critical temperature and pressure of the gas, respectively. In this study, the assumptions used for the calculation of the dynamic pressure are:

(i) Gas temperature is assumed constant (T_g);
(ii) The biocomposite sheet temperature is assumed constant ($T_{sheet} = T_g$);
(iii) At every moment, the pressure between the sheet and the mold is assumed constant (ΔP);
(iv) The contact between the biocomposite sheet and the mold is assumed to be a sticky contact as the polymer cools and stiffens rapidly during the sheet/mold contact.

For a reference state in volume (V_0) and number of molds (n_0), Equation (11) becomes:

$$P_0 = \frac{n_0RT_g}{V_0 - bn_0} - \frac{n_0^2 a}{V_0[V_0 + bn_0]\sqrt{T_g}} \tag{13}$$

The dynamic pressure, responsible for inflating the thermoplastic membrane, represents the difference between the internal pressure, induced by the introduction of n(t) mole of gas, and the initial pressure P_0:

$$\Delta P(t) = P(t) - P_0 \tag{14}$$

Equation (15) describes, over time, the internal pressure induced by the fluidic charge (air). This pressure, in turn, is responsible for the inflation of the membrane (work). It follows that the following relation expresses the virtual external work in terms of closed volume [24]:

$$\delta W^{ext} = \Delta P(t)\, \delta V \tag{15}$$

6.4. Analysis of Reliability of Experimental Tests Characterization on Thermoforming

For the study, we considered the thermoforming of a circular PMSQ–HDPE membrane, similar to the one used in experiments for free blowing (with a radius of 4 cm and a thickness of 1.5 mm). For the applied load, we considered a non-linear airflow as shown in Figure 9. The geometries of the mold and the composite sheet discretized by triangular membrane elements are shown in Figure 8. The material temperature was assumed constant at 130 °C. The rheological parameters of the Christensen behavior law are given in Table 2 (relating to the DMA test) and Table 1 (relating to the free biaxial test).

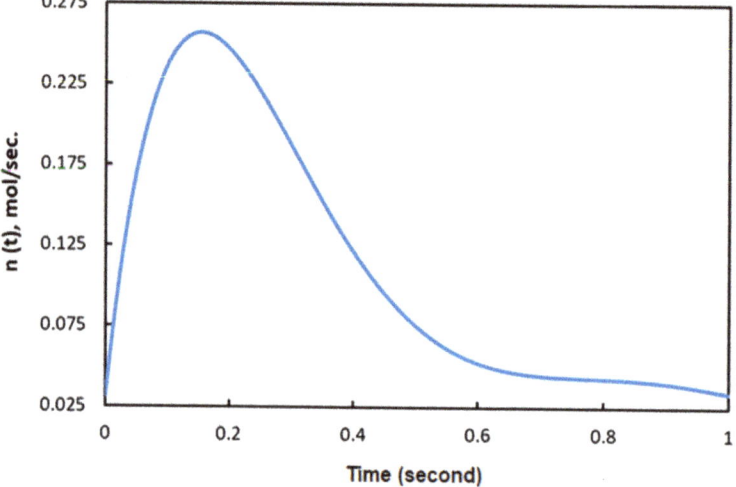

Figure 9. Airflow versus time.

In the following sections of the study, the viscoelastic behavior of the PMSQ-HDPE material relative to the biaxial and DMA tests will be referred to as MB and MD respectively. Figure 10 shows the evolution of the pressure, generated by the airflow, with the volumes for MB and MD. Figure 11 shows the evolution, over time, of its volumes. According to this figure, and contrary to MB, we can see that MD resisted inflation and was unable to ensure its shaping by thermoforming for the treated example, which involved large deformations. To clarify this situation, we have presented in Figure 12a comparison between the two models MB and MD with respect to the principal extensions λ_3 and the von Mises stress at 0.0143, 0.0293, 0.0443, and 0.0593 s. It can be seen that the action of the air flow on the MD membrane induced, on the one hand, much higher von Mises stresses than those on MB and, on the other hand, a lower stretch. In Table 4, the critical values of the von Mises stresses as well as the principal stretching for the MB and MD models have been provided. Therefore, the MD material is not a candidate for thermoforming and blowing thin, hollow parts that typically induce large deformations. To illustrate this behavior for the MD model, we have presented views of the von Mises constraints at 0.0143, 0.0293, and 0.0443 s in Figure 13, and in Figure 14, we have presented a view of the distribution of its constraints in the proximity of the critical time of 0.0593 s.

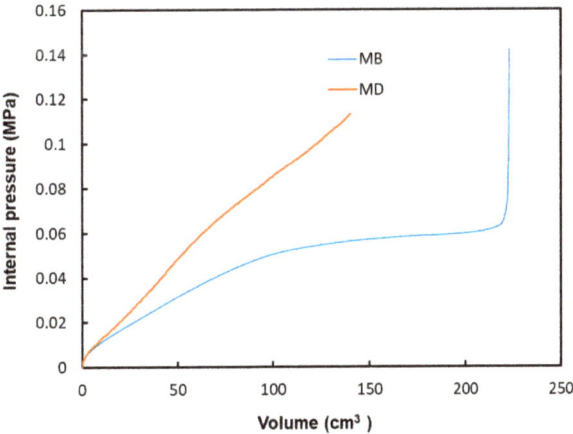

Figure 10. Internal pressure versus volume of PMSQ–HDPE.

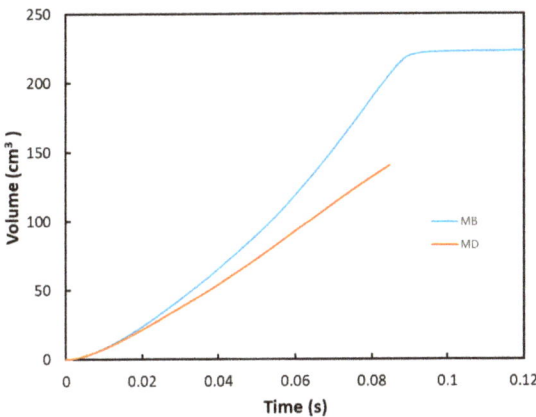

Figure 11. Volume evolution with time of PMSQ–HDPE.

Table 4. Critical values of von Mises stress and principal stretch λ_3.

Time (s)	Von Mises Stress MPa		Principal Stretch λ_3	
	MB Model	MD Model	MB Model	MD Model
0.0143	0.04923	0.1116	0.9676	0.9788
0.0293	0.2298	0.4866	0.8503	0.8736
0.0443	0.6185	0.9467	0.6222	0.6966
0.0593	1.158	1.658	0.408	0.5307

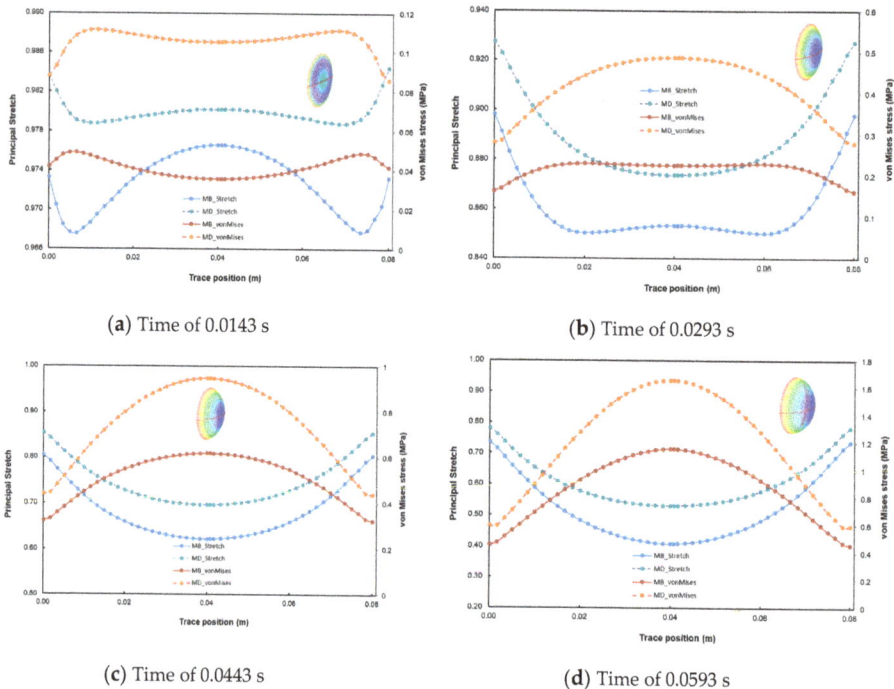

Figure 12. Transient evolution of the principal stretch and von Mises stresses at different instants: (**a**) time = 0.0143 s, (**b**) time = 0.0291 s, (**c**) time = 0.0443 s and (**d**) time = 0.0593 s.

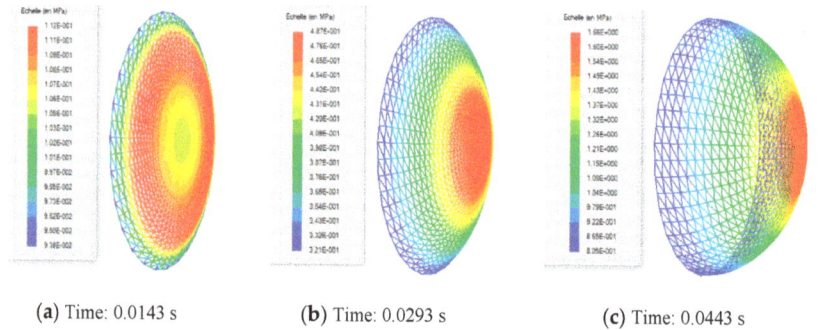

Figure 13. MD model: von Mises stresses induced in PMSQ–HDPE at times of 0.0143 (**a**), 0.0293 (**b**), and 0.0443 s (**c**).

It should be pointed out that after the critical time of 0.06 s, the pressure loading had no significant effect on the deformation of the MD membrane, but it had a considerable effect on the stresses. To this effect, we have presented in Figure 15, within the mold, a view of the final shape of the membrane (including the contact nodes), and a view of the von Mises stresses. The MD became quasi-rigid.

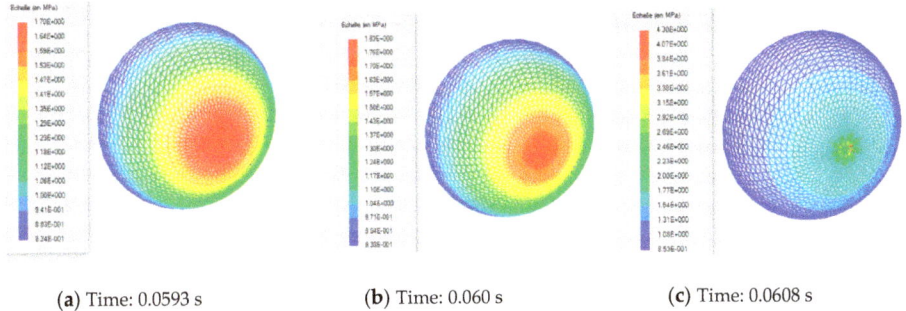

(a) Time: 0.0593 s   (b) Time: 0.060 s   (c) Time: 0.0608 s

Figure 14. Von Mises constraints induced in the HDPE–PMSQ with the MD model in the time-critical neighborhood 0.0593 s.

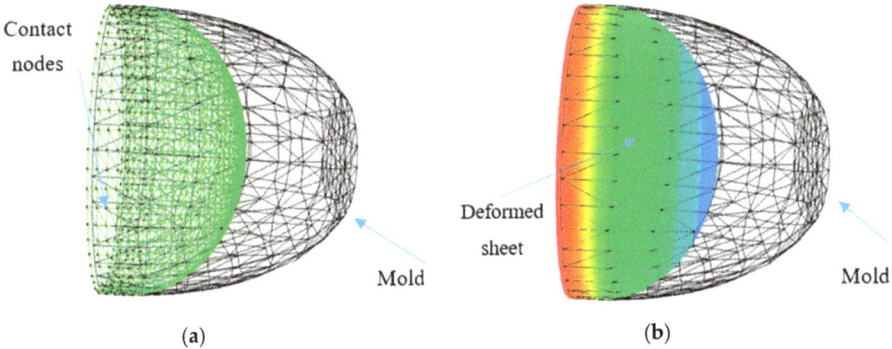

(a)   (b)

Figure 15. Shape of the deformed MD and principal stretch λ_3 in the time-critical neighborhood 0.06 s. (**a**) Final shape of the deformed MD. (**b**) Distribution of the principal extensions λ_3 in the final deformed MD.

For the rest of the study, only the MB material is considered. Figure 16 illustrates the evolution of the nodes, which were in contact with the mold during the forming process, represented by the black dots.

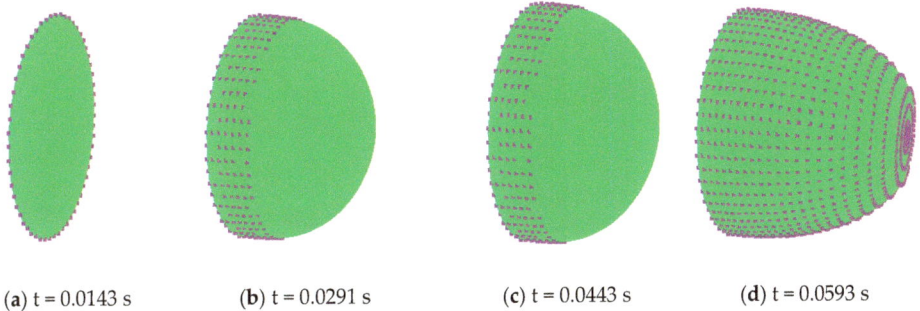

(a) t = 0.0143 s   (b) t = 0.0291 s   (c) t = 0.0443 s   (d) t = 0.0593 s

Figure 16. Evolution of the distribution of the contact nodes with the mold at at different instants: (**a**) time = 0.0143 s, (**b**) time = 0.0291 s, (**c**) time = 0.0443 s and (**d**) time = 0.0593 s.

When simulating the thermoforming of a thin part, it is important to predict the thickness and stress distributions in the molded part. In fact, the predictions of the residual stresses and dimensional stability of the final shape of the molded part are closely related to the estimated stresses. In addition, the effect of localized thinning of the deformed membrane is usually accompanied by an increase in the Cauchy stresses (or actual stresses). For this purpose, we have presented in Figure 17 the von Mises stress distribution and the main extensions on the trace of the thermoformed part. The maximum value of the von Mises constraint is of the order 5.4 MPa and is located at the positions 2.5 and 4.5 m. For this critical stress value, the principal stretch λ_3 is 0.093. In Figure 18, we have presented an overview of the von Mises stresses (Figure 18a) and the main extensions (Figure 18b) induced in the thermoformed part.

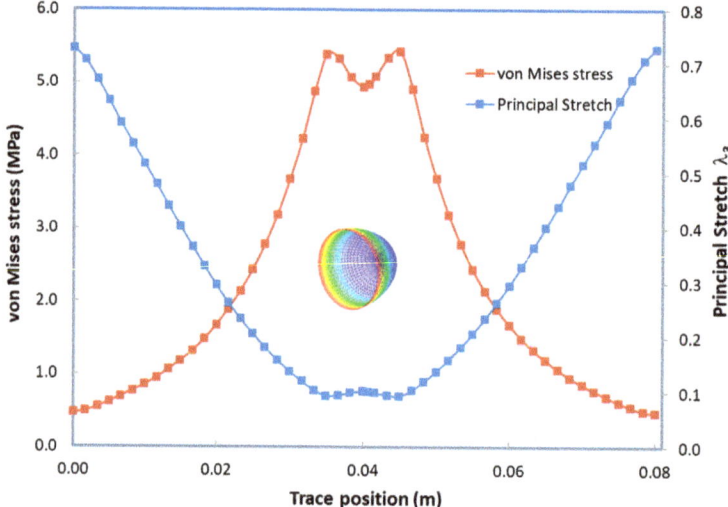

Figure 17. Von Mises stress and stretch ratio, λ_3, on the half plane of symmetry at the end of the forming cycle.

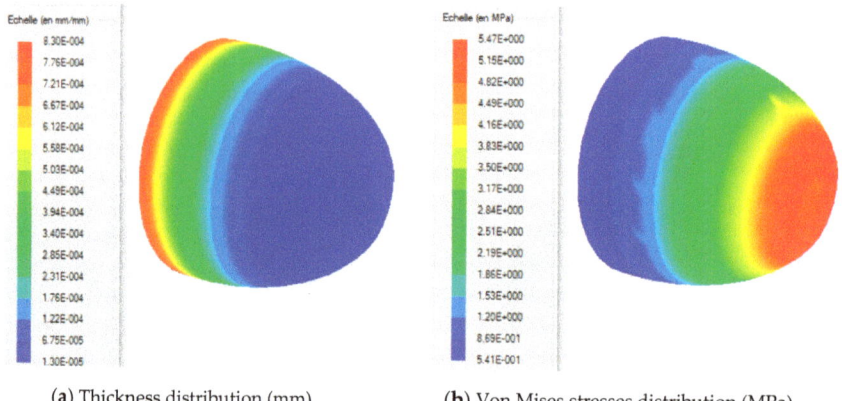

(**a**) Thickness distribution (mm). (**b**) Von Mises stresses distribution (MPa).

Figure 18. Thickness and von Mises stresses distributions in the thermoformed part: (**a**) Thickness distribution and (**b**) Von Mises stresses distribution.

In light of the results presented above, the following remarks can be made about the use of experimental tests for viscoelastic identification relative to Christensen's model:

- The experimental test used for the construction of the constitutive behavior law of polymers plays a key role on the qualities of the results;

- The results obtained by the mechanical blowing test, which induces deformation modes similar to those encountered in thermoforming, seem to be the most appropriate;
- The construction of viscoelastic laws from DMA is more suitable for small deformations for thermoforming applications;
- The choice to use the finite element method with a pressure load, which is derived from a thermodynamic law, is judicious for the integrated analysis in large deformations of the forming of a thin part;
- Experimental temperature can improve the quality of viscoelastic identification for thermoforming applications. The material becomes softer.

7. Conclusions

The study was conducted on the reliability of the experimental method for viscoelastic identification of a nanocomposite reinforced with Polymethylsilsesquioxane nanoparticles (PMSQ–HDPE). To do so, two tests of different nature were used. One was based on free inflation of the membrane and the other on a dynamic mechanical test (DMA). The experiments were carried out at a temperature of 130 °C. The material constants for Christensen's model were determined by the least squares optimization. The comparative study of viscoelastic behavior of PMSQ–HDPE shows that the biaxial test is more appropriate for the construction of a behavior law for applications in thermoforming. Concerning the viscoelastic identification obtained from the rheological data of the DMA, it does not seem to be able to represent the thermoforming of a part which requires large deformations.

Following this study, comparative studies between the DMA and the free blowing should be carried out at temperatures above 130 °C for viscoelastic identification. This will make it possible to characterize the effect of temperature on the reliability of the tests in thermoforming.

Author Contributions: Conceptualization F.E., H.K. and A.I.; methodology, F.E., K.Z. and A.B.; software, ThermoForm (house software developed by F.E.); validation, K.Z. and F.E.; formal analysis, F.E., H.K. and A.I.; data curation, A.B., K.Z. and F.E.; writing—original draft preparation, F.E., H.K. and A.I.; writing—review and editing, F.E.; supervision, F.E.; project administration, F.E. All authors have read and agreed to the published version of the manuscript.

Funding: This research received no external funding.

Conflicts of Interest: The authors declare no conflict of interest.

References

1. Chyan, Y.; Shiu-Wan, H. Modeling and Optimization of a Plastic Thermoforming Process. *J. Reinf. Plast. Compos.* **2004**, *23*, 109–121.
2. Derdouri, A.; Erchiqui, F.; Bendada, A.; Verron, E.; Peseux, B. Viscoelastic behaviour of polymer membrane under inflation. *20000-XIII Int. Congr. Rheol.* **2000**, *3*, 394–396.
3. Toth, G.; Nagy, D.; Bata, A.; Belina, K. Determination of polymer melts flow-activation energy a afunction of wide range shear rate. *IOP Conf. Ser. J. Phys.* **2018**, *1045*, 012040. [CrossRef]
4. Buckley, C.P.; Bucknell, C.B. *Principles of Polymer Engineering*, 2nd ed.; Oxford University Press: New York, NY, USA, 2011.
5. Christensen, R.M. A Nonlinear Theory of Viscoelasticity for Application to Elastomers. *J. Appl. Mech. ASME Trans.* **1980**, *47*, 762–768. [CrossRef]
6. Bernstein, B.; Kearsley, E.A.; Zapas, L.J. A study of stress relaxation with finite strain. *Trans. Soc. Rheol.* **1963**, *7*, 391–410. [CrossRef]
7. Lodge, A.S. Elastic Liquids; An Introductory Vector Treatment Of Finite-strain Polymer Rheology. *J. Am. Chem. Soc.* **1964**, *86*, 5056.
8. Engelmann, S. *Advanced Thermoforming: Methods, Machines and Materials, Applications and Automation*; Wiley Series on Polymer Engineering and Technology; John & Son Inc.-Wiley: Hoboken, NJ, USA, 2012; ISBN 978-0-470-49920-7.

9. Janhui, H.; Wujun, C.; Fan, P.; Gao, J.; Fang, G.; Cao, Z.; Peng, F. Uniaxial tensile tests and dynamic mechanical analysis of satin weave reinforced epoxy shape memory polymer composite. *Polym. Test.* **2017**, *64*, 235–241.
10. Jerabek, M.; Major, Z.; Lang, R.W. Uniaxial compression testing of polymeric materials. *Polym. Test.* **2012**, *29*, 302–309. [CrossRef]
11. Daiyan, H.; Andreassen, E.; Grytten, F.; Osnes, H. Shear Testing of Polypropylene Materials Analysed by Digital Image Correlation and Numerical Simulations. *Exp. Mech.* **2012**, *52*, 1355–1369. [CrossRef]
12. Erchiqui, F.; Ozdemir, Z.; Souli, M.; Ezaidi, H.; Dituba-Ngoma, G. Neuronal networks approach for characterization of viscoelastic polymers. *Can. J. Chem. Eng.* **2011**, *89*, 1303–1310. [CrossRef]
13. Souli, M.; Erchiqui, F. Experimental and Numerical Investigation of Instructions for Hyperelastic Membrane Inflation Using Fluid Structure Coupling. *Comput. Model. Eng. Sci.* **2011**, *77*, 183–200.
14. Potter, S.; Graves, J.; Drach, B.; Leahy, T.; Hammel, C.; Feng, Y.; Baker, A.; Sacks, M.S. A Novel Small-Specimen Planar Biaxial Testing System with Full In-Plane Deformation Control. *J. Biomech. Eng.* **2018**, *140*, 0510011–05100118. [CrossRef] [PubMed]
15. Meissner, J.; Raible, T.; Stephenson, S.E. Rotary clamp in uniaxial and biaxial rheometry of polymer melts. *J. Rheol.* **1981**, *25*, 1–28. [CrossRef]
16. Benjeddou, A.; Jankovich Hadhri, T. Determination of the parameters of Ogden's law using biaxial data and Levenberg-Marquardt-Fletcher algorithm. *J. Elastomers Plast.* **1993**, *25*, 224–248. [CrossRef]
17. Marquardt, D. An Algorithm for the Least-Squares Estimation of Non-linear Parameters. *SIAM J. Appl. Math.* **1963**, *11*, 431–441. [CrossRef]
18. Baatti, A.; Erchiqui, F.; Godard, F.; Bussières, D.; Bébin, P. A two-step Sol-Gel method to synthesize ladder polymethylsilsesquioxane nanoparticles. *Adv. Powder Technol.* **2017**, *28*, 1038–1046. [CrossRef]
19. Baatti, A.; Erchiqui, F.; Godard, F.; Bussières, D.; Bébin, P. DMA analysis, thermal study and morphology of polymethylsilsesquioxane nanoparticles-reinforced HDPE nanocomposite. *J. Therm. Anal. Calorim.* **2020**, *139*, 789–797. [CrossRef]
20. Ben Aoun, N.; Erchiqui, F.; Mrad, H.; Dituba-Ngoma, G.; Godard, F. Viscoelastic characterization of high-density polyethylene membranes under the combined effect of the temperature and the gravity for thermoforming applications. *Polym. Eng. Sci.* **2020**, *60*, 2676. [CrossRef]
21. Erchiqui, F.; Gakwaya, A.; Rachik, M. Dynamic finite element analysis of nonlinear isotropic hyperelastic and viscoelastic materials for thermoforming applications. *Polym. Eng. Sci.* **2005**, *45*, 125–134. [CrossRef]
22. Courant, R.; Friedrichs, K.; Lewy, H. On the partial difference equations of mathematical physics. *IBM J. Res. Dev.* **1967**, *11*, 215–234. [CrossRef]
23. Redlich, O.; Kwong, J.N.S. V-An equation of state. *Chem. Rev.* **1949**, *44*, 233. [CrossRef] [PubMed]
24. Erchiqui, F. A New hybrid approach using the explicit dynamic finite element method and thermodynamic law for the analysis of the thermoforming and blow molding processes for polymer materials. *Polym. Eng. Sci.* **2006**, *46*, 1554. [CrossRef]

Publisher's Note: MDPI stays neutral with regard to jurisdictional claims in published maps and institutional affiliations.

© 2020 by the authors. Licensee MDPI, Basel, Switzerland. This article is an open access article distributed under the terms and conditions of the Creative Commons Attribution (CC BY) license (http://creativecommons.org/licenses/by/4.0/).

Article

Study of the Properties of a Biodegradable Polymer Filled with Different Wood Flour Particles

Francisco Parres [1,*], Miguel Angel Peydro [1], David Juarez [1], Marina P. Arrieta [2,3] and Miguel Aldas [4]

1. Departamento de Ingeniería Mecánica y Materiales, Universitat Politècnica de València, Plz Ferrándiz y Carbonell, s/n, 03801 Alcoy, Spain; mpeydro@mcm.upv.es (M.A.P.); djuarez@mcm.upv.es (D.J.)
2. Departamento de Ingeniería Química y del Medio Ambiente, Escuela Politécnica Superior de Ingenieros Industriales, Universidad Politécnica de Madrid (ETSII-UPM), Calle José Gutierrez Abascal 2, 28006 Madrid, Spain; m.arrieta@upm.com
3. Grupo de Investigación, Polímeros, Caracterización y Aplicaciones (POLCA), 28006 Madrid, Spain
4. Departamento de Ciencia de Alimentos y Biotecnología, Facultad de Ingeniería Química y Agroindustria, Escuela Politécnica Nacional, Quito 170517, Ecuador; miguel.aldas@epn.edu.ec
* Correspondence: fraparga@dimm.upv.es; Tel.: +34-966-528-570

Received: 28 November 2020; Accepted: 11 December 2020; Published: 13 December 2020

Abstract: Lignocellulosic wood flour particles with three different sizes were used to reinforce Solanyl® type bioplastic in three compositions (10, 20, and 30 wt.%) and further processed by melt-extrusion and injection molding to simulate industrial conditions. The wood flour particles were morphologically and granulometric analyzed to evaluate their use as reinforcing filler. The Fuller method on wood flour particles was successfully applied and the obtained results were subsequently corroborated by the mechanical characterization. The rheological studies allowed observing how the viscosity was affected by the addition of wood flour and to recover information about the processing conditions of the biocomposites. Results suggest that all particles can be employed in extrusion processes (shear rate less than 1000 s^{-1}). However, under injection molding conditions, biocomposites with high percentages of wood flour or excessively large particles may cause an increase in defective injected-parts due to obstruction of the gate in the mold. From a processing point of view and based on the biocomposites performance, the best combination resulted in Solanyl® type biopolymer reinforced with wood flour particles loaded up to 20 wt.% of small and medium particles size. The obtained biocomposites are of interest for injected molding parts for several industrial applications.

Keywords: biopolymer; Solanyl®, wood flour; Lignocel®, lignocellulosic particles; fuller method; mechanical properties; rheological characterization

---

1. Introduction

The consumption of polymers has been steadily increasing since they first appeared. It would be difficult to imagine life today without these materials. Thus, the world plastic production is nowadays about 350 million tons per year [1]. The most significant problem with traditional polymers is the amount of plastic waste, particularly those coming from short-term applications such as packaging. Although a high amount of plastic wastes is recycled every year, there still is a great problem regarding the large number of plastic wastes that finishes in landfills.

Despite the existence of photodegradable and biodegradable polymers, the design of many landfills is not efficient [2]. For this reason, recycling, reuse, composting, and incineration seem to be the most environmentally and feasible way to reduce plastic waste after their useful life. There is a special scientific interest in the use of more sustainable polymers in the packaging field (i.e., biopolymers and/or reusable packaging) to reduce the disposal of plastic waste in landfills. Moreover, it should be considered that landfills should be adapted to current plastic wastes [3,4]. Biopolyesters have emerged as the most promising sustainable polymers to replace traditional thermoplastics used for packaging materials, mainly due to their easy processability because of their available processing technologies at the industrial level (i.e., melt blending, extrusion, injection molding, etc.) [5,6]. Additionally, great efforts have been made in rheological [7] and mechanical properties [3] research in biodegradable polymers. These properties are essential for plastic processing manufacturing and their further application in the industrial sector.

Solanyl® is a bio-based material based on reclaimed potato starch [8], the main component of which is the poly(lactic acid) (PLA), as is further confirmed by the DSC technique in the results section of this work. PLA is nowadays, undoubtedly, the most attractive biodegradable polymer for rigid and flexible packaging applications, due to its many advantages such as availability in the market, ease of processing, economic competitiveness, high transparency, and environmentally friendly characteristics [5,9]. However, PLA possesses some disadvantages such as poor thermal, mechanical, and barrier properties [10,11]. Thus, considerable academic and industrial efforts have been focused on improving PLA performance, such as copolymerization or blending with others biopolyesters such as poly(ε-caprolactone) (PCL) and/or poly(hydroxy alkanoates) (PHAs) [12–14], mainly focused on increasing its crystallinity for extending the industrial applications. Although many PLA-based biodegradable polymers are currently commercialized, they are not widely used at the industrial level, mainly due to the substantially higher price of biopolymers compared to the more conventional petroleum-derived and/or non-degradable polymers. On the other hand, there are many ways to reduce the price of a material. First, and perhaps most difficult, is the use of cheaper raw materials. Second, there is the possibility of generating a more widespread use of material, in which case demand and production would increase, while costs would naturally decrease. A third option is to introduce a reinforcing material or fillers into the polymeric matrix by developing composites and/or nanocomposites [15]. This approach is probably the most currently scalable option in the industrial sector. Polymeric composites have been used to reduce the cost of producing plastic materials. However, to guarantee the green character of the final polymeric formulation, the filler materials should be also biobased and/or biodegradable. Thus, during the last decades, many research works have been focused on the development of biocomposites and bionanocomposites, for improving the different properties of the materials.

In this context, several authors have analyzed the effect of different particles (i.e., chitosan, cellulose nanocrystal [15,16], clay and silver nanoparticles, organomontmorillonite/graphene [17], kraft lignin [18] and TiO_2 [19], among others, on different properties of biopolymers [20–23]. Additionally, others have analyzed the degradation process of biopolymers in detail using Taylor Dispersion Analysis (TDA) [24]. Nowadays, the use of residues as raw material sources for lignocellulosic derivatives production is considered to be one of the most promising strategies for reducing industrial waste [25]. Lignocellulose derivatives can be obtained from agricultural by-products of several industries [26,27]. They seem to be optimal reinforcing fillers for biopolymeric matrices since they are biobased, biodegradable, lightweight, stiff, non-abrasiveness to the processing equipment, and highly abundant in nature at low cost [15,16]. Although nanocellulose has been recognized in recent years as one of the most interesting reinforcement for biopolymers due to its extraordinary performance [15,16,25,27], the extraction process from agricultural sources requires bleaching treatments, basic treatments, and further the isolation of cellulose nanocrystals from cellulose fibers by acid hydrolysis [27]. Thus, nanocellulose production at an industrial scale is still in development, and it is not cost-effective for the plastic processing industry. In this regard, the wood industry produces a wide range of by-products which can be used to gain value as derivatives in many fields, for example, bark on biofuel [28], in agricultural fields as an alimentary complement for cows [29], animal bedding, for wood-smoking meat; and in materials field as an absorbent material and as reinforcing fillers for the development of composites. Wood flour has been studied and analyzed as a filler in different materials. It is common to use coupling agents to improve interactions between particles and polymers, these results show improving the adhesion between particle and matrix, and thermal stability [30,31].

On this basis, the present work aims to study the influence of wood flour particle size, at the microscale level, on the properties of biodegradable composite materials based on Solanyl® type bioplastic. Wood flour with different granulometry, ranging from 70 to 1100 μm, was assessed as filler. Commercial wood flour particles with three different sizes were selected: Lignocel® CB120 (70–150 μm), Lignocel® BK 40-90 (300–500 μm), and Lignocel® Grade 9 (800–1100 μm). The lignocellulose particles were added at three different loadings (10, 20, and 30 wt.%) to reinforce the Solanyl® type matrix. The materials were processed by melt-extrusion followed by injection molding processes to simulate the industrial processing conditions. The obtained formulations were assessed by rheological, mechanical, and thermal characterization to study the processability and performance of Solanyl®/Lignocel® formulations, the influences of the size of lignocellulose particles as well as the suitability of these materials for industrial production in the plastic sector.

2. Materials and Methods

2.1. Materials

Solanyl® C1201 (IMCD España Especialidades Quimicas, S.A., Barcelona, Spain) and three different wood flour particles, Lignocel® CB120 (0.07–0.15 mm), Lignocel® BK 40-90 (0.30–0.50 mm) and Lignocel® Grade 9 (0.80–1.10 mm) (Rettenmaier Iberica S.L. y Cia. S. Com, Grupo JRS GmbH, Barcelona, Spain), were used in the study. The characteristics of the wood flour particles are reported in Table 1.

Table 1. Physical properties of Lignocel® fillers.

	CB120	BK 40-90	Grade 9
Color	Yellow	yellow	yellow
Structure	Fibrous	cubic	cubic
Particle range (mm)	0.07–0.15	0.30–0.50	0.80–1.10

2.2. Granulometric Analysis

The particle size distribution of wood flour was determined by sieving. For this purpose, different sieves (2000 μm, 1000 μm, 500 μm, 250 μm, 125 μm, 63 μm, and bottom) were employed. The sieving machine used was CISA SIEVE SHAKER, model RP09 (Barcelona, Spain) with a sample of 200 g. The particle size distribution was calculated following the Fuller and Thompson method; Equation (1):

$$y = 100 \left(\frac{d}{D}\right)^e \quad (1)$$

where:

d is the sieve size (in mm) being considered (32, 16, 8, 4, 2, 1, 0.5, 0.25, 0.125, 0.063). d value is always minor or equal to D value.

D is maximum particle size (in mm). D value is equal to size mesh with cumulative retained less to 15%

e is the parameter that adjusts the curve. e value is 0.5 to Fuller and Thompson method.

2.3. Optical Measurements

Morphological characteristics (length and diameter) of different wood flour particles were analyzed using a compact stereomicroscope OLIMPUS model SZX7 with a Galilean optical system at 0.8×–5.6× zoom range.

2.4. Sample Preparation

The biopolymer and different microparticles of wood were mixed mechanically varying the filler contents in increments of 10% from 0% to 30% (wt.%). Later, the different samples were extruded in a twin-screw co-rotating extruder, using the temperatures recommended by Solanyl® C1201 (Zone 1: 135 °C–Zone 2: 150 °C–Zone 3: 160 °C–Die temperature: 165 °C) as a reference and at rotating speed of 60 rpm. Finally, a Meteor 270/75 injector supplied by Mateu-Sole®, (Barcelona, Spain) was used to obtain the samples for thermal and mechanical characterization.

2.5. Rheological Characterization

A Thermo Haake Rheoflixer MT1 capillary rheometer (Thermo Fisher Scientific, Karlsruhe, Germany) and current standard ISO 11443 was used to study the viscosity variations of all samples. The conditions of the tests were: temperature 160 °C, shearing speeds 100, 200, 500, 1000, 2000, 5000 and 10,000 s^{-1}; diameter of die 1 mm, relation length/diameter 10, 20 and 30.

2.6. Mechanical Properties Measurement

Tensile and impact tests were used to evaluate the mechanical performance of Solanyl® type reinforced with lignocellulose particle formulations. ELIB 30 electro-mechanical universal testing machine by Ibertest (S.A.E. Ibertest, Madrid, Spain) and Charpy impact machine (S.A.E. Ibertest, Madrid, Spain) were employed. UNE-EN-ISO 527 and ISO 179 standards were followed for tensile and impact tests, respectively. Five samples of each formulation were used in both tests and the mean and standard deviation of the measurements are reported.

2.7. Scanning Electron Microscopy (SEM) Measurements

The surface of the samples was metalized using gold sputter coating under vacuum conditions. Afterward, the different samples were analyzed in a ZEISS Ultra 55 Scanning Electronic Microscope (Carl Zeiss Iberia, Madrid, Spain).

2.8. Thermal Characterization

DSC Mettler-Toledo 821 equipment (Mettler-Toledo Inc., Schwerzenbach, Switzerland) allowed analysis of the different thermic transitions of all samples. The weight of the samples was 7–8 mg. A first heating (isothermal 5 min at 25 °C, and heating from 25 °C to 220 °C at 5 °C min^{-1}) was completed, followed by a cooling process (from 220 °C to 30 °C at 5 °C min^{-1}) and a second heating process (from 30 °C to 220 °C at 5 °C min^{-1}). The tests were performed in a nitrogen environment (flow rate of 30 mL min^{-1}).

The crystallinity of the PLA can be calculated using the following equation [32]:

$$\chi_c = \frac{\Delta H_m - \Delta H_{cc}}{\Delta H_m^0} \times \frac{100}{w} \quad (2)$$

where:

χ_c is the degree of crystallinity in %
ΔH_m is the melting enthalpy in J g^{-1}
ΔH_{cc} is the cold crystallization enthalpy in J g^{-1}
ΔH_m^0 is the calculated melting enthalpy of purely crystalline PLA, 93.7 J g^{-1}
w is the weight fraction of the PLA sample.

Thermogravimetric Analysis, TGA, was carried out using a Mettler-Toledo TGA/SDTA 851 (Mettler-Toledo Inc., Schwerzenbach, Switzerland) with an initial temperature of 50 °C and final temperature of 800 °C, using a 20 °C min^{-1} heating rate. The analysis was carried out using a nitrogen atmosphere (20 mL min^{-1}) in samples of 5 mg approximately.

To determine the Heat Deflection Temperature, a Vicat/HDT model Deflex 687-A2 (Metrotec, S.A, San Sebastian, Spain) was used. The oil used for the softening was silicon Dow Corning 200 Fluid 100 CS. The ISO 75 standard (B method) was followed employing a heating rate of 120 °C h^{-1} on samples sizing 80 × 10 × 4 mm^3.

3. Results

3.1. Wood Flour Particles Characterization

The wood flour particles were characterized by optical microscopy, where it was possible to verify the presence of two types of particles, long and short, as shown in Figure 1. In the case of Lignocel® CB120, fillers resulted in yellow particles with a fibrous structure. Meanwhile, Lignocel® BK-40-90 and Lignocel® Grade 9 resulted in long particles with cubic structure. This structure will further directly influence the processability as well as the final properties of Solanyl® type-based composites.

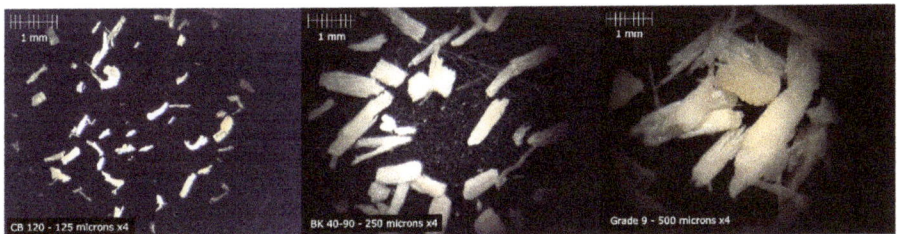

Figure 1. From left to right, micrographs of CB 120 retained on the sieve of 125 microns, BK 40-90 retained on the sieve of 250 microns and Grade 9 retained on the sieve of 500 microns.

A granulometric study of all Lignocel® particles with different sizes was carried out. The results are shown in Figure 2a, where it is possible to see the weight retained values in each of the sieves used. CB 120 sample possesses the smallest particles, and it is mostly composed of particles with sizes between 63 and 250 microns. Among the sizes, 125-micron particles are the most frequent. On the other hand, the BK 40-90 sample shows size particles between 250 and 1000 microns. Finally, 93% of particles of Grade 9 showed a size between 500 and 2000 microns, emphasizing 1000-micron particles, with 28%.

Mechanical properties of the composites depend on the size and particle distribution but are not the only factors, since the morphology of the particles is also very important. Since the optical micrographs showed significant differences in all wood flour particles (as shown in Figure 1), it was necessary to carry out a statistical study regarding the relationship between length and width (L/W) of the particles. The statistical analysis of different samples shows that the relation length/width decrease in bigger size samples and increases in smaller particles (CB 120), as shown in Figure 2b.

Another variable on which the mechanical properties of the composites material depend is the interaction of the phases (matrix phase (biopolymer) and disperse phase (wood flour)). Therefore, wood flour should be homogeneously dispersed among the Solanyl® type bioplastic to obtain composites with improved properties. It is very important to have an optimal size distribution, and the Fuller and Thompson method is most often used due to its high simplicity. Table 2 shows the values obtained for the analysis of this methodology.

From the obtained results, it is possible to establish that the D value for the samples is 0.5 mm for CB 120, 1 mm for BK 40-90 and 2 mm for Grade 9. The graphical representation of cumulative retained values versus sieve allowed observing differences between ideal particle size distribution and experimental values of wood flour (Figure 3). Finally, from the wood flour particles analysis, it can be concluded that the CB 120 sample is the most suitable particle for incorporation into the polymer matrix.

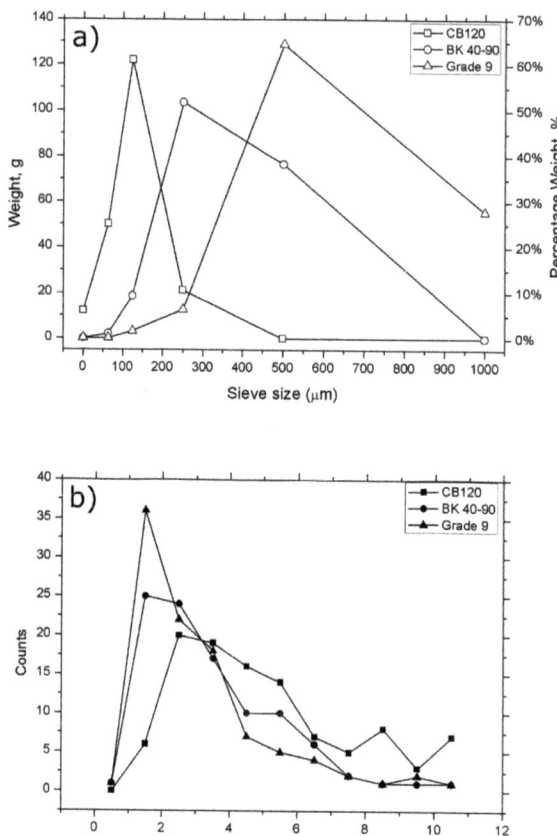

Figure 2. (**a**) Histogram of the weight distribution for different wood flour particles, and (**b**) histogram of the relationship between the length/width of different wood flour particles.

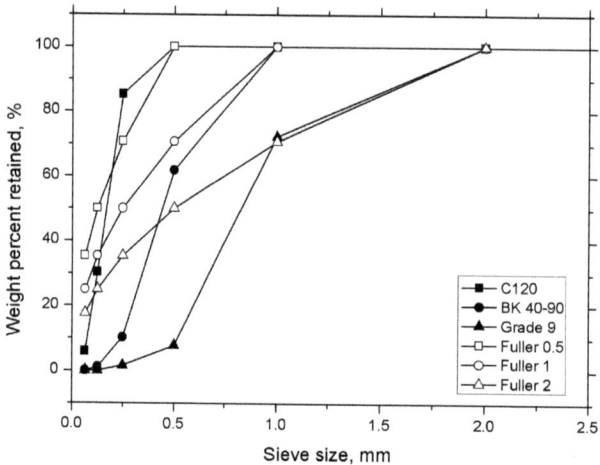

Figure 3. Representation of ideal particle size distribution (Fuller D = 0.5, D = 1, and D = 2) and wood flour size distribution.

Table 2. Weight values retained (g), cumulative retained (%), and passed of different particles CB 120, BK 40-90 and Grade 9.

Mesh Size	CB 120				BK 40-90				Grade 9			
	Weight Retained	Cumulative Retained		Total % Passing	Weight Retained	Cumulative Retained		Total % Passing	Weight Retained	Cumulative Retained		Total % Passing
mm	g	g	%	%	g	g	%	%	g	g	%	%
32	0	0	0	100	0	0	0	100	0	0	0	100
16	0	0	0	100	0	0	0	100	0	0	0	100
8	0	0	0	100	0	0	0	100	0	0	0	100
4	0	0	0	100	0	0	0	100	0	0	0	100
2	0	0	0	100	0	0	0	100	0	0	0	100
1	0	0	0	100	0	0	0	100	55.5	55.5	27.75	72.25
0.5	31	31	15.12	84.88	76.5	76.5	38.15	61.85	129	184.5	92.25	7.75
0.25	112	143	69.75	30.25	103.5	180	89.77	10.23	12.5	197	98.5	1.5
0.125	50	193	94.14	5.86	18.5	198.5	99.00	1	3	200	100	0
0.063	12	205	100	0	2	200.5	100	0	0	200	100	0
bottom					0	200.5			0	200	100	0
Total	205				200.5				200			

3.2. Biocomposites Characterization

Once the lignocellulosic particles were characterized, they were used to obtain biocomposites using Solanyl® as the biopolymeric matrix, in 10, 20, and 30 wt.% loading. The Solanyl®/Lignocel-based composites were successfully processed by the melt-blending process followed by the injection molding process to simulate the industrial condition and, thus, assessing the possibility for their scalable production at the industrial level.

3.2.1. Rheological Characterization

During the melt-extrusion and injection molding process, polymers and particularly biopolymers undergo degradation due to the strong shear stresses that act in the viscous molten polymer [33,34]. Thermal and mechanical characterization of material is extremely important, as this provides information on the variation produced by the introduction of filler, thermic treatments, ultraviolet radiation, and thermic cycles. However, if the material is going to be used at an industrial level, it is also important to establish the rheological characteristics to establish temperature and pressure parameters.

The use of capillary rheology requires the application of a range of corrections. There is a difference in the pressure between the radius of the deposit and the radius of the nozzle and the Bagley correction allows this value to be accurately obtained. The Rabinowitch correction allows the shear rate gradient to be corrected regarding the relationship between polymers and Newtonian fluids. After applying the Bagley and Rabinowitch corrections, the rheological curves show a slight increase in viscosity as the different wood particles are introduced (wood flour fillers). This behavior has been observed by other authors using polymers matrix. The viscosity of the PLA-based polymeric formulations increased with the addition of microparticles [35], while wood flour also showed the ability to increase the viscosity of polymeric matrix (i.e., polypropylene, PP) with increasing wood flour content. These studies showed that the interaction between polymer and particle produces an increase in viscosity of the end product material [36].

The same result can be observed in the evolution of values of viscosity in wood flour biocomposites, as shown in Table 3.

Table 3. Viscosity values for Solanyl® composites with a varying wood flour fraction, C120, BK 40-90, and Grade 9.

Material	% of filler	\multicolumn{7}{c}{Viscosity, Pa s}						
Shear rate (s^{-1})		100	200	500	1000	2000	5000	10,000
Virgin	0	3061.25	1730.12	815.25	463.87	236.75	90.38	49.92
CB 120	10	3244.37	1900.46	842.38	500.87	250.78	95.31	52.85
	20	3300.25	2000.65	974.27	520.51	251.09	95.91	59.22
	30	3492.52	2101.56	985.37	543.18	280.06	103.15	63.23
BK 40-90	10	3408.25	2109.37	1022.62	532.46	256.84	110.56	60.87
	20	3662.51	2271.56	1095.25	587.18	295.46	120.87	70.23
	30	3720.21	2500.32	1200.54	610.87	306.43	126.32	75.98
Grade 9	10	3304.37	2036.87	1021.62	565.87	291.90	110.28	60.67
	20	3400.25	2286.56	1109.62	598.56	309.81	118.95	65.99
	30	3563.12	2500.65	1200.87	620.15	325.46	120.38	70.15

Increasing viscosity is not the only effect of adding wood flour to polymer matrices. When the behavior of the material subjected to high temperatures is analyzed in detail, the pressure shows a gradual increase as the shear rate increases from 100 to 10,000 s^{-1}, as shown in Figure 4.

Rheological characterization is obtained from the constant pressure values for a given shear speed; subsequently, the application of various corrections (Bagley and Rabinowitch) makes it possible to obtain the rheological curves of the material analyzed.

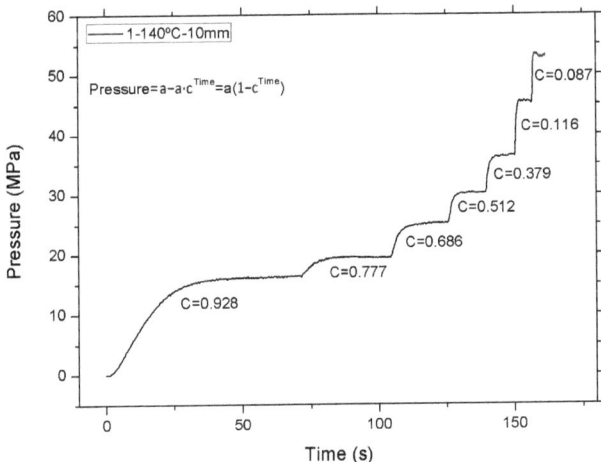

Figure 4. Evolution of the pressure versus time of the Solanyl® at 140 °C with a nozzle length of 10 mm.

The variation in pressure versus time can be defined by the following equation:

$$Pressure_t = P_a(1 - C^t) \tag{3}$$

P_a is the pressure value when it has been stabilized for a particular shear rate in MPa.

C is the speed with which the pressure is achieved (dimensionless, always less than 1, with values close to zero suggesting that the rate at which the pressure increases is faster).

t is the time in s.

The trend of C value is zero as the shear speed increases. This behavior can be observed in virgin material regardless of the type of nozzle used in the capillary rheometer. In this case, the values of C remain very close to each other for every shear speed, as shown in Figure 5.

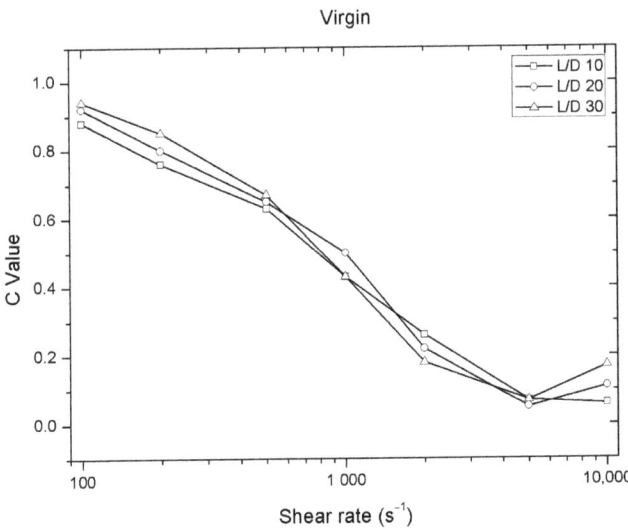

Figure 5. Evolution of C values versus shear rate for different nozzles length of the Solanyl® at 140 °C.

The value of C is indicative of the rate at the pressure increases. This value can be affected by the presence of particles in the polymer matrix. C values of CB 120 and virgin polymer are very similar, but in the case of the smaller particle size (CB 120), the dispersion of C values increase when 2000 s^{-1} of shear rate is employed. This dispersion is greater in samples with 30 wt.% of wood flour and it is more evident in 1000 s^{-1} of shear rate, as can be seen in Figure 6.

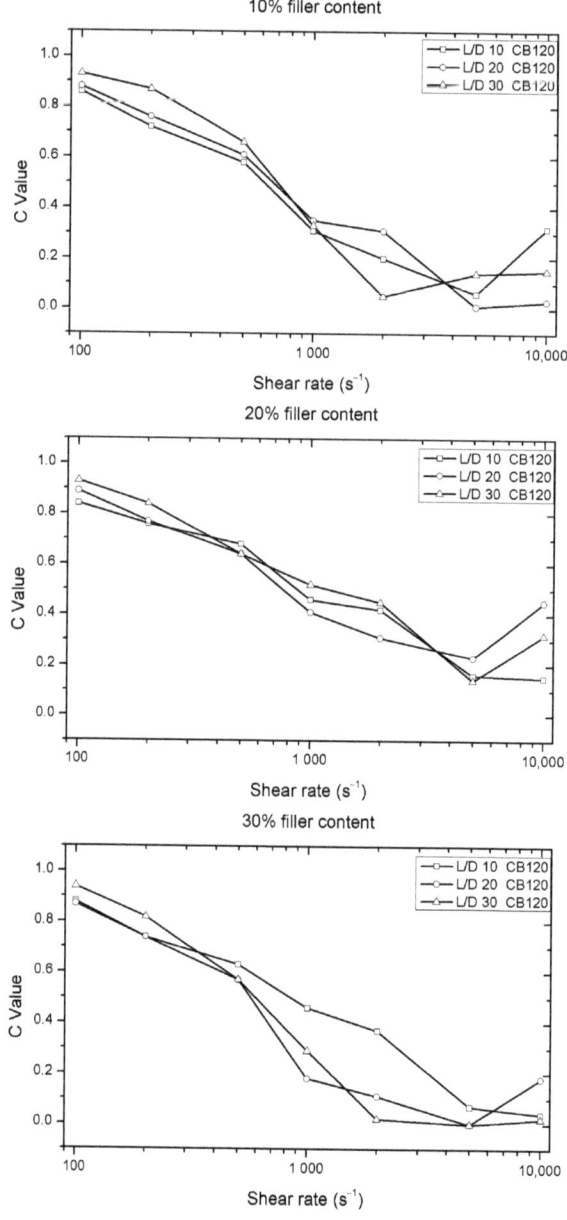

Figure 6. Evolution of C values versus shear rate for different nozzles length of the Solanyl® with different percentages of fillers (CB 120).

On the other hand, BK 40-90 wood flour sample shows a behavior similar to CB 120. In this case, it is possible to see C value dispersion in the shear speed of 500 s^{-1}, as seen in Figure 7. This behavior is due to the morphological properties of this sample, where 50% of the samples show lengths greater than 1 mm. Thus, the use of this sample can increase the probability of nozzle blockage.

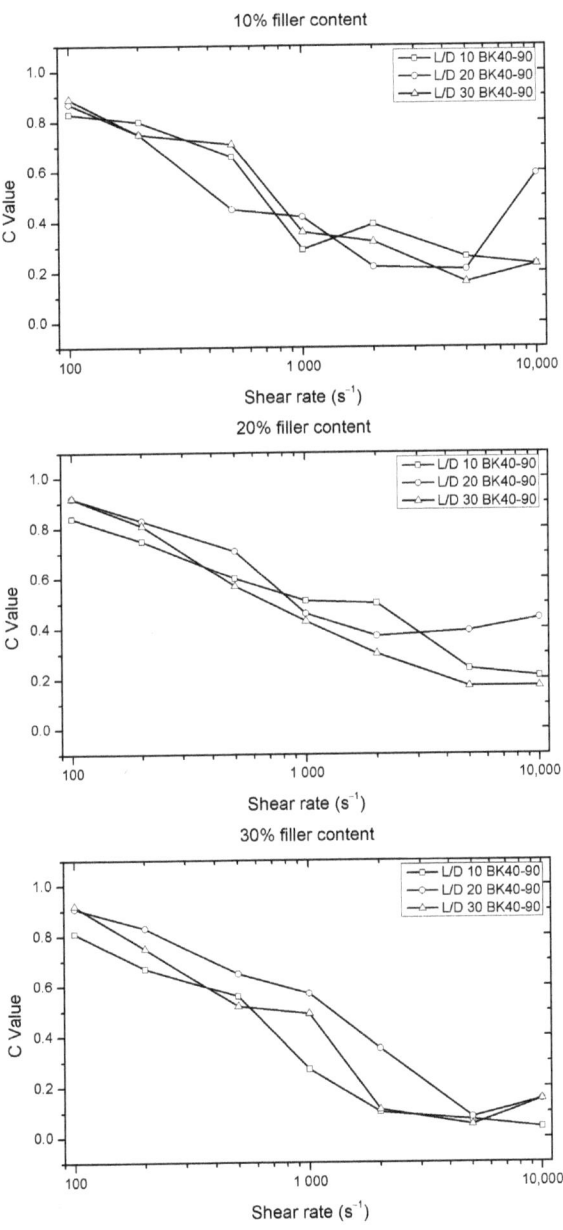

Figure 7. Evolution of C values versus shear rate for different nozzle lengths of the Solanyl® with different percentages of fillers (BK 40-90).

Finally, in the case of the largest particles (Grade 9), the dispersion of the C values is much greater in samples with wood flour contents of 20 and 30 wt.%. Moreover, this phenomenon appears at relatively low shear rates, as shown in Figure 8.

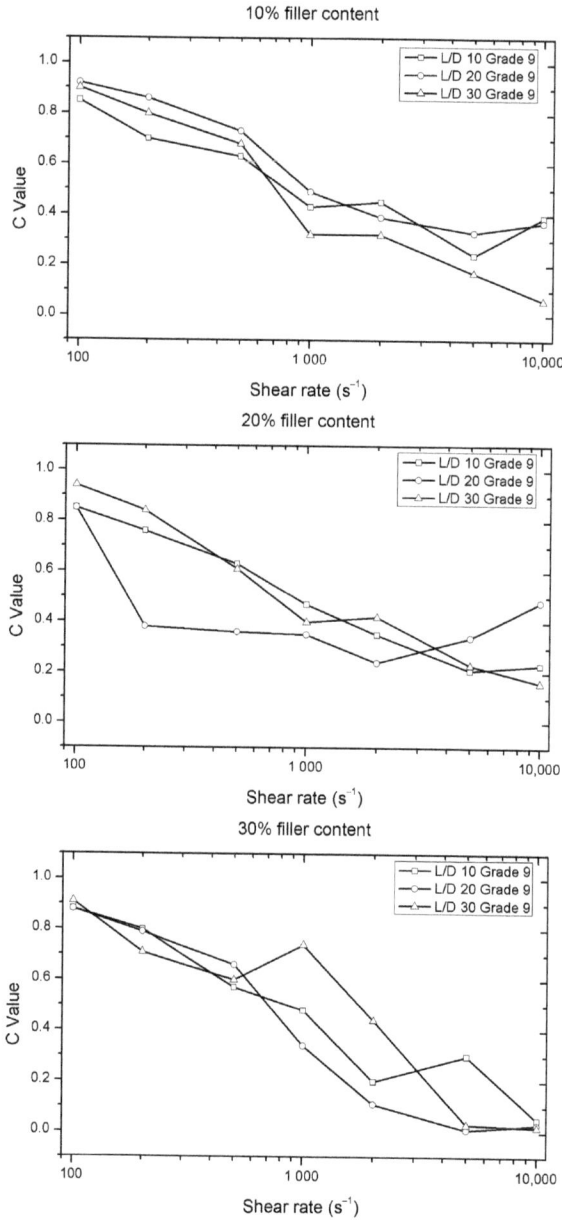

Figure 8. Evolution of C values versus shear rate for different nozzles length of the Solanyl® with different percentages of fillers (Grade 9).

In summary, the presence of wood flour not only affects viscosity, but also means that the pressure increases much more quickly as greater percentages of filler are added. This behavior is repeated for all three types of particles used.

The variation in C values is attributed to the obstruction of the nozzle during the sample processing using the capillary rheometer. The fact that the pressure increases faster does not imply differences in rheological behavior, but it can cause certain problems in the manufacture of parts at an industrial level, especially in injection molding processes. The manufacture of parts by injection requires molds, which have sprues, runners, gates, and pieces. In sprues and runners, large enough sections are used to allow the material to flow without problems, whereas the reduction of the section in the case of gates is quite important and they are susceptible to obstruction by the presence of particles in the polymeric matrix.

3.2.2. Mechanical Properties

Injection molded polymers for rigid applications (e.g., packaging, domestic appliance, toys, etc.) are required for high mechanical performance to overcome the strong shear stresses during processing in order to successfully obtain injected molded parts, as well as to offer good performance during service [33]. At the engineering level, the application of a material is often determined by its mechanical properties. Tensile strength, elongation at break, impact strength, and Young's modulus should be analyzed.

Tensile strength is one of the most important properties; it is the characteristic that decides the maximum load that a material can support. Sometimes, the incorporation of filler may have a negative effect if there is no interaction between the matrix and the particle, as a consequence of non-homogeneous dispersion of the particle into the polymeric matrix. When the level of interaction is low, it can lead to a general decrease in mechanical properties. Nevertheless, when the interaction is good, due to the good dispersion of particles that reach a homogeneous distribution into the polymeric matrix, it allows an improvement of the overall mechanical performance.

Young's modulus of all biocomposites increased with the filler percentage, up to 20 wt.% (Figure 9a). The tensile strength showed a similar tendency when smaller fillers were used (Lignocel® CB120 and BK 40-90) and they showed a slight improvement compared to the filler Grade 9, which has the largest sized particles. This last filler was unable to significantly increase the tensile strength of the Solanyl® matrix (Figure 9b).

Regarding the elongation at break, for filler contents of 20 and 30 wt.%, a slight decrease is seen with Lignocel® BK 40-90 and Lignocel® Grade 9 particles. In contrast, the samples with smaller sized particles (Lignocel® CB120) showed an increase in elongation at break up to 20% of wood flour content. However, these increasing tendencies slightly fall for samples with the highest content of 30 wt.% of this wood flour filler (Figure 9c).

Although the values for tensile strength and elongation at break were slightly improved in comparison with neat Solanyl® type (polymeric matrix without filler), particularly with the smallest wood flour particles (Lignocel® CB120), the application of high impact creates very different results for polymers with and without filler. In the samples with smaller sized particles (Lignocel® CB120), the impact strength was practically the same in the formulations and the matrix (Figure 9d). Nevertheless, when the particle sizes are intermediate or large, the impact strength increases as the percentage of filler increases, achieving values up to 140% for Lignocel® BK 40-90 and 180% for the biggest sized particles (Lignocel® Grade 9). This maximum is produced in samples with a filler content of 20 wt.%. Once this content is surpassed, the matrix collapses and there is a huge decrease in the impact strength values. Other authors have observed increases in Young's modulus depending on wood flour content. In contrast, the strain and Izod impact strength decreased [37].

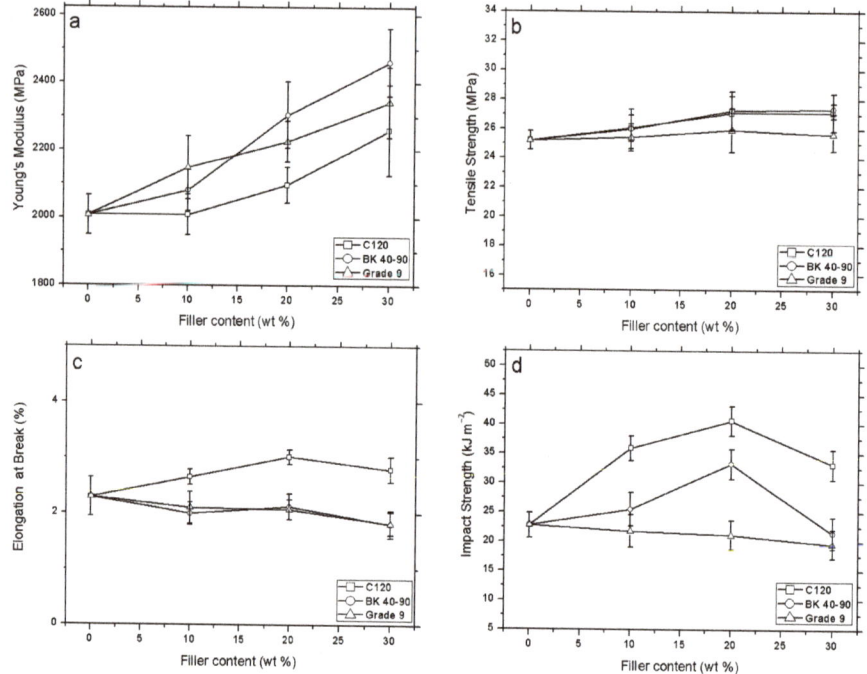

Figure 9. Mechanical properties of different samples: (**a**) Young's modulus, (**b**) tensile strength, (**c**) elongation at break, (**d**) impact strength.

3.2.3. SEM Studies

The effect of the addition of wood flour particles with different granulometry on the microstructure of Solanyl® type bioplastic was evaluated through SEM analysis, focusing on the dispersion of lignocellulose particles into the polymeric matrix, as well as on the interaction between wood flour particles with the polymeric matrix, to better understand the rheological and mechanical performance of the biocomposites. Considering the fibrous structure of lignocellulose fillers, the orientation into the polymeric matrix was also considered.

In general, it is possible to observe oriented particles with a certain degree of inclination. Paying attention to the particle/matrix interphase, somewhat symptoms of lack of adhesion between the particle and the matrix phase are observed with the increasing amount of particle loading. Figure 10, shows the fracture surface of the produced Solanyl®/Lignocel biocomposites loaded with the smallest particles (CB 120, particles size between 63 and 250 microns) at different loadings of 10 wt.% (Figure 10a,b), 20 wt.% (Figure 10c,d) and 30 wt.% (Figure 10e,f). A clear homogeneous distribution of lignocellulose fillers in the polymeric matrix was observed, confirming the good processing ability of the smallest lignocellulose particles used here. Given their length and flexibility, it seems that CB 120 particles at 10 wt.% were well dispersed in the polymeric matrix, while non-oriented fibers were observed. Good adhesion between lignocellulose particles and the polymer matrix was observed, thus making the breaking process difficult. At higher loadings (20 wt.% Figure 10c,d) and particularly at 30 wt.% (Figure 10e,f), it seems that the interaction of the particles with the polymeric matrix is relatively low. Two phenomena can be observed depending on the orientation of the fibers. On one side, very thin fibers with a considerable length are partially attached to the polymeric matrix, while on the other side, small cavities are observed. This last phenomenon occurs when the orientation of the fiber is perpendicular to the working section.

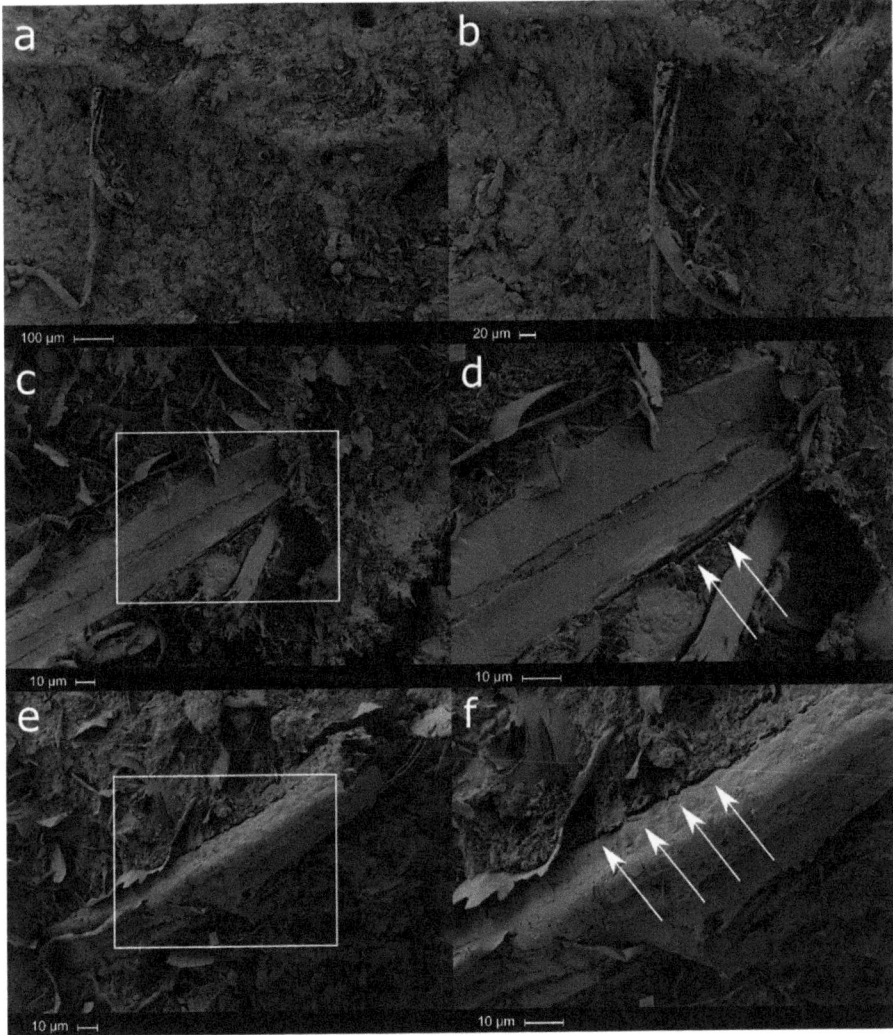

Figure 10. SEM micrographs of smallest particles, CB120—10 wt.% wood flour: (**a**) ×100, (**b**) ×200; CB120—20 wt.% wood flour: (**c**) ×500, (**d**) ×1000; and CB120—30 wt.% wood flour: (**e**) ×500, (**f**) ×1000.

Medium-sized particle (BK 40-90)-loaded biocomposites were characterized by partially maintaining the structure of the wood (Figure 11). In Figure 11a,b, the SEM images of Solanyl® loaded with 10 wt.% of medium-sized particles (BK 40-90) are shown. Lignocellulosic particles appeared to be aligned with a certain degree of crushing, probably due to the wood grinding process, as well as due to the processing of biocomposites by melt-extrusion and injection molding. Although there is a homogeneous distribution of medium-sized particles, it can be observed that there are partially embedded lamellar particles into the polymeric matrix with evidence of fiber pull-out. Moreover, a clear crack is observed along the surface until it meets a particle, but borders it without great difficulty in fracturing.

Figure 11. SEM micrographs of medium size particles—10 wt.% wood flour: (**a**) ×100, (**b**) ×200; 20 wt.% wood flour: (**c**) ×50, (**d**) ×200, (**e**) ×200, (**f**) ×200; 20 wt.% wood flour: (**g**) ×50; and 30 wt.% wood flour: (**h**) ×50.

Increasing the loading amount of medium-sized particles, lamellar particles tend to agglomerate, and the degree of crushing, produced either by the wood grinding process or by the shear stress suffered during biocomposite processing, became more evident (Figure 11c–f).

As occurred with the smallest particles, the medium-sized particles showed different orientations into the polymeric matrix, with more evidence of fiber pull-out, suggesting a lower degree of adhesion between the wood flour particles and the polymeric matrix (Figure 11g–h).

Finally, the larger lignocellulose particles used present the lamellar structure of wood, with lengths of approximately 1 mm, and thus following the obtained weight distribution (Figure 12). Regarding the orientation of the fibers in the polymeric matrix, different orientations were found. Indeed, it is expected that large particles flow slowly in the polymer molten state, making their orientation difficult, as is evident from the near-surface region (Figure 12). Moreover, at the interface, fiber debonding, as well as fiber pull-out phenomena, are evident.

Figure 12. SEM micrographs of largest sized particles—10 wt.% of wood flour ×50.

Instead, in the internal areas, it seems that the fibers acquire a certain degree of inclination. During the breaking process, the particle remains adhered to the polymeric matrix on one of its faces. Meanwhile, the other face is completely clear. This phenomenon was observed in all biocomposites prepared with the largest sized particles (Figure 13).

As a general conclusion, as the particle size increases, non-optimal adhesion and debonding phenomena are observed. A homogeneous and better dispersion was observed for the smallest particles used leading to the better mechanical response obtained.

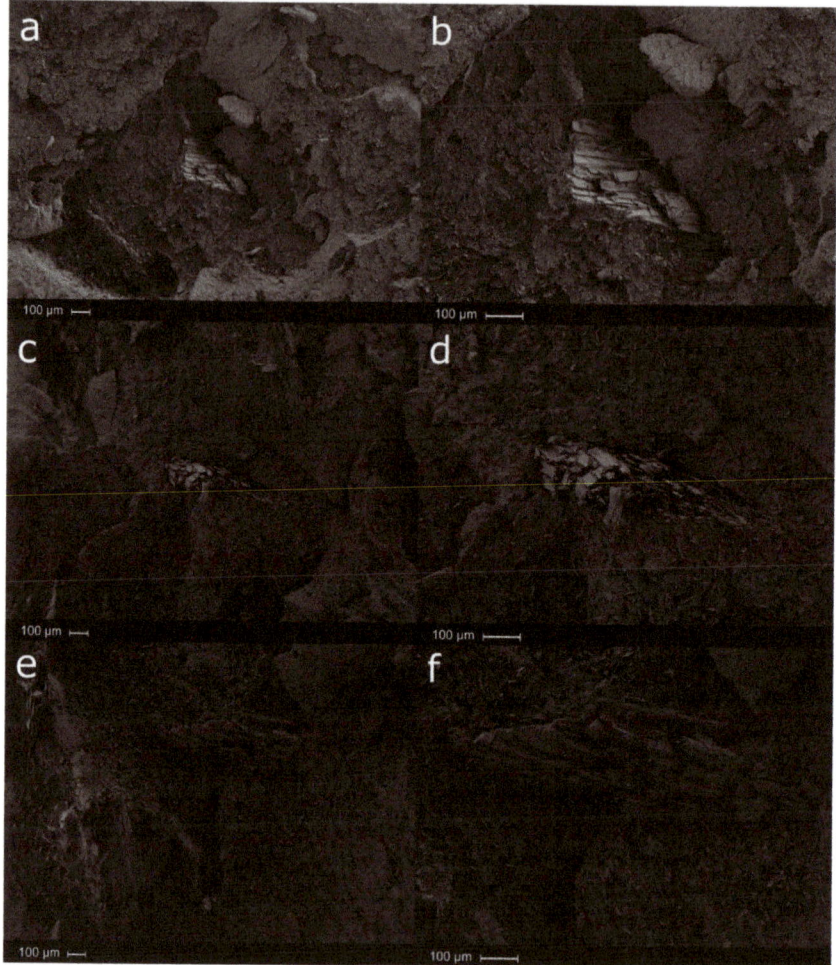

Figure 13. SEM micrographs of largest sized particles, Grade 9—10 wt.% wood flour: (**a**) ×50, (**b**) ×100; Grade 9—20 wt.% wood flour: (**c**) ×50, (**d**) ×100; and Grade 9—30 wt.% wood flour: (**e**) ×50, (**f**) ×100.

3.2.4. Thermal Properties

Differential scanning calorimetry provides a range of information, such as glass transition temperature and crystallization temperature. Moreover, it allows mixtures of polymers to be detected. The DSC analysis (Figure 14a) shows three types of thermal curves and two endothermic reactions, one at around 60 °C, most probably due to the presence of PCL [38] in the compound, and another one at around 160 °C, which may correspond to the PLA phase (PLA) [11]. Furthermore, between 70 °C and 90 °C, an exothermic peak appears, which is related to the PLA cold crystallization. Moreover, the thermogravimetric curve (TGA) allows an understanding of the relationship between the matrix and dispersed phases. The maximum degradation rate of neat Solanyl® C1201 is produced at 342 °C and 462 °C, corresponding to the thermal degradation of PLA and PCL phases, respectively (Figure 14b) [39]. The TGA analysis also revealed the presence of a starch-based material, as stated by the producer [8], since there is about a 5% mass loss at low temperatures (lower than 100 °C), as shown in Figure 14b [40].

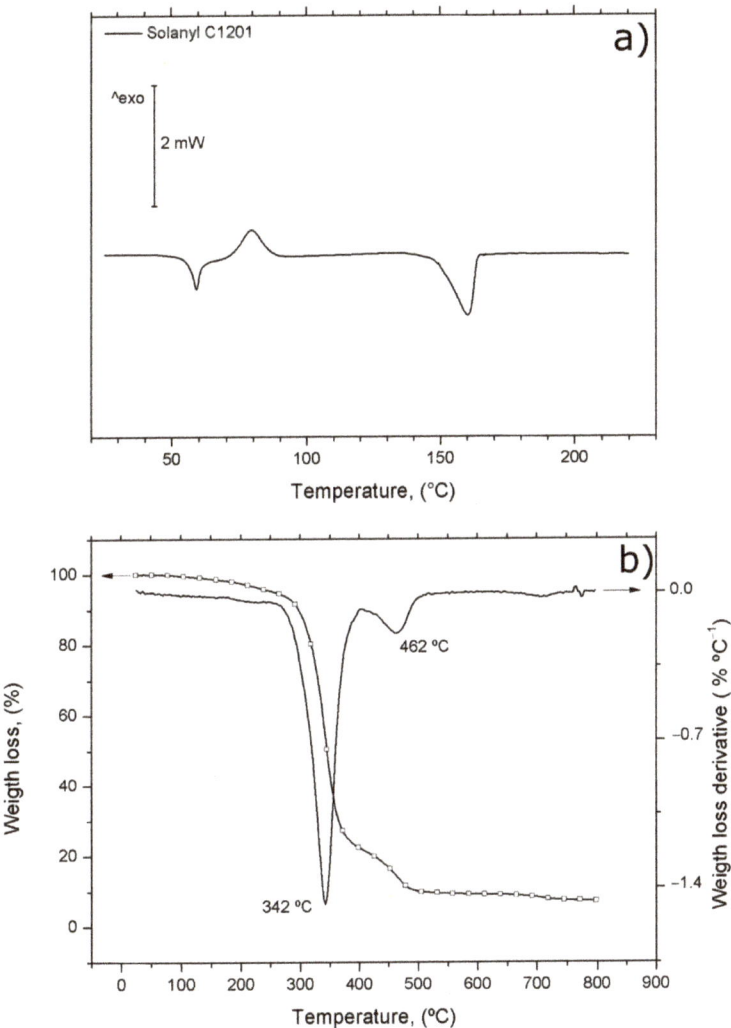

Figure 14. (a) DSC curve analysis of Solanyl® C1201, (b) TGA and DTGA curves of Solanyl® C1201.

The presence of two phases in the neat polymeric matrix of Solanyl® C1201 was further corroborated by SEM images which, at ×500, clearly shows the presence of both phases. This is relevant because the interaction between the two phases determines the mechanical properties of the material. The dispersed phase can be characterized as flakes embedded in the matrix phase (Figure 15) and the presence of the flakes confirm the presence of a starch-based material [41–43].

Understanding the matrix phases is extremely important when analyzing the behavior of a material, as are these phases that define the properties of the material. The dispersed phase, if it is homogeneously dispersed in the polymeric matrix, will further improve the overall performance of the final material. It can be seen that the first rise is much greater than the second, indicating that the matrix phase is composed of a higher amount of PLA (around 80%) and around 20% of PCL (Figure 14b). This data are again extremely important, as the mechanical properties are governed by the interaction between the PLA and the PCL matrices, as well as between both polymeric matrices and the lignocellulosic additive particles. Furthermore, the Solanyl® C1201 biopolymer has a degree of crystallinity (as shown in DSC results in Figure 14a). This is significant, as crystallinity is often an important factor in deciding the mechanical behavior of a polymer. Any variation in this characteristic will have repercussions on the mechanical properties of the final material. Wood flour was added to increase the crystallinity of the Solanyl® type bioplastic.

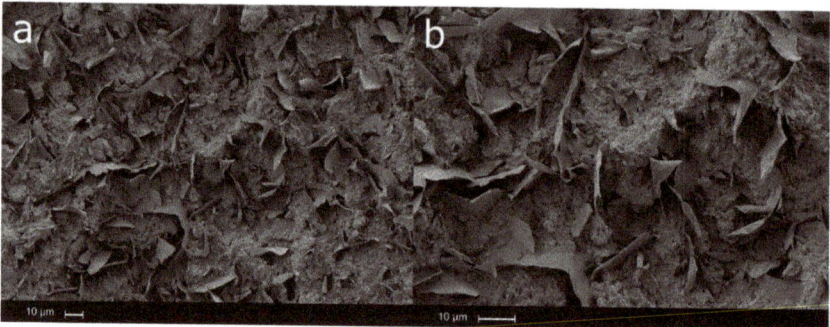

Figure 15. SEM micrograph of Solanyl® C1201: (**a**) ×500 and (**b**) ×1000.

Table 4 shows the DSC thermal parameters and the evolution of crystallinity in the Solanyl® type biocomposites reinforced with Lignocel® particles from both the first and second heating DSC scans. In the first heating scan, the cold crystallization exothermic peak was observed in all formulations, showing that the injection molding processing conditions did not produce the crystallization of PLA, and thus it crystallizes during DSC heating. The degree of crystallinity in general increases with an increasing amount of lignocellulosic particles, particularly in small (Lignocel® CB 120) and middle-sized particles (Lignocel® BK 40-90), showing the ability of wood flour to act as a nucleating agent. However, the crystallinity from the first heating cannot be taken as representative as the cooling conditions varied from sample to sample. To eliminate the effect of the cooling rate during processing, and thus the thermal history, a second DSC heating was carried out. The DSC cooling conditions applied made it possible to crystallize PLA, and thus it did not crystallize during the second heating scan. In this case, the crystallinity values do indeed better represent the effect of the presence of lignocellulose particles in the samples.

It is worth highlighting the fact that crystallinity increases depending on the presence of lignocellulosic filler regardless of the particle size, although the increase is very slight for samples with the larger sized particles, Lignocel® Grade 9. This trend in the crystallinity values was observed by other authors in composites based on PLA with pinewood flour composites [44,45].

In both cases, crystallinity increases depending on the quantity of filler present. Moreover, the use of blends in polymers led to significant advances in this field, including the creation of materials with properties that the individual materials do not have. The difficulty, however, is that on certain occasions, thermal transitions are hidden by the presence of another polymer. This is the case with the glass transition temperature (T_g) of PLA, which is concealed by the melting of PCL. The difficulty to determine the T_g during the DSC heating scans due to its overlapping with the melting peak of PCL frequently occurs in PLA/PCL-based formulations [12–14].

Finally, the heat deflection temperature (HDT) values remain virtually unchanged regardless of the type and percentage of filler used, with a slight increase only in the samples with 30 wt.% filler, as shown in Table 5. This indicates good interaction between polymer and filler [46,47].

Table 4. Enthalpy values of different samples and the calculated crystallinity.

Mixture	Tcc, °C	Tm, °C	First Heating at 5 °C min^{-1}		Crystallinity χ_c, (%)
			Normalized Enthalpy		
			Cold (J g^{-1})	Hot (J g^{-1})	
PLA-PCL	78.85	160.17	6.43	14.31	10.51
10—CB 120	78.77	161.86	6.40	14.70	12.31
20—CB 120	79.72	161.91	4.70	15.43	17.90
30—CB 120	79.83	160.58	4.42	16.80	23.61
10—BK 40-90	79.94	161.28	3.60	12.15	12.67
20—BK 40-90	80.35	163.03	5.84	14.18	13.90
30—BK 40-90	80.36	162.74	5.17	14.85	18.45
10—Grade 9	82.50	161.64	5.92	13.27	10.89
20—Grade 9	79.77	162.23	4.77	13.61	14.73
30—Grade 9	79.39	160.28	4.23	12.56	15.88
			Second Heating at 5 °C min^{-1}		
Mixture	Tcc, °C	Tm, °C	Normalized Cold (J g^{-1})	Enthalpy Hot (J g^{-1})	Crystallinity χ_c, (%)
PLA-PCL	-	156.86	-	12.62	16.84
10—CB 120	-	158.21	-	12.17	18.04
20—CB 120	-	158.19	-	12.62	21.05
30—CB 120	-	158.35	-	14.11	26.89
10—BK 40-90	-	157.71	-	12.43	18.43
20—BK 40-90	-	159.88	-	13.66	22.78
30—BK 40-90	-	159.42	-	14.28	27.22
10—Grade 9	-	158.16	-	10.66	15.80
20—Grade 9	-	158.58	-	12.16	20.28
30—Grade 9	-	157.23	-	11.12	21.19

Table 5. HDT values (°C) of different samples with wood flour.

Filler Content wt.%	CB 120	BK 40-90	Grade 9
0	53.8	53.8	53.8
10	53.8	54.2	54.4
20	54.4	54.6	54.8
30	54.8	55.0	55.3

4. Conclusions

The present study focused on the influence of three types of wood flour, Lignocel® C120 (70–150 µm), BK 40-90 (300–500 µm) and Grade 9 (800–1100 µm), on the rheological, mechanical, and thermal properties of Solanyl® type bioplastic. The main novelty of this study is the analysis of the pressure variation during the rheological study of the prepared samples. The granulometric and morphological study of different wood flour samples and the application of the Fuller-Thompson method allowed us to determine which sample presented better characteristics for improving the properties of the polymeric mixture (polymer—wood flour) to obtain biocomposites with improved performance. The rheological study revealed an increase in viscosity as the filler percentage increases. The viscosity data are especially relevant for use at the industrial level, particularly by extrusion and injection molding processes. Despite the low viscosity variation as the percentage of filler increased, a more detailed analysis showed that pressures rose more quickly, which can be attributed to the obstruction of the nozzle by the presence of wood flour particles. This fact could cause an increase in defective pieces due to the obstruction of the gate inside the mold. All these results indicate that the material can be processed by extrusion.

The tensile test results showed that, in general, Young's modulus gave increasing values for all types of particles. Similarly, the tensile strength increased with an increasing amount of wood flour small-sized (CB 120) and medium-sized particles (BK 40-90) in the formulation. The largest particles (Grade 9) did not produce significant changes in the tensile strength values. The elongation at break showed little variation with the addition of different percentages of medium-sized particles (BK 40-90) and the largest particles (Grade 9) to the Solanyl® type bioplastic. Nevertheless, the incorporation of the smallest lignocellulosic microparticles (CB 120) increased the elongation at break. The impact strength was only affected by the incorporation of small and medium-sized particles (CB 120 and BK 40-90) showing a maximum value for biocomposites with 20 wt.% wood flour filler. SEM images revealed good interaction between particles and the Solanyl® type polymeric matrix, which positively affects the mechanical behavior of the biocomposites. On the other hand, an increase in crystallinity leads to an increase in the mechanical properties, visible through the impact strength and Young's modulus. Of all the samples analyzed, those with 10 or 20 wt.% filler of the smallest particles (CB120) and medium-sized particles (BK 40-90) showed the best combination for processing by injection molding as well as leading to biocomposites with improved properties and, thus, that are of interest for sustainable industrial applications.

Author Contributions: Conceptualization, F.P., M.A.P. and D.J.; Data curation, F.P. and M.A.; Formal analysis, F.P., M.A.P. and D.J.; Funding acquisition, F.P.; Investigation, F.P., M.A.P., D.J., M.P.A. and M.A.; Methodology, F.P., M.P.A. and M.A.; Project administration, F.P. and M.A.P.; Resources, F.P.; Supervision, F.P. and M.A.P.; Validation, F.P., M.A.P. and M.P.A.; Visualization, F.P. and M.A.; Writing—original draft, F.P.; Writing—review & editing, F.P., M.P.A. and M.A. All authors have read and agreed to the published version of the manuscript.

Funding: This research received no external funding.

Conflicts of Interest: The authors declare have no conflict of interest with Solanyl® biodegradables. This research received no external funding.

Data Availability: All relevant data (mechanical properties, SEM micrographs, Thermal Analysis curves) are available via e-mail (fraparga@dimm.upv.es).

References

1. Plastics Europe Market Research Group (PEMRG). Plastics-the Facts 2019 An Analysis of European Plastics Production, Demand and Waste Data. *Plast. Facts* **2019**, *1*, 1–42.
2. Aldas, M.; Paladines, A.; Valle, V.; Pazmiño, M.; Quiroz, F. Effect of the Prodegradant-Additive Plastics Incorporated on the Polyethylene Recycling. *Int. J. Polym. Sci.* **2018**, *2018*, 1–10. [CrossRef]
3. Kolek, Z. Recycled Polymers from Food Packaging in Relation to Environmental Protection. *Polish J. Environ. Stud.* **2001**, *10*, 73–76.
4. Aldas, M.; Valle, V.; Aguilar, J.; Pavon, C.; Santos, R.; Luna, M. Ionizing Radiation as Adjuvant for the Abiotic Degradation of Plastic Bags Containing Pro-oxidant Additives. *J. Appl. Polym. Sci.* **2021**, *138*, 49664. [CrossRef]
5. Arrieta, M.P.; Samper, M.D.; Aldas, M.; López, J. On the Use of PLA-PHB Blends for Sustainable Food Packaging Applications. *Materials* **2017**, *10*, 1008. [CrossRef] [PubMed]
6. Pavon, C.; Aldas, M.; De La Rosa-Ramírez, H.; López-Martínez, J.; Arrieta, M.P. Improvement of PBAT Processability and Mechanical Performance by Blending with Pine Resin Derivatives for Injection Moulding Rigid Packaging with Enhanced Hydrophobicity. *Polymers* **2020**, *12*, 2891. [CrossRef]
7. Owen, A.J.; Jones, R.A.L. Rheology of a Simultaneously Phase-Separating and Gelling Biopolymer Mixture. *Macromolecules* **1998**, *31*, 7336–7339. [CrossRef]
8. What is Solanyl Biopolymers? Available online: https://www.solanylbiopolymers.com/about-solanyl.html (accessed on 27 November 2020).

9. Auras, R.; Harte, B.; Selke, S. An Overview of Polylactides as Packaging Materials. *Macromol. Biosci.* **2004**, *4*, 835–864. [CrossRef]
10. Villegas, C.; Arrieta, M.P.; Rojas, A.; Torres, A.; Faba, S.; Toledo, M.J.; Gutierrez, M.A.; Zavalla, E.; Romero, J.; Galotto, M.J.; et al. PLA/Organoclay Bionanocomposites Impregnated with Thymol and Cinnamaldehyde by Supercritical Impregnation for Active and Sustainable Food Packaging. *Compos. Part B Eng.* **2019**, *176*, 107336. [CrossRef]
11. De La Rosa-Ramírez, H.; Aldas, M.; Ferri, J.M.; López-Martínez, J.; Samper, M.D. Modification of Poly (Lactic Acid) through the Incorporation of Gum Rosin and Gum Rosin Derivative: Mechanical Performance and Hydrophobicity. *J. Appl. Polym. Sci.* **2020**, *137*, 1–15. [CrossRef]
12. Arrieta, M.P.; Samper, M.D.; Jiménez-López, M.; Aldas, M.; López, J. Combined Effect of Linseed Oil and Gum Rosin as Natural Additives for PVC. *Ind. Crops Prod.* **2017**, *99*, 196–204. [CrossRef]
13. Rhim, J.-W.; Park, H.-M.; Ha, C.-S. Bio-Nanocomposites for Food Packaging Applications. *Prog. Polym. Sci.* **2013**, *38*, 1629–1652. [CrossRef]
14. Sessini, V.; Navarro-Baena, I.; Arrieta, M.P.; Dominici, F.; López, D.; Torre, L.; Kenny, J.M.; Dubois, P.; Raquez, J.-M.; Peponi, L. Effect of the Addition of Polyester-Grafted-Cellulose Nanocrystals on the Shape Memory Properties of Biodegradable PLA/PCL Nanocomposites. *Polym. Degrad. Stab.* **2018**, *152*, 126–138. [CrossRef]
15. Fortunati, E.; Peltzer, M.; Armentano, I.; Torre, L.; Jimenez, A.; Kenny, J.M. Effects of Modified Cellulose Nanocrystals on the Barrier and Migration Properties of PLA Nano-Biocomposites. *Carbohydr. Polym.* **2012**, *90*, 948–956. [CrossRef] [PubMed]
16. Arrieta, M.P.; Fortunati, E.; Dominici, F.; López, J.; Kenny, J.M.M. Bionanocomposite Films Based on Plasticized PLA–PHB/Cellulose Nanocrystal Blends. *Carbohydr. Polym.* **2015**, *121*, 265–275. [CrossRef] [PubMed]
17. Bouakaz, B.S.; Pillin, I.; Habi, A.; Grohens, Y. Synergy between Fillers in Organomontmorillonite/Graphene-PLA Nanocomposites. *Appl. Clay Sci.* **2015**, *116*, 69–77. [CrossRef]
18. Gordobil, O.; Delucis, R.; Egues, I.; Labidi, J. Kraft Lignin as Filler in PLA to Improve Ductility and Thermal Properties. *Ind. Crops Prod.* **2015**, *72*, 46–53. [CrossRef]
19. Mofokeng, J.P.; Luyt, A.S. Morphology and Thermal Degradation Studies of Melt-Mixed Poly(Lactic Acid) (PLA)/Poly(Epsilon-Caprolactone) (PCL) Biodegradable Polymer Blend Nanocomposites with TiO_2 as Filler. *Polym. Test.* **2015**, *45*, 93–100. [CrossRef]
20. Aguilar, R.; Nakamatsu, J.; Ramírez, E.; Elgegren, M.; Ayarza, J.; Kim, S.; Pando, M.A.; Ortega-San-Martin, L. The Potential Use of Chitosan as a Biopolymer Additive for Enhanced Mechanical Properties and Water Resistance of Earthen Construction. *Constr. Build. Mater.* **2016**, *114*, 625–637. [CrossRef]
21. Bagheriasl, D.; Carreau, P.J.; Riedl, B.; Dubois, C.; Hamad, W.Y. Shear Rheology of Polylactide (PLA)-Cellulose Nanocrystal (CNC) Nanocomposites. *Cellulose* **2016**, *23*, 1885–1897. [CrossRef]
22. John, M.J. Biopolymer Blends Based on Polylactic Acid and Polyhydroxy Butyrate-Co-Valerate: Effect of Clay on Mechanical and Thermal Properties. *Polym. Compos.* **2015**, *36*, 2042–2050. [CrossRef]
23. Vanamudan, A.; Sudhakar, P.P. Biopolymer Capped Silver Nanoparticles with Potential for Multifaceted Applications. *Int. J. Biol. Macromol.* **2016**, *86*, 262–268. [CrossRef] [PubMed]
24. Chamieh, J.; Biron, J.P.; Cipelletti, L.; Cottet, H. Monitoring Biopolymer Degradation by Taylor Dispersion Analysis. *Biomacromolecules* **2015**, *16*, 3945–3951. [CrossRef] [PubMed]
25. Arrieta, M.P.; Peponi, L.; López, D.; Fernández-García, M. Recovery of Yerba Mate (Ilex Paraguariensis) Residue for the Development of PLA-Based Bionanocomposite Films. *Ind. Crops Prod.* **2018**, *111*, 317–328. [CrossRef]
26. Luzi, F.; Puglia, D.; Sarasini, F.; Tirillò, J.; Maffei, G.; Zuorro, A.; Lavecchia, R.; Kenny, J.M.; Torre, L. Valorization and Extraction of Cellulose Nanocrystals from North African Grass: Ampelodesmos Mauritanicus (Diss). *Carbohydr. Polym.* **2019**, *209*, 328–337. [CrossRef]
27. Berglund, L.; Noel, M.; Aitomaki, Y.; Oman, T.; Oksman, K. Production Potential of Cellulose Nanofibers from Industrial Residues: Efficiency and Nanofiber Characteristics. *Ind. Crops Prod.* **2016**, *92*, 84–92. [CrossRef]
28. Nosek, R.; Holubcik, M.; Jandacka, J. The Impact of Bark Content of Wood Biomass on Biofuel Properties. *Bioresources* **2016**, *11*, 44–53. [CrossRef]
29. Kuzmina, I. Influence the Food Additive of Fermented Cedar Elfin Wood on Biochemical Indices of Cows Blood in Magadan Region. *Zootekhniya* **2015**, *6*, 6–8.

30. Rollinson, A.N.; Williams, O. Experiments on Torrefied Wood Pellet: Study by Gasification and Characterization for Waste Biomass to Energy Applications. *R. Soc. Open Sci.* **2016**, *3*, 150578. [CrossRef]
31. Samarzija-Jovanovic, S.; Jovanovic, V.; Petkovic, B.; Dekic, V.; Markovic, G.; Zekovic, I.; Marinovic-Cincovic, M. Nanosilica and Wood Flour-Modified Urea-Formaldehyde Composites. *J. Thermoplast. Compos. Mater.* **2016**, *29*, 656–669. [CrossRef]
32. Pawlak, F.; Aldas, M.; López-Martínez, J.; Samper, M.D. Effect of Different Compatibilizers on Injection-Molded Green Fiber-Reinforced Polymers Based on Poly(Lactic Acid)-Maleinized Linseed Oil System and Sheep Wool. *Polymers* **2019**, *11*, 1514. [CrossRef] [PubMed]
33. Aldas, M.; Pavon, C.; López-Martínez, J.; Arrieta, M.P. Pine Resin Derivatives as Sustainable Additives to Improve the Mechanical and Thermal Properties of Injected Moulded Thermoplastic Starch. *Appl. Sci.* **2020**, *10*, 2561. [CrossRef]
34. Sarasini, F.; Puglia, D.; Fortunati, E.; Kenny, J.M.; Santulli, C. Effect of Fiber Surface Treatments on Thermo-Mechanical Behavior of Poly(Lactic Acid)/Phormium Tenax Composites. *J. Polym. Environ.* **2013**, *21*, 881–891. [CrossRef]
35. Arrieta, M.P.; Diez Garcia, A.; Lopez, D.; Fiori, S.; Peponi, L. Antioxidant Bilayers Based on PHBV and Plasticized Electrospun PLA-PHB Fibers Encapsulating Catechin. *Nanomaterials* **2019**, *9*, 346. [CrossRef]
36. Lewandowski, K.; Piszczek, K.; Zajchowski, S.; Mirowski, J. Rheological Properties of Wood Polymer Composites at High Shear Rates. *Polym. Test.* **2016**, *51*, 58–62. [CrossRef]
37. Nitz, H.; Semke, H.; Landers, R.; Mulhaupt, R. Reactive Extrusion of Polycaprolactone Compounds Containing Wood Flour and Lignin. *J. Appl. Polym. Sci.* **2001**, *81*, 1972–1984. [CrossRef]
38. Pavon, C.; Aldas, M.; López-Martínez, J.; Ferrándiz, S. New Materials for 3D-Printing Based on Polycaprolactone with Gum Rosin and Beeswax as Additives. *Polymers* **2020**, *12*, 334. [CrossRef]
39. Mofokeng, J.P.; Kelnar, I.; Luyt, A.S. Effect of Layered Silicates on the Thermal Stability of PCL/PLA Microfibrillar Composites. *Polym. Test.* **2016**, *50*, 9–14. [CrossRef]
40. Bulatović, V.O.; Mandić, V.; Kučić Grgić, D.; Ivančić, A. Biodegradable Polymer Blends Based on Thermoplastic Starch. *J. Polym. Environ.* **2020**, 1–17. [CrossRef]
41. Ferri, J.M.; Garcia-Garcia, D.; Sánchez-Nacher, L.; Fenollar, O.; Balart, R. The Effect of Maleinized Linseed Oil (MLO) on Mechanical Performance of Poly(Lactic Acid)-Thermoplastic Starch (PLA-TPS) Blends. *Carbohydr. Polym.* **2016**, *147*, 60–68. [CrossRef]
42. Aldas, M.; Ferri, J.M.; Lopez-Martinez, J.; Samper, M.D.; Arrieta, M.P. Effect of Pine Resin Derivatives on the Structural, Thermal, and Mechanical Properties of Mater-Bi Type Bioplastic. *J. Appl. Polym. Sci.* **2020**, *137*, 48236. [CrossRef]
43. Aldas, M.; Rayón, E.; López-Martínez, J.; Arrieta, M.P. A Deeper Microscopic Study of the Interaction between Gum Rosin Derivatives and a Mater-Bi Type Bioplastic. *Polymer* **2020**, *12*, 226. [CrossRef] [PubMed]
44. Kelnar, I.; Kratochvil, J.; Kapralkova, L. Crystallization and Thermal Properties of Melt-Drawn PCL/PLA Microfibrillar Composites. *J. Therm. Anal. Calorim.* **2016**, *124*, 799–805. [CrossRef]
45. Pilla, S.; Gong, S.; O'Neill, E.; Rowell, R.M.; Krzysik, A.M. Polylactide-Pine Wood Flour Composites. *Polym. Eng. Sci.* **2008**, *48*, 578–587. [CrossRef]
46. Chuayjuljit, S.; Kongthan, J.; Chaiwutthinan, P.; Boonmahitthisud, A. Poly(Vinyl Chloride)/Poly(Butylene Succinate)/Wood Flour Composites: Physical Properties and Biodegradability. *Polym. Compos.* **2018**, *39*, 1543–1552. [CrossRef]
47. Zhang, Y.; Yu, C.; Chu, P.K.; Lv, F.; Zhang, C.; Ji, J.; Zhang, R.; Wang, H. Mechanical and Thermal Properties of Basalt Fiber Reinforced Poly(Butylene Succinate) Composites. *Mater. Chem. Phys.* **2012**, *133*, 845–849. [CrossRef]

Publisher's Note: MDPI stays neutral with regard to jurisdictional claims in published maps and institutional affiliations.

© 2020 by the authors. Licensee MDPI, Basel, Switzerland. This article is an open access article distributed under the terms and conditions of the Creative Commons Attribution (CC BY) license (http://creativecommons.org/licenses/by/4.0/).

Article

Effect of Material Parameter of Viscoelastic Giesekus Fluids on Extensional Properties in Spinline and Draw Resonance Instability in Isothermal Melt Spinning Process

Geunyeop Park [1,†], Jangho Yun [2,†], Changhoon Lee [1] and Hyun Wook Jung [1,*]

1 Department of Chemical and Biological Engineering, Korea University, Seoul 02841, Korea; gypark@grtrkr.korea.ac.kr (G.P.); forza@grtrkr.korea.ac.kr (C.L.)
2 Hyundai Oilbank, Central Technology Research Institute, Yongin 16891, Korea; jjangho99yun@gmail.com
* Correspondence: hwjung@grtrkr.korea.ac.kr; Tel.: +82-2-3290-3306
† These authors contributed equally to this study.

Abstract: The draw resonance instability of viscoelastic Giesekus fluids was studied by correlating the spinline extensional features and transit times of several kinematic waves in an isothermal melt spinning process. The critical drawdown ratios were critically dependent on the Deborah number (De, the ratio of material relaxation time to process time) and a single material parameter (α_G) of the Giesekus fluid. In the intermediate range of α_G, the stability status changed distinctively with increasing De, i.e., the spinning system was initially stabilized and subsequently destabilized, as De increases. In this α_G regime, the level of velocity and extensional-thickening rheological property in the spinline became gradually enhanced at low De and weakened at high De. The draw resonance onsets for different values of α_G were determined precisely using a simple indicator composed of several kinematic waves traveling the entire spinline and period of oscillation. The change in transit times of kinematic waves for varying De adequately reflected the effect of α_G on the change in stability.

Keywords: viscoelastic spinning; draw resonance; kinematic waves; extensional deformation; stability indicator; Giesekus fluid

1. Introduction

Fiber spinning is one of the representative extensional deformation polymer processes, fabricating highly oriented fibers with large drawdown ratios ($r = V_L/V_0$) of velocities at the take-up (V_L) and spinneret (V_0) positions in the spinline (Figure 1a) [1]. The product quality and processability of the fibers are greatly influenced by the rheological properties of the polymeric filaments and spinline conditions. The most important concern to ensure the uniformity of the fibers is the stability of the spinning flow in the spinline from the spinneret to the take-up positions. The well-known instability in the spinning flow is draw resonance that is characterized by self-sustained periodic oscillations of state variables such as fiber diameter and spinline tension (Figure 1b), when the drawdown ratio exceeds the critical value. This was first observed by Christensen [2] and Miller [3]. Subsequently, various theoretical and experimental developments on this phenomenon were reported in melt spinning processes with various complex fluids such as Newtonian, viscous, and viscoelastic ones [4,5] using linear stability analysis [6–10], direct transient responses [11], kinematic traveling waves [12–15], bifurcation theory [16,17], and experimental observations [18,19].

The investigation of basic spinning flow as a uniaxial extensional deformation process has been a well-known classical topic in the last four to five decades in academia and industries. The linear stability method, focusing on the theoretical and numerical aspects of the draw resonance, is very useful for determining critical onsets in various spinning processes by transforming nonlinear governing equations into eigen-systems linearized by infinitesimal disturbances based on steady states. The nonlinear periodic oscillations of state variables beyond the onsets, exhibiting limit cycles (Figure 1c), were elucidated using direct transient simulations. Hyun and coworkers [12–15] tried to address the fundamental physics behind draw resonance by incorporating kinematic waves penetrating the entire spinline as the stability indicator. Draw resonance was found to be a type of supercritical Hopf bifurcation using the bifurcation theory [16,17]. Recently, Kwon et al. [20] determined draw resonance onsets precisely using the transfer function method under the constant force boundary condition that rendered the system constantly stable.

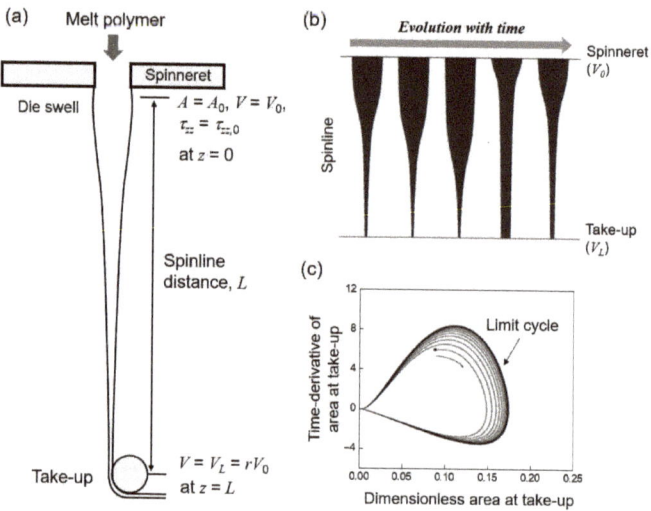

Figure 1. Schematics of (**a**) melt spinning process with spinline conditions, (**b**) periodic oscillation of filament during draw resonance, (**c**) limit cycle of spinline area at take-up position over the critical drawdown ratio.

The exact value of the critical drawdown ratio for Newtonian fluids is known to be 20.218 for an isothermal spinning flow without any secondary forces. Based on this value, various stability windows for generalized Newtonian and viscoelastic fluids were established, depending on their material properties. The shear-thinning nature (when the power-law index, n, is less than one) makes the system less stable [21]. Viscoelasticity results in dichotomous behavior of the onsets with respect to the Deborah number (De), defined as $\lambda V_0/L$ (a dimensionless number representing the ratio of a material relaxation time to a characteristic time for the deformation process, where λ is the material relaxation time and L is the spinline length), stabilizing for an extensional-thickening fluid such as low-density polyethylene (LDPE) and destabilizing for an extensional-thinning fluid such as high-density polyethylene (HDPE) with De. For instance, White–Metzner [8,22] and Phan-Thien and Tanner (PTT) fluids [23] showed distinct dichotomous stability curves with respect to De, depending on each material parameter characterizing the extensional feature in their fluid models. In this study, we attempted to explain the nature of the stability curves in the spinning process of Giesekus fluids—initially stabilizing and subsequently destabilizing pattern with respect to De in the intermediate range of the material parameter. It is important to take into consideration the relationship between the stability window and spinline extensional characteristics for viscoelastic Giesekus fluids, which demonstrate an unusual dependence of the spinning stability on De at the intermediate values of material parameter. The Giesekus fluid model [24,25] is a prominent constitutive equation that reflects the realistic viscoelastic features of polymeric liquids and successfully predicts the material functions in extensional as well shear flows using a singlematerial parameter [26]. This fluid was reliably implemented to investigate polymer extensional deformation processes such as fiber spinning [27], film casting [28,29], and film blowing [30].

Various theoretical approaches were considered in this study to examine the changes in the stability curves with respect to the material parameter of Giesekus fluids in the isothermal spinning process without cooling, including steady velocity profiles and extensional deformation properties in the spinline, and kinematic waves traveling along the entire spinline.

2. Simulation Methods

2.1. Governing Equations of Spinning Flows

Simplified one-dimensional governing equations for the isothermal spinning flow of Giesekus fluids are given here under the following assumptions: (1) The equation set neglects radial stress and all secondary forces such as inertia, gravity, surface tension, and air drag. Including them will not change the fundamental aspects described here. (2) The origin at the maximum position of the die swell excludes the pre-shear history in the nozzle. (3) The fiber is slender with uniform properties in the cross-section [15,20].

Equation of continuity (EOC):

$$\frac{\partial a}{\partial t} + \frac{\partial (av)}{\partial x} = 0 \qquad (1)$$

where $a = \frac{A}{A_0}$, $v = \frac{V}{V_0}$, $x = \frac{z}{L}$, $t = \frac{t^* V_0}{L}$.

Equation of motion (EOM):

$$\frac{\partial}{\partial x}(a\tau) = 0 \qquad (2)$$

where $\tau = \frac{2\tau_{zz} L}{\eta V_0}$.

Constitutive Equation (CE, Giesekus model):

$$\tau + De\left[\frac{\partial \tau}{\partial t} + v\frac{\partial \tau}{\partial x} - 2\tau\frac{\partial v}{\partial x}\right] + 2\alpha_G \tau^2 De = \frac{\partial v}{\partial x} \qquad (3)$$

Boundary conditions:

$$a = 1,\ v = 1,\ \tau = \tau_0 \text{ at } x = 0, \text{ and } v = r \text{ at } x = 1 \qquad (4)$$

The aforementioned equations are non-dimensionalized using the following dimensionless variables; a denotes the dimensionless spinline cross-sectional area of A, v is the dimensionless spinline velocity of V, t is the dimensionless time of t^*, x is the dimensionless spatial coordinate of z, τ is the dimensionless axial stress of τ_{zz}, De is the Deborah number, and r is the drawdown ratio. α_G represents a material parameter portraying the extensional behavior of the Giesekus fluid. The subscripts 0 and L represent the spinneret and take-up positions, respectively.

The steady velocity profiles along the spinline and corresponding apparent extensional properties were solved using the 4th-order Runge–Kutta method, ensuring the acceptable level of numerical accuracy.

2.2. Linear Stability Analysis of Steady Flows

The governing Equations (1)–(3) were linearized using the following perturbation variables based on steady states for constructing the linear eigen-systems.

$$a(t,x) = a_s(x) + \alpha(x)e^{\Omega t},\ v(t,x) = v_s(x) + \beta(x)e^{\Omega t},\ \tau(t,x) = \tau_s(x) + \gamma(x)e^{\Omega t} \qquad (5)$$

where, the subscript s denotes the steady state and Ω is the complex eigenvalue. α, β, and γ are infinitesimal perturbed quantities (i.e., eigenvectors).

Linearized EOC:

$$\Omega\alpha = \left(\frac{v_s'}{v_s^2}\right)\beta - \left(\frac{1}{v_s}\right)\beta' - (v_s')\alpha - (v_s)\alpha' \qquad (6)$$

Linearized EOM:

$$0 = -\left(\frac{v_s'}{v_s^2}\right)\gamma + \left(\frac{1}{v_s}\right)\gamma' + (\tau_s')\alpha + (\tau_s)\alpha' \qquad (7)$$

Linearized CE:

$$\Omega\gamma = -\tau_s'\beta + \left(2\tau_s + \frac{1}{De}\right)\beta' + \left(2v_s' - \frac{1}{De} - 4\alpha_G\tau_s\right)\gamma - v_s\gamma' \qquad (8)$$

Boundary conditions:

$$\alpha(0) = \beta(0) = 0 \text{ at } x = 0 \text{ and } \beta(1) = 0 \text{ at } x = 1 \qquad (9)$$

The boundary conditions (given by Equation (9)) indicate that the flow rate at the spinneret and velocity at the take-up position are unperturbed under constant velocity operation. The prime (′) symbol signifies the derivative with respect to x. The critical drawdown ratios, when the real part of the first normal mode is zero, are obtained by solving the eigenvalues from the linear eigen-system, $\Omega \underline{My} = \underline{Ay}$, where $\underline{y} = \begin{bmatrix} \alpha, \beta, \gamma \end{bmatrix}$ via the shift-invert method [15]. They are plotted with respect to De for different values of α_G.

2.3. Simple Stability Indicator Using Traveling Times of Kinematic Waves

Draw resonance is a hydrodynamic instability that can be physically figured out by several kinematic waves penetrating the entire spinline from the spinneret to the take-up position [13,14]. The stability criterion (Equation (10)) comprising the unity-throughput wave, maximum/minimum area wave, and period of oscillation was confirmed to correctly interpret the draw resonance dynamics for various fluid systems [4,5].

$$(t_L)_1 + (t_L)_2 + \frac{T}{2} \overset{>}{\underset{<}{=}} (\theta_L)_1 + (\theta_L)_2 \text{ for } r \overset{>}{\underset{<}{=}} r_c \qquad (10)$$

where $(t_L)_1$ and $(t_L)_2$ represent the dimensionless traveling times of the unity-throughput waves, $(\theta_L)_1$ and $(\theta_L)_2$ are the dimensionless traveling times of the maximum and minimum cross-sectional area waves, and T is the dimensionless period of oscillation.

In the case of $r < r_c$, the left-hand side (LHS; designated as the required time) of Equation (10) becomes larger than the right-hand side (RHS; designated as the allowed time), implying that the oscillation cannot be sustained due to insufficient time for inducing the draw resonance [12,13]. For $r = r_c$, both sides of Equation (10) are identical, triggering the draw resonance. When $r > r_c$, the magnitude of both sides is reversed and periodic oscillation by draw resonance is continuously maintained. Although the real traveling times and period data must be acquired by the direct transient simulation of nonlinear governing equations, it was shown that the simplified version of the stability indicator from the linear stability analysis could reliably determine the onsets [15]. In this study, we tried to identify the size change on both sides of Equation (10) using linear eigen-mode data.

3. Results and Discussion

3.1. Neutral Stability Curves with Respect to De for Giesekus Fluids with Different α_G Values

As illustrated in Figure 2, the critical drawdown ratios for Giesekus fluids were determined from the first normal mode of the linearized spinning systems, when De and α_G were varied. The extensional properties (extensional-thickening or -thinning) of fluids played a significant role in the stability of various spinning processes. Unlike other viscoelastic models such as White–Metzner and PTT fluids [8,31], the Giesekus fluid exhibited interesting stability patterns depending on the value of α_G. Three stability patterns can be shown in Figure 2. When α_G was less than about 0.01, De rendered the system more stable implying that the critical drawdown ratio increased with increasing De and a secondary stable region at high draw ratios was observed, as in the typical extensional-thickening case (e.g., LDPE) of an upper-convected Maxwell (UCM) fluid with $\alpha_G = 0$. It is noted that the PTT fluid under extensional-thickening conditions did not show the secondary stable region [31].

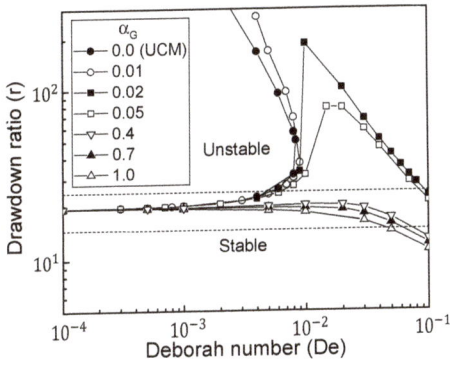

Figure 2. Stability windows of Giesekus fluids for various α_G values. Here, α_G represents a material parameter of the Giesekus fluid. If $\alpha_G = 0$, it is identical to the upper-convected Maxwell (UCM) model.

When α_G was in the intermediate range 0.01–0.4, the system was first stabilized and then destabilized as De increased, distinguishing it from other viscoelastic fluids. The extensional deformation features in the spinline, as described in the next section, qualitatively explain the trend in the stability of the Giesekus fluid for intermediate value of α_G. A value of α_G greater than about 0.4 made the system less stable to disturbances with increasing De; this was frequently observed in extensional-thinning fluids such as HDPE.

3.2. Steady Extensional Properties of Giesekus Fluids in the Spinline

The aforementioned neutral stability curves for various values of α_G are basically associated with the extensional behavior and properties in the spinline. First, the steady velocity profiles around the onsets for varying De in three fluid cases ((a) $\alpha_G = 0.01$, $r = 25$; (b) $\alpha_G = 0.05$, $r = 25$; and (c) $\alpha_G = 0.7$, $r = 15$) were compared and are shown in Figure 3. The extensional-thickening fluid with $\alpha_G = 0.01$ showed a higher velocity level in the same spinline position, as De increased (Figure 3a), requiring less residence time along the spinline with increasing De. However, the spinline velocity profiles in Figure 3c for the case of the extensional-thinning fluid with $\alpha_G = 0.7$ were the opposite with respect to the previous case, i.e., gradually decreasing velocity level in the same spinline position with increasing De. For intermediate value of α_G, the dependence of the steady velocity profiles on De was different around the first and second drawdown ratio onsets at low and high De regions, exhibiting a higher spinline velocity level near the first onset and subsequent lower velocity level around the second onset, as De increased (Figure 3b).

The extension rate ($\dot{\varepsilon}$) and apparent extensional viscosity (η_E) were evaluated from the steady velocity and tensile stress profiles in the spinline, respectively, under the spinning conditions shown in Figure 3. The apparent extensional viscosity rapidly increased with increasing De for the extensional-thickening case with $\alpha_G = 0.01$, as shown in Figure 4a. It must be noted that the non-zero value of α_G in this case effectively prevents the infinite growth of tensile stress [32]. In the case of intermediate value of α_G (Figure 4b), the growth rate of the extensional viscosity with increasing De is analogous to the case of $\alpha_G = 0.01$ around the lower onset, but not significant around the higher onset, as compared to that shown in Figure 4a. The extensional-thinning fluid (Figure 4c) showed decreasing extensional viscosity with respect to the extension rate and lower viscosity level with increasing De, as expected.

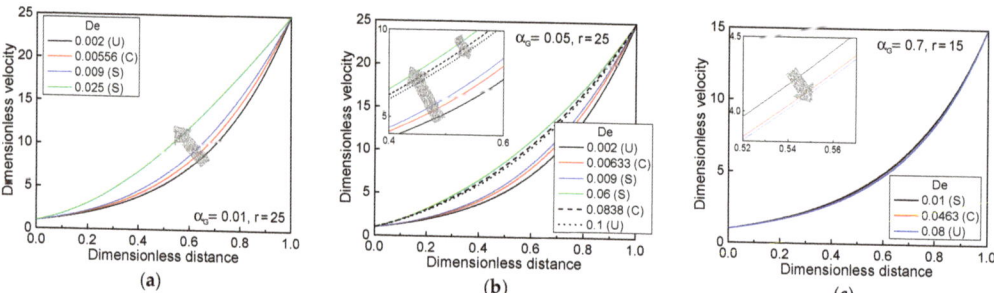

Figure 3. Dimensionless steady spinline velocity profiles under several material parameter conditions: (**a**) $\alpha_G = 0.01$, $r = 25$, (**b**) $\alpha_G = 0.05$, $r = 25$, and (**c**) $\alpha_G = 0.7$, $r = 15$. S, C, and U in the box indicate stable, critical, and unstable states, respectively. The arrow indicates the direction in which De increases.

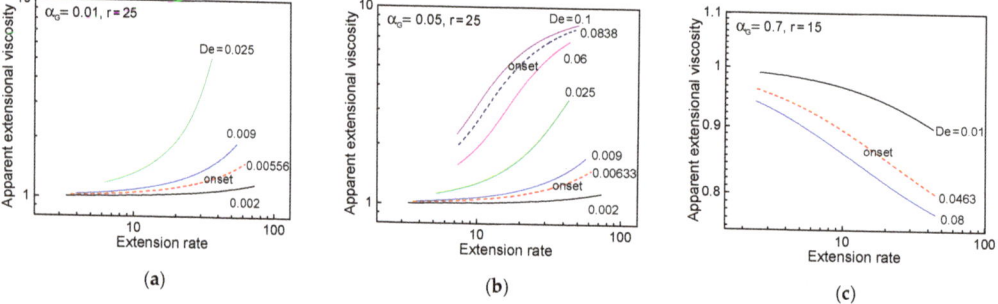

Figure 4. Apparent extensional viscosity in the spinline with respect to extension rate under several material parameter conditions: (**a**) $\alpha_G = 0.01$, $r = 25$, (**b**) $\alpha_G = 0.05$, $r = 25$, and (**c**) $\alpha_G = 0.7$, $r = 15$.

3.3. Transit Times of Kinematic Waves for Different Giesekus Fluids

Figure 5 displays the changes in the LHS and RHS times (composed of the traveling times of two kinematics waves and period of oscillation) of the simple stability indicator (Equation (10)) with respect to De, to confirm the draw resonance onsets for three Giesekus fluids with $\alpha_G = 0.01, 0.05,$ and 0.7. As De increased, the magnitude of the LHS and RHS times crossed exactly at critical points. For instance, the extensional-thickening case for $\alpha_G = 0.01$ was stable after a critical $De = 0.00556$ at $r = 25$; the intermediate case for $\alpha_G = 0.05$ was stable only in the range of $De = 0.00633$–0.0838 at $r = 25$; the extensional-thinning case for $\alpha_G = 0.7$ became unstable after a critical $De = 0.0463$ at $r = 15$. It was observed that these onsets were identical to those shown in Figure 1 obtained from the linear stability analysis.

Figure 6 compares each traveling time of the kinematic waves and period of oscillation with respect to De for the three fluid cases illustrated in Figure 5. Interestingly, the traveling time of the maximum or minimum cross-sectional area wave showed a slight upward turn after the second critical point at a higher drawdown ratio, as shown in Figure 6b, as qualitatively described in the steady velocity profiles.

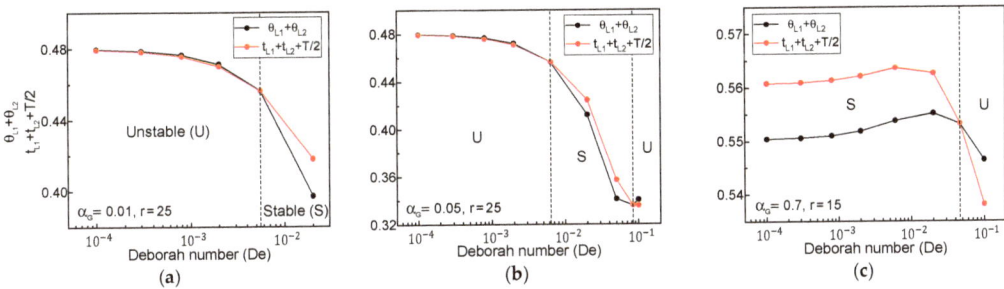

Figure 5. Changes in time scales of left-hand side (LHS) and right-hand side (RHS) of the simple indicator: (**a**) $\alpha_G = 0.01$, $r = 25$, (**b**) $\alpha_G = 0.05$, $r = 25$, and (**c**) $\alpha_G = 0.7$, $r = 15$. Here, θ_{L1} and θ_{L2} represent the dimensionless traveling times of the maximum and minimum cross-sectional area waves; t_{L1} and t_{L2} are the dimensionless traveling times of the unity-throughput waves; T is the dimensionless period of oscillation.

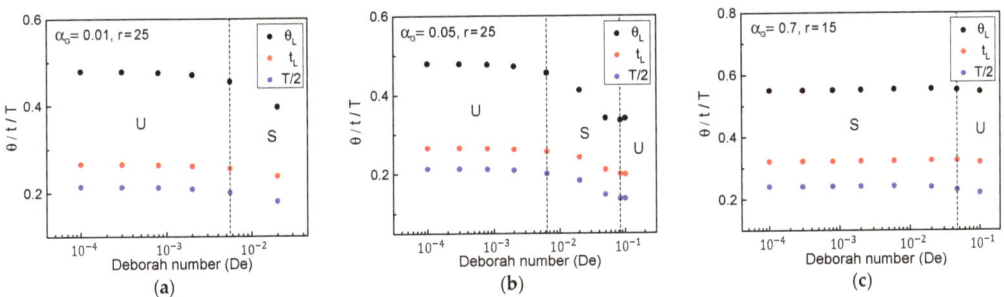

Figure 6. Traveling times of unity-throughput wave (t_L) and maximum/minimum cross-sectional area wave (θ_L), and period of oscillation (T) for (**a**) $\alpha_G = 0.01$, $r = 25$, (**b**) $\alpha_G = 0.05$, $r = 25$, and (**c**) $\alpha_G = 0.7$, $r = 15$.

4. Conclusions

Neutral stability curves for Giesekus fluids were established in the melt spinning processes. The material parameter α_G in this fluid model suitably depicted the extensional-thickening (stabilizing effect of De) and extensional-thinning (destabilizing effect of De) properties of viscoelastic fluids in extensional deformation processes. In the intermediate range of values of α_G (approximately $0.01 < \alpha_G < 0.4$), the effect of De on the stability was unusual—the system was stabilized in the low and medium De regions and then destabilized in the high De region. This tendency may be qualitatively interpreted by extensional flow characteristics in the spinline. When De increased at a fixed drawdown ratio (e.g., $r = 25$ condition applied in this study) in the intermediate α_G region, the system, starting from the unstable state, became stable after the first onset for low or medium De, resulting in a higher level of spinline velocity and strain-hardening viscosity. It became unstable again beyond the second onset for high De, yielding a lowered spinline velocity and an insignificant strain-hardening feature. A combination of transit times of the kinematic waves penetrating the entire spinline and period of oscillation, i.e., the simple indicator from the linear stability analysis, predicted well the draw resonance onsets under different De and α_G conditions. It was confirmed that these transit times of kinematic waves for varying De adequately reflected the dependence of change in stability on the values of α_G.

Author Contributions: Conceptualization, G.P., J.Y. and H.W.J.; methodology, G.P. and J.Y.; validation, C.L. and H.W.J.; formal analysis, G.P. and C.L.; investigation, G.P. and H.W.J.; resources, J.Y. and C.L.; data curation, G.P.; writing—original draft preparation, G.P. and H.W.J.; writing—review and editing, G.P. and H.W.J.; supervision, H.W.J.; project administration, H.W.J.; funding acquisition, H.W.J. All authors have read and agreed to the published version of the manuscript.

Funding: This research was supported by the Ministry of Trade, Industry & Energy (MOTIE, Korea) under the Industrial Technology Innovation Program (Grant No. 20004044).

Institutional Review Board Statement: Not applicable.

Informed Consent Statement: Not applicable.

Data Availability Statement: The data presented in this study are available on request from the corresponding author.

Conflicts of Interest: The authors declare no conflict of interest.

References

1. Denn, M.M. Continuous drawing of liquids to form fibers. *Annu. Rev. Fluid Mech.* **1980**, *12*, 365–387. [CrossRef]
2. Christensen, R.E. Extrusion coating of polypropylene. *SPE J.* **1962**, *18*, 751–755.
3. Miller, J.C. Swelling behavior in extrusion. *Polym. Eng. Sci.* **1963**, *3*, 134–137. [CrossRef]
4. Jung, H.W.; Hyun, J.C. Instabilities in extensional deformation polymer processing. In *Rheology Reviews, 2006*; Binding, D.M., Walters, K., Eds.; The British Society of Rheology: Aberystwyth, UK, 2006; pp. 131–164.
5. Jung, H.W.; Hyun, J.C. Fiber spinning and film blowing instabilities. In *Polymer Processing Instabilities: Control and understanding*, 1st ed.; Hatzikiriakos, S.G., Migler, K.B., Eds.; Marcel Dekker: New York, NY, USA, 2005; pp. 321–381.
6. Gelder, D. The stability of fiber drawing processes. *Ind. Eng. Chem. Fundam.* **1971**, *10*, 534–535. [CrossRef]
7. Fisher, R.J.; Denn, M.M. Finite-amplitude stability and draw resonance in isothermal melt spinning. *Chem. Eng. Sci.* **1975**, *30*, 1129–1134. [CrossRef]
8. Jung, H.W.; Hyun, J.C. Stability of isothermal spinning of viscoelastic fluids. *Korean J. Chem. Eng.* **1999**, *16*, 325–330. [CrossRef]
9. Götz, T.; Perera, S.S.N. Stability analysis of the melt spinning process with respect to parameters. *ZAMM-Z. Angew. Math. Mech.* **2009**, *89*, 874–880. [CrossRef]
10. Bechert, M.; Scheid, B. Combined influence of inertia, gravity, and surface tension on the linear stability of Newtonian fiber spinning. *Phys. Rev. Fluids* **2017**, *2*, 1–19. [CrossRef]
11. Kase, S.; Matsuo, T. Studies on melt spinning. II. Steady-state and transient solutions of fundamental equations compared with experimental results. *J. Appl. Polym. Sci.* **1967**, *11*, 251–287. [CrossRef]
12. Hyun, J.C. Theory of draw resonance: Part I. Newtonian fluid & Part II Power-law and Maxwell fluids. *AIChE J.* **1978**, *24*, 418–426.
13. Kim, B.M.; Hyun, J.C.; Oh, J.S.; Lee, S.J. Kinematic waves in the isothermal melt spinning of Newtonian fluids. *AIChE J.* **1996**, *42*, 3164–3169. [CrossRef]
14. Jung, H.W.; Song, H.S.; Hyun, J.C. Draw resonance and kinematic waves in viscoelastic isothermal spinning. *AIChE J.* **2000**, *46*, 2106–2111. [CrossRef]
15. Lee, J.S.; Jung, H.W.; Hyun, J.C.; Scriven, L.E. Simple indicator of draw resonance instability in melt spinning processes. *AIChE J.* **2005**, *51*, 2869–2874. [CrossRef]
16. Schultz, W.W.; Zebib, A.; Davis, S.H.; Yee, L. Nonlinear stability of Newtonian fibres. *J. Fluid Mech.* **1984**, *149*, 455–475. [CrossRef]
17. Yun, J.H.; Shin, D.M.; Lee, J.S.; Jung, H.W.; Hyun, J.C. Direct calculation of limit cycles of draw resonance and their stability in spinning process. *Nihon Reoroji Gakkaishi*. **2008**, *36*, 133–136. [CrossRef]
18. Bergonzoni, A.; DiCresce, A.J. The phenomenon of draw resonance in polymeric melts. Part I—Qualitative view, Part II—Correlation to molecular parameters. *Polym. Eng. Sci.* **1966**, *6*, 45–59. [CrossRef]
19. Demay, Y.; Agassant, J.F. Experimental study of draw resonance in fiber spinning. *J. Non-Newton. Fluid Mech.* **1985**, *18*, 187–198. [CrossRef]
20. Kwon, I.; Chun, M.S.; Jung, H.W.; Hyun, J.C. Determination of draw resonance onsets in tension-controlled viscoelastic spinning process using transient frequency response method. *J. Non-Newton. Fluid Mech.* **2016**, *228*, 31–37. [CrossRef]
21. Pearson, J.R.A.; Shah, Y.T. On the stability of isothermal and nonisothermal fiber spinning of power-law fluids. *Ind. Eng. Chem. Fundam.* **1974**, *13*, 134–138. [CrossRef]
22. Lee, J.S.; Jung, H.W.; Kim, S.H.; Hyun, J.C. Effect of fluid viscoelasticity on the draw resonance dynamics of melt Spinning. *J. Non-Newton. Fluid Mech.* **2001**, *99*, 159–166. [CrossRef]
23. Lee, J.S.; Jung, H.W.; Hyun, J.C. Melt spinning dynamics of Phan-Thien Tanner fluids. *Korea-Aust. Rheol. J.* **2000**, *12*, 119–124.
24. Giesekus, H. A simple constitutive equation for polymer fluids based on the concept of deformation-dependent tensorial mobility. *J. Non-Newton. Fluid Mech.* **1982**, *11*, 69–109. [CrossRef]
25. Giesekus, H. A unified approach to a variety of constitutive models for polymer fluids based on the concept of configuration-dependent molecular mobility. *Rheol. Acta.* **1982**, *11*, 366–375. [CrossRef]
26. Khan, S.A.; Larson, R.G. Comparison of simple constitutive equations for polymer melts in shear and biaxial and uniaxial extensions. *J. Rheol.* **1987**, *31*, 207–234. [CrossRef]
27. Dhadwal, R. Numerical study of effect of inertia on stability of fibre spinning. *Int. J. Appl. Comput. Math.* **2016**, *2*, 699–711. [CrossRef]
28. Iyengar, V.R.; Co, A. Film casting of a modified Giesekus fluid: Stability analysis. *Chem. Eng. Sci.* **1996**, *51*, 1417–1430. [CrossRef]

29. Pis-Lopez, M.E.; Co, A. Multilayer film casting of modified Giesekus fluids Part2. Linear stability analysis. *J. Non-Newton. Fluid Mech.* **1996**, *66*, 95–114. [CrossRef]
30. Doufas, A.K.; McHugh, A.J. Simulation of film blowing including flow-induced crystallization. *J. Rheol.* **2001**, *45*, 1085–1104. [CrossRef]
31. Chang, J.C.; Denn, M.M. Sensitivity of the stability of isothermal melt spinning to rheological constitutive assumptions. In *Rheology Vol 3: Application*, 1st ed.; Astarita, G., Marrucci, G., Nicolais, L., Eds.; Plenum Press: New York, NY, USA, 1980; pp. 9–13.
32. Papanastasiou, T.C.; Macosko, C.W.; Scriven, L.E. Fiber spinning of viscoelastic liquid. *AIChE J.* **1987**, *33*, 834–842. [CrossRef]

Article

Fabrication of Drug-Eluting Nano-Hydroxylapatite Filled Polycaprolactone Nanocomposites Using Solution-Extrusion 3D Printing Technique

Pang-Yun Chou [1,2], Ying-Chao Chou [3], Yu-Hsuan Lai [1], Yu-Ting Lin [1], Chia-Jung Lu [1] and Shih-Jung Liu [1,3,*]

[1] Department of Mechanical Engineering, Chang Gung University, Taoyuan 33302, Taiwan; m7406@cgmh.org.tw (P.-Y.C.); s9409612563@gmail.com (Y.-H.L.); yutinna9876@mail.cgu.edu.tw (Y.-T.L.); happy2231017@mail.cgu.edu.tw (C.-J.L.)
[2] Department of Plastic and Reconstructive Surgery and Craniofacial Research Center, Chang Gung Memorial Hospital, Taoyuan 33305, Taiwan
[3] Department of Orthopedic Surgery, Bone and Joint Research Center, Chang Gung Memorial Hospital-Linkou, Taoyuan 33305, Taiwan; enjoycu@ms22.hinet.net
* Correspondence: shihjung@mail.cgu.edu.tw; Tel.: +886-3-2118166; Fax: +886-3-2118558

Citation: Chou, P.-Y.; Chou, Y.-C.; Lai, Y.-H.; Lin, Y.-T.; Lu, C.-J.; Liu, S.-J. Fabrication of Drug-Eluting Nano-Hydroxylapatite Filled Polycaprolactone Nanocomposites Using Solution-Extrusion 3D Printing Technique. *Polymers* **2021**, *13*, 318. https://doi.org/10.3390/polym13030318

Academic Editors: Khalid Lamnawar and Abderrahim Maazouz
Received: 4 January 2021
Accepted: 18 January 2021
Published: 20 January 2021

Publisher's Note: MDPI stays neutral with regard to jurisdictional claims in published maps and institutional affiliations.

Copyright: © 2021 by the authors. Licensee MDPI, Basel, Switzerland. This article is an open access article distributed under the terms and conditions of the Creative Commons Attribution (CC BY) license (https://creativecommons.org/licenses/by/4.0/).

Abstract: Polycaprolactone/nano-hydroxylapatite (PCL/nHA) nanocomposites have found use in tissue engineering and drug delivery owing to their good biocompatibility with these types of applications in addition to their mechanical characteristics. Three-dimensional (3D) printing of PCL/nHA nanocomposites persists as a defiance mostly because of the lack of commercial filaments for the conventional fused deposition modeling (FDM) method. In addition, as the composites are prepared using FDM for the purpose of delivering pharmaceuticals, thermal energy can destroy the embedded drugs and biomolecules. In this report, we investigated 3D printing of PCL/nHA using a lab-developed solution-extrusion printer, which consists of an extrusion feeder, a syringe with a dispensing nozzle, a collection table, and a command port. The effects of distinct printing variables on the mechanical properties of nanocomposites were investigated. Drug-eluting nanocomposite screws were also prepared using solution-extrusion 3D printing. The empirical outcomes suggest that the tensile properties of the 3D-printed PCL/nHA nanocomposites increased with the PCL/nHA-to-dichloromethane (DCM) ratio, fill density, and print orientation but decreased with an increase in the moving speed of the dispensing tip. Furthermore, printed drug-eluting PCL/nHA screws eluted high levels of antimicrobial vancomycin and ceftazidime over a 14-day period. Solution-extrusion 3D printing demonstrated excellent capabilities for fabricating drug-loaded implants for various medical applications.

Keywords: polycaprolactone; nano-hydroxylapatite; 3D printing; solution extrusion; process optimization; drug release

1. Introduction

Degenerative pathologies, injuries, and trauma can harm bone tissues and lead to the requirement for therapies that facilitate their repair, replacement, or regeneration of the tissue. Scaffolds made of distinct biomaterials have been used as a substitute for bone regeneration. Bioceramic such as nano-hydroxyapatite (nHA) is known to promote cell proliferation and osteoconduction, and has been widely used as a bone graft substitute due to its good biocompatible and osteoconductive properties [1–3]. However, nHA possesses low mechanical properties because of its brittleness. Polycaprolactone (PCL), on the other hand, is a degradable polymer widely researched for use in long-term implants and controlled drug release applications [4]. nHA-filled PCL nanocomposites can be a good candidate as a synthetic alternative for bone tissue engineering and drug delivery, mainly owing to their excellent biocompatible and mechanical properties [5].

The 3D printing technique [6,7] is a novel technique for making unusual or complex component geometries that might be difficult to fabricate via other processes. The 3D printing process enables and facilitates the manufacture of moderate to mass numbers of parts that can be specifically customized. The technologies provide new opportunities with regard to the manufacturing paradigm and fabrication possibilities with substantially reduced times. New designs require only a short time to market, and customer demands can be fulfilled more rapidly.

Among various 3D printing methods, FDM is the most extensively used technique [8]. The procedure extrudes hot polymer melts from a nozzle and paves them on a collection table for product development. The extruding nozzle shifts in a horizontal position to form a single layer at a time after which the extruding nozzle moves consecutively in a vertical position so as to make a fresh layer. A computer is generally used to command the migration of the extrusion nozzle until a 3D-printed part is acquired.

The 3D printing of virgin PCL and composites [9,10] continues to be a defiance, largely because the filaments for FDM are generally limited [11,12]. Guerra and Ciurana [13] proposed a fused filament fabrication (FFF) printing device to print 3D stents out of PCL. Jhao et al. [14] explored hydroxyapatite/PCL scaffold printing using a lab-exploited melt-differential FDM printing facility. Hollander et al. [15] used FDM-printed PCL grafts to transport micronized indomethacin. Visscher et al. [16] integrated FDM and a salt leaching process for manufacturing a PCL scaffold of micro-/macro-porosity. In spite of these efforts, all developments rely on the FDM scheme that extrudes hot melted polymer during the printing process. The use of polymer melt extrusion, however, leads to some restrictions, specifically as the manufactured product is used for the intent of pharmaceutical delivery. Compounding or mixing drugs and hot polymer melt in the 3D extrusion printing process can damage or degenerate the pharmaceuticals [17].

One solution for coping with this concern is the use of the solution-extrusion 3D printing technique [18], which integrates a fluid-delivery unit and an automatized three-axis migration device for extrusion printing. To print a part, PCL, fillers, and solvent are primarily compounded and extruded from a feeding system consisting of a syringe equipped with a delivery nozzle. The nozzle is controlled and shifted by a computer/microprocessor. Once the solvent becomes volatile, the solution that is expelled from the delivery nozzle solidifies and forms successive layers to create 3D products of needed shapes.

This study exploited drug-eluting nHA-filled PCL nanocomposites using the solution-extrusion 3D printing technique so as to avoid the deactivation of embedded pharmaceuticals. An empirical study was completed to assess the effect of distinct printing variables on the tensile properties of 3D-printed PCL/nHA nanomaterials. The tensile strength of post-printed PCL/nHA specimens was estimated using a tensile test machine, and the morphological structure was examined via a field emission scanning electron microscope (FESEM) and a projector microscope. In addition, vancomycin- and ceftazidime-loaded PCL/nHA screws were also prepared using solution-extrusion 3D printing. Printed screws were assessed by differential scanning calorimeter (DSC) and Fourier-transform infrared (FTIR) spectroscopy. The elution characteristics of incorporated drugs were also evaluated via a high-performance liquid chromatography (HPLC).

2. Materials and Methods
2.1. Materials

PCL (M_n: 80,000 Da), nHA (<200 nm, M_w: 502.31 g/mol), and DCM were used for 3D printing, and vancomycin hydrochloride and ceftazidime hydrate were used as pharmaceuticals. All of them were acquired from Sigma-Aldrich (St. Louis, MO, USA).

2.2. Experimental Methods

The 3D printing experiments were completed on a lab-made device (Figure 1A), which consists of an extrusion feeder, steering step motors, a syringe with a delivering nozzle (inner diameter: 180 μm), a collection table, and a control port connected to a computer. An open control Cura code was used to monitor the entire printing course.

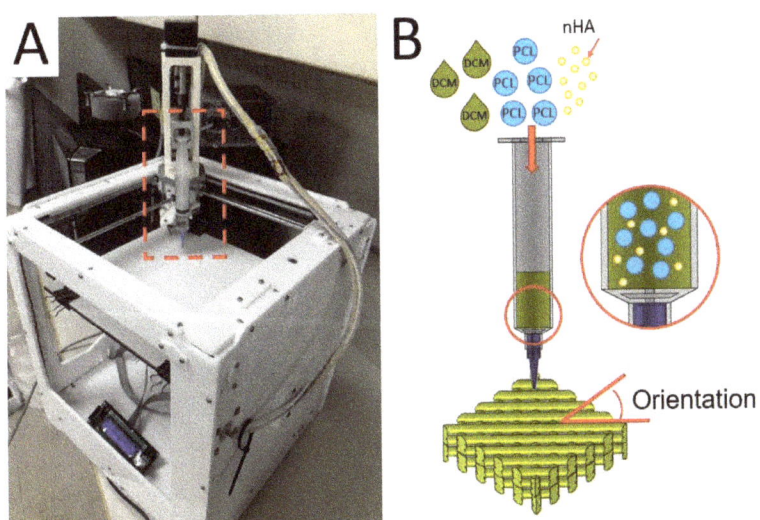

Figure 1. (**A**) Image of the solution-extrusion three-dimensional (3D) printer and (**B**) the solution printing of polycaprolactone/nano-hydroxylapatite (PCL/nHA) composites with desired orientation.

To print nanocomposite parts, PCL (2.5 g) and nHA (0.133 g) were fixed with DCM based on distinct weight-to-volume ratios and stirred for 3 h. The mixed solution was then added to the extruding feeder for printing (Figure 1B). In the 3D printing process, the delivery nozzle was actuated by a computer-commanded step motor, thus extruding and layering the PCL/nHA solution on the collection table. As soon as the solvent evaporated, strips of PCL/nHA (about 0.2 mm in thickness) were laid on the table in successive layers.

2.3. Processing Variables

The effect of distinct processing variables on the tensile strengths of printed PCL/nHA specimens was investigated. Four variables were chosen: (1) PCL/nHA-to-DCM ratio, (2) fill density, (3) orientation of the extruded strips (Figure 1B), and (4) moving speed of the delivering nozzle. A few test trials were first completed to identify the ranges of processing values able to successfully print the nanocomposite specimens. The ratios of PCL/nHA to DCM were 2.5 g/0.133 g:5.8 mL, 2.5 g/0.133 g:6.0 mL, 2.5 g/0.133 g:6.2 mL, and 2.5 g/0.133 g:6.4 mL (w/v). Meanwhile, the fill densities were 50%, 55%, 60%, and 65%. The shifting speeds of the delivering nozzle spanned from 30 to 60 mm/s. The orientations of extruded strips were 45°, 60°, 75°, and 90°. The variables and variable values used for the experiment are listed in Table 1. After printing, the specimens were placed in an oven at room temperature for 72 h to completely vaporize the solvents.

Table 1. Variables utilized for the three-dimensional (3D) printing of PCL/nHA parts.

Variable	A: PCL/nHA to DCM Ratio (w/v)	B: Fill Density (%)	C: Print Speed (mm/s)	D: Print Orientation
Level 1	2.5 g/0.133g:5.8 mL	50	30	45°
Level 2	**2.5 g/0.133g:6.0 mL**	**55**	40	60°
Level 3	2.5 g/0.133g:6.2 mL	60	50	75°
Level 4	2.5 g/0.133g:6.4 mL	65	60	90°

A dumbbell geometry (Figure 2A) was used to 3D print the test parts. The code used to control the migration of the dispensing nozzle was established using commercial software from Solidworks (Waltham, MA, USA). Post-printing, the tensile strengths of prepared PLC/nHA nanocomposites were assessed with a Lloyd test machine (Ameteck, Berwyn, PA, USA). The stretching rate for the specimens was 50 mm/min. The tensile strengths were calculated with the equation.

$$\text{Strength (MPa)} = \text{Maximum load (N)}/\text{Part cross} - \text{sectional area (mm}^2) \quad (1)$$

As shown in Table 1, one variable was varied every time while keeping the others constant (bold ones). The influence of every variable on the tensile strengths of the printed samples could be assessed. The experiment was repeated three times (N = 3) for every specimen.

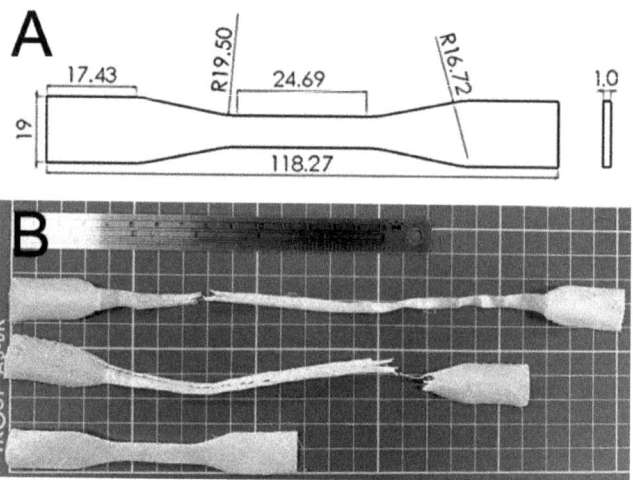

Figure 2. (**A**) Layout and dimensions of the samples and (**B**) ruptured specimens.

2.4. Microscopic Examinations

The morphological structure of the 3D-printed samples was examined by a JEOL Model JSM-7500F FESEM (Tokyo, Japan) and an APEX-2010 profile projector (Taipei, Taiwan).

2.5. Printing of Drug-Eluting Screws

To demonstrate the capability of solution-extrusion 3D printing in fabricating drug-loaded implants, the optimum processing conditions obtained in previous sections were used to print the drug-eluting PCL/nHA screws. PCL (2500 mg), nHA (132 mg), and vancomycin and ceftazidime (312.5 mg each) were mixed with 6 mL of DCM. The solution was then used to print the screws.

2.6. Fourier-Transform Infrared Assay

The spectra of virgin PCL, nHA, PCL/nHA, and drug-loaded PCL/nHA were assessed employing a Nicolet iS5 Fourier-transform infrared (FTIR) spectrometer assay (Thermo Fisher, Waltham, MA, USA). The samples were first compressed as KBr discs for the assay, which was conducted at a resolution of 4 cm^{-1} and 32 scans. The spectra of the assay ranged from 400 to 4000 cm^{-1}.

2.7. Differential Scanning Calorimeter Assay

The thermal properties of virgin PCL, PCL/nHA, and drug-loaded PCL/nHA were estimated by a TA-DSC25 differential scanning calorimeter (New Castle, DE, USA). The scanning temperature ranged from 30 to 350 °C while the specimens were heated at 10 °C/min.

2.8. In Vitro Release of PLC/nHA Screws

The elution patterns of vancomycin/ceftazidime from the drug-loaded PLC/nHA screws were assessed using an in vitro elution method [19]. Screws were put in glass tubes (one screw in each tube, N = 3) and filled with 1 mL of phosphate buffer solution (0.15 mol/L, pH 7.4). The tubes were kept in an isothermal oven at 37 °C for 24 h until the eluent was gathered and assayed. New phosphate buffer solution (1 mL) was added to the tubes for the next 24 h time interval. The process was duplicated for 14 days. The drug levels in the gathered eluents were characterized with Hitachi L-2200R multi-solvent high-performance liquid chromatography (HPLC) (Tokyo, Japan).

3. Results

3.1. Effects of Processing Parameters on Mechanical Strengths

PCL/nHA specimens were satisfactorily printed using solution-extrusion 3D printing. Figure 2B displays the fractured PCL/nHA nanocomposites post-tensile test. All specimens exhibited good ductile properties.

Figure 3 displays the tensile characteristics of 3D-printed PCL and PCL/nHA specimens. As expected, nHA-filled PCL parts showed superior tensile strengths to virgin PCL parts [18]. The results in Figure 3 also suggest that the maximum tensile strengths of PLC/nHA specimens improved as the concentration of DCM increased. Composite samples printed using a PCL/nHA:DCM ratio of 2.5 g/0.133g:6.4 mL showed the most superior mechanical strength, whereas printed specimens with a ratio of 2.5 g/0.133g:5.8 mL displayed the most inferior mechanical properties. Figure 4a,b shows the surface images of printed nanocomposite parts from the profile projector and SEM. The abundant solvent during the printing process helped promote healing at the extruded strip interfaces not only in the same layer, but also across layers. The tensile strengths increased accordingly.

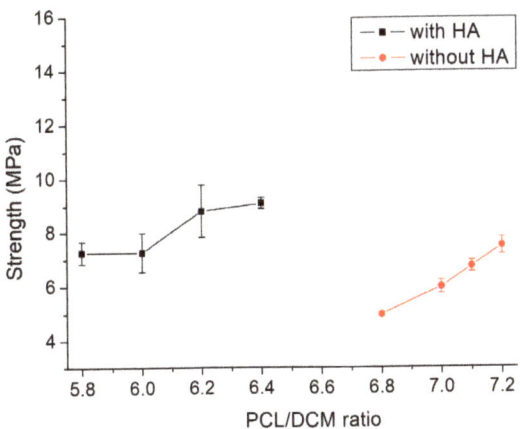

Figure 3. Influence of the polycaprolactone/dichloromethane (PCL/DCM) ratio on the tensile strengths of 3D-printed virgin PCL and PCL/nHA nanocomposite specimens.

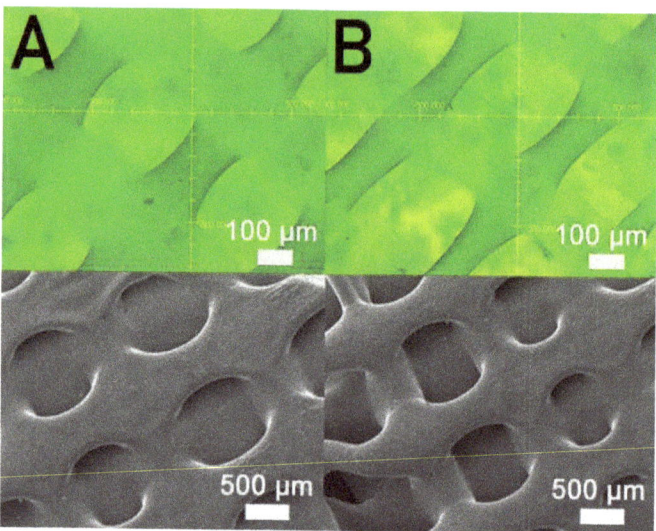

Figure 4. Profile projector (**top**) and scanning electron microscopy (SEM) (**bottom**) micro-photos of nanocomposite specimens printed with PCL/DCM ratios of (**A**) 2.5 g/6.4 mL and (**B**) 2.5 g/5.8 mL.

Figure 5 shows the measured strengths of PCL/nHA specimens printed with different fill densities. The estimated ultimate tensile raised with the fill density. Figure 6 shows the micro-images of specimen surfaces printed with 65% and 50% fill densities. Imperfectly healed pores were observed on the surfaces of the printed specimens. Superior healing can be noted in Figure 6A (65% fill density) compared to Figure 6B (50% fill density). Since the specimens made with 65% fill density possessed the smallest pore sizes, they showed superior mechanical strengths. Meanwhile, nanocomposites printed with a lower fill density exhibited pores of greater sizes, resulting in stress concentrations in the tensile test process. Composite parts therefore possessed inferior mechanical properties.

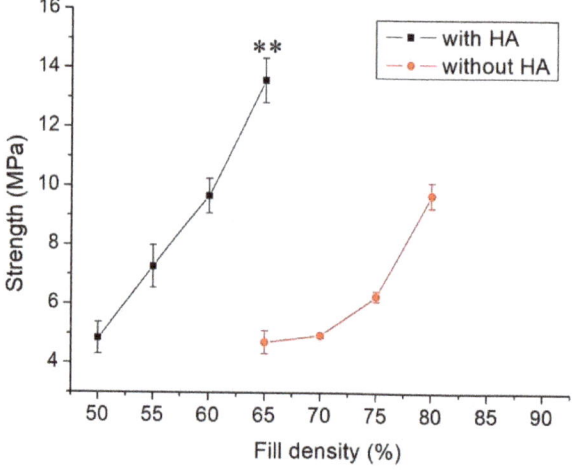

Figure 5. Effect of fill density on the tensile strengths of 3D-printed virgin PCL and PCL/nHA nanocomposite specimens (** $p < 0.01$, virgin PCL vs. PCL/nHA).

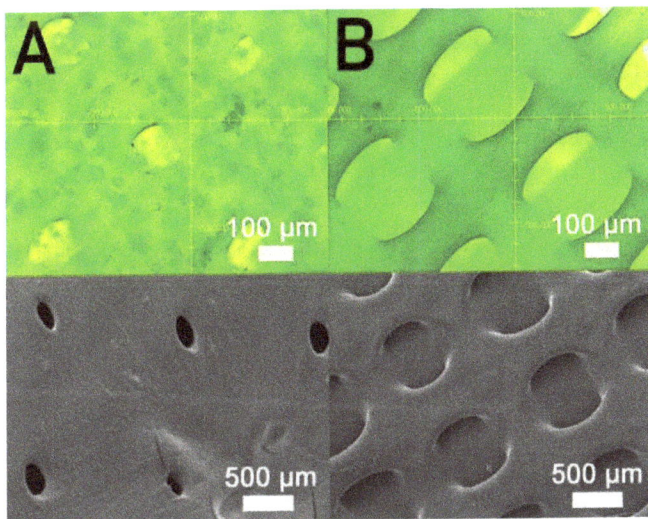

Figure 6. Profile projector (**top**) and SEM (**bottom**) micro-photos of nanocomposite specimens printed with fill densities of (**A**) 65% and (**B**) 50%.

Figure 7 displays the influence of print speed on the ultimate strength of the printed nanocomposites. The ultimate strength decreased in general as the tip speed increased. Nanocomposite parts printed at a speed of 30 mm/s showed the highest strengths, and samples printed at a print speed of 60 mm/s displayed the most inferior strengths. Figure 8A,B shows the images of part surfaces printed using 30 and 60 mm/s, respectively. PCL/nHA nanocomposite parts printed at a speed of 60 mm/s possessed bigger pores than those printed with 30 mm/s. Small orifices were observed on the surface of 60 mm/s printed parts, which mainly resulted from incomplete healing. Manufactured 60 mm/s parts thus illustrated inferior mechanical strengths.

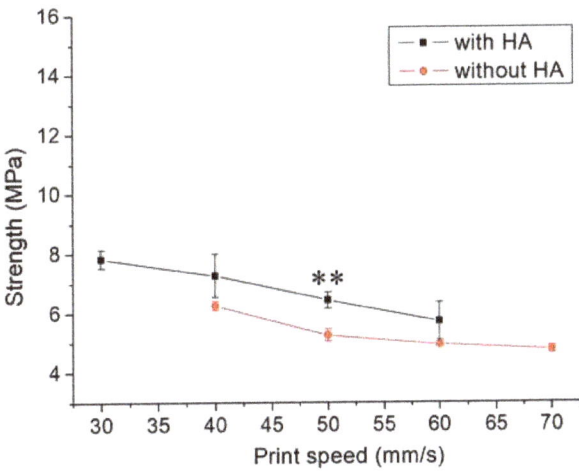

Figure 7. Influence of print speed on the tensile strengths of 3D-printed virgin PCL and PCL/nHA nanocomposite specimens (** $p < 0.01$, virgin PCL vs. PCL/nHA).

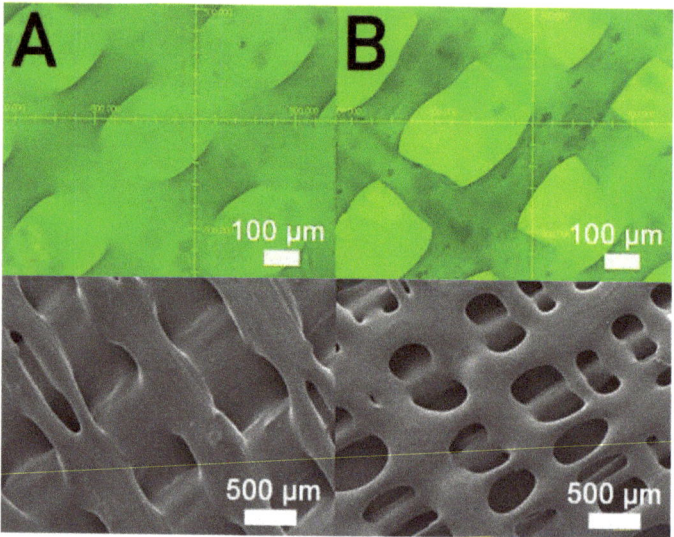

Figure 8. Profile projector (**top**) and SEM (**bottom**) micro-photos of nanocomposite specimens printed with nozzle speeds of (**A**) 30 mm/s and (**B**) 60 mm/s.

Figure 9 indicates the maximum tensile strength of these samples printed using different orientations of 45°, 60°, 75°, and 90°, suggesting that the nanocomposite specimen printed at a 90° orientation showed the most superior tensile strength, and those at 45° led to the least mechanical strengths. Figure 10A,B shows the images of part surfaces printed with orientations of 90° and 45°. Clearly, 45° printing led to less healing and larger pores on the parts' surfaces. As these specimens are exposed to foreign loads, stress can happen and result in ruptures. Printed part strengths thus diminish under these loads.

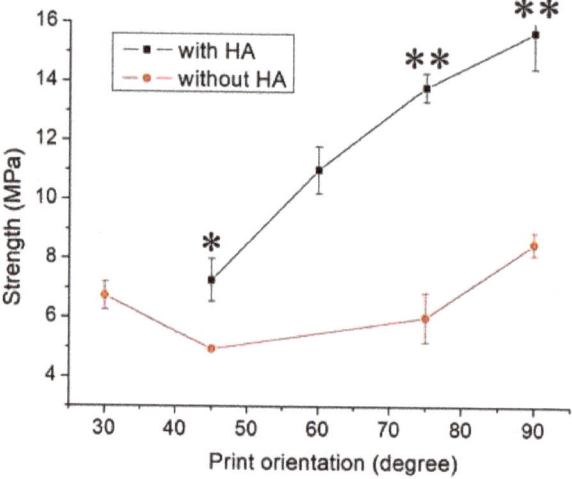

Figure 9. Influence of print orientation on the tensile strengths of 3D-printed virgin PCL and PCL/nHA nanocomposite specimens (* $p < 0.05$; ** $p < 0.01$, virgin PCL vs. PCL/nHA).

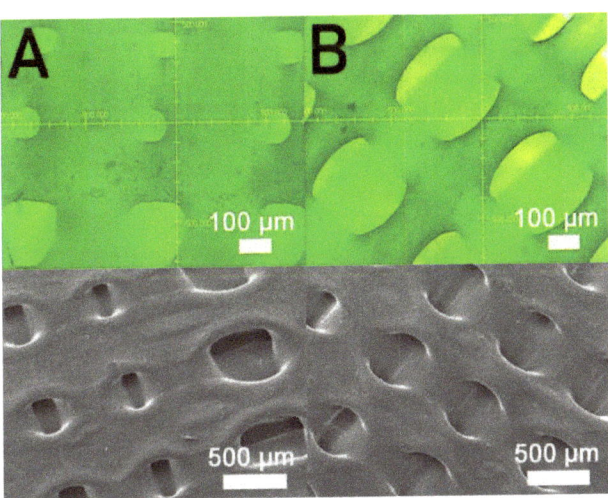

Figure 10. Profile projector (**top**) and SEM (**bottom**) micro-photos of nanocomposite specimens printed with orientations of (**A**) 90° and (**B**) 45°.

3.2. Drug Release from Printed Implants

Drug-eluting screws were prepared using the optimum processing conditions obtained in Section 3.1, namely a PCL/DCM ratio of 2.5 g/0.133 g:6.4 mL, a fill density of 65%, a nozzle shifting speed of 30 mm/s, and a 90° printing orientation. The ultimate strength and elastic modulus thus obtained were 15.67 ± 1.22 and 37.49 ± 1.36 MPa, respectively.

Figure 11 displays the Fourier-transform infrared (FTIR) spectra of pure PCL, PCL/nHA, and drug-loaded PCL/nHA screws. The vibration peak of PO_4^{3-} near 1040 cm^{-1} for nHA diminished after the material was incorporated into the PCL matrix [20]. The peaks at 1724 and 1635 cm^{-1} may be attributed to the C=O and C=C bonds, respectively, of the incorporated drugs. The vibration peak at 2942 cm^{-1}, corresponding to a CH_2 bond, was promoted due to the addition of vancomycin. Additionally, the vibration at 3340 cm^{-1} resulted from the N–H bond of the anti-microbial agents [21,22].

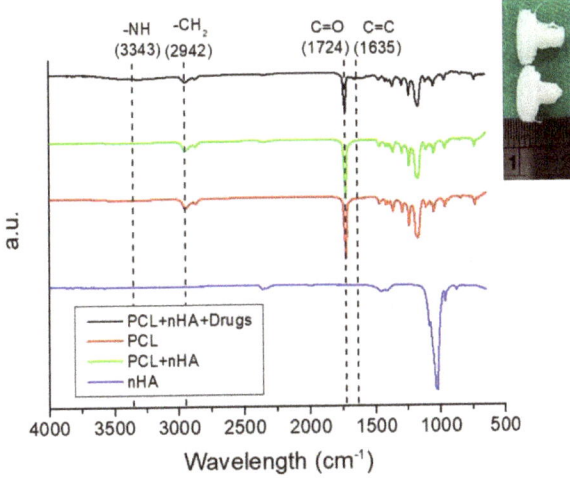

Figure 11. Fourier-transform infrared (FTIR) spectra of virgin PCL, nHA, PCL/nHA, and drug-loaded PCL/nHA nanocomposites (**upper right** is a photo of 3D-printed drug loaded PLC/nHA screws).

The thermal properties of virgin PCL, PCL/nHA, and drug-loaded PCL/nHA screws were assessed, and the results are displayed in Figure 12. Although the incorporation of nHA caused a slight increase in the melting point of pure PCL from 62.78 to 65.02 °C, the addition of drugs tended to reduce the melting temperature of 3D-printed PCL/nHA screws to 57.75 °C.

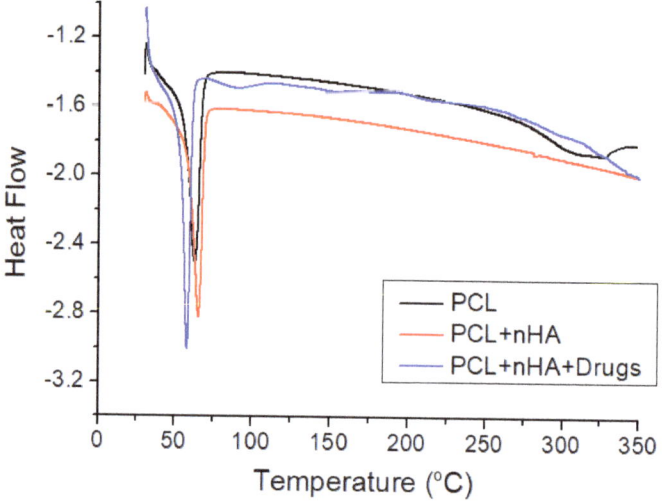

Figure 12. Differential calorimetry scanning (DSC) curves of virgin PCL, PCL/nHA, and drug-loaded PCL/nHA composites.

The release of antibiotics from the 3D-printed screws was characterized. Figure 13 displays the daily and cumulative releases of vancomycin and ceftazidime. A burst release was noticed for the anti-microbial agents at day 1, after which a steady and diminishing elution of pharmaceuticals was observed. The drug-loaded PCL/nHA screws could elute high levels of vancomycin and ceftazidime (higher than the minimum inhibitory concentrations) for more than 14 days. Antibiotic levels were maintained at a high level after the 3D solution-extrusion printing procedure, demonstrating that the 3D printing did not inactivate the antimicrobial agents during the fabrication process.

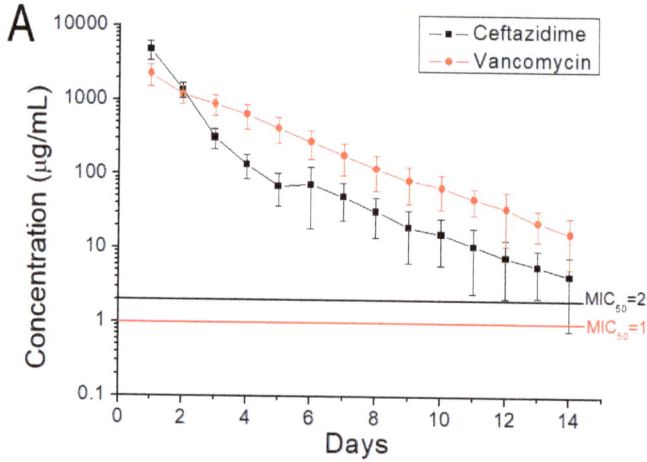

Figure 13. *Cont.*

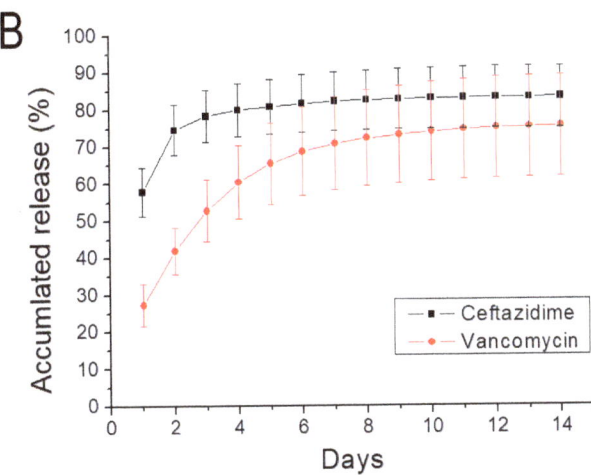

Figure 13. (**A**) Daily and (**B**) cumulative release of vancomycin and ceftazidime from PCL/nHA screws.

4. Discussion

This study explored the effect of distinct printing variables on mechanical strengths of printed PCL/nHA materials. PCL pertains to a biodegradable polymer material that resorbs gradually through hydrolysis [23], whereas HA is a primary component for hard tissues, including bone and teeth, and has been widely used for bone repair, bone augmentation, and implant coatings [24]. The mechanical properties of a material represent its response to an externally applied load, and are one of the most important basic characteristics of a good design. Two factors might have influenced the mechanical strengths of our 3D-printed nanocomposites and helped decide the success of the composites for a specific application. The primary factor pertained to the healing/sealing of extruded materials, and the next factor was the morphological structure of printed products.

The first factor that affected the final product property was the healing and chain entanglement at the inter-boundaries of polymeric strips. In 3D printing, solvent vaporization and molecular chain diffusion occur at the solidification stage of polymer materials. Promotion of molecular entanglement at the interface of extruded strips is required and accounts for the eventual tensile characteristics of 3D-printed samples [25,26]. The optimum status tends to be a semi-dilute and moderately entangled mode that arises at the critical entanglement concentration [27]. The period is a crossover point between a semi-diluted unentangled mode and a semi-diluted moderately entangled mode. Promoted entanglement and the associated strips sealing at the interfaces are thus expected. With regard to the second factor, a solution-extrusion 3D-printed part may possess irregular morphology owing to imperfectly healed pores or flaws that result in stress concentrations as exposed to foreign loads. This process may in turn cause deterioration of the tensile strengths of printed nanocomposites.

The empirical outcomes in Figure 3 indicate that the measured tensile strengths of 3D-printed nanocomposites increased with the volume fraction of the solvent. The nanocomposites prepared with the PCL/nHA-to-DCM ratio of 2.5 g/0.133g:6.4 mL exhibited the greatest tensile strengths. When a small amount of DCM was used, the solvent may have evaporated too fast, resulting in lack of time for chain entanglement at the interfaces of extruded strips. An abundant concentration of DCM kept the polymers within a semi-diluted status for a longer period of time and promoted strip healing/sealing at the inter-boundaries, either within the current layer or across distinct layers (Figure 4). Printed materials therefore illustrate superior strengths.

Figure 5 shows that the estimated tensile strengths of nanocomposites increased as the fill density was increased. Fill density represents the quantity of polymer nanocomposites used in manufacturing a specimen. A greater fill density represents potent polymeric materials inside printed specimens, thereby leading to a more superior part. Additionally, fulfilling the specimens with extra materials also resulted in pores of smaller sizes (Figure 6). Printed product quality increased accordingly.

Figure 7 implies that mechanical strength declined when print speed decreased. Following the solution extrusion from the delivering nozzle, the solvent started to vaporize. When the print speed was excessively high, not enough time was allowed for molecular entanglement of the polymers across the strip interfaces (Figure 8). Additionally, the solvent may not have had sufficient time to vaporize and may then have diffused and disintegrated the surrounding strips. Chain entanglement and the relevant part strengths diminished accordingly.

Nanocomposite specimens printed with a 45° orientation demonstrated the most inferior part strength, and 90° printed parts showed the highest strengths (Figure 9). This finding might be due to the fact that the 90° oriented strips restricted the quantity of solvent that vaporized quickly. Abundant time was allowed for the polymer solution to achieve chain entanglement across strip boundaries (Figure 10A). The tensile properties of 3D-printed samples thereby increased. Additionally, the printed 45° orientated nanocomposite parts presented a greater number of pores (Figure 10B). When stretched by the external tensile forces, these large-size pores possessed greater chances of undergoing damage from stress. Printed parts therefore displayed the most inferior tensile properties.

Finally, this study successfully developed antimicrobial agent-loaded PCL/nHA screws using the solution of extrusion 3D printing technology. After 3D printing, some drugs may have remained at the surface of the screws, thus leading to a burst release when the screws were submerged in a PBS solution. After the burst release, the release mechanism was mainly controlled by channel diffusion. When the loading of antibiotics was low, the pharmaceuticals would have been separated in the polymer matrix. The drugs may not have been capable of penetrating the matrix at an effective rate. When the pharmaceutical loading was further increased, the drugs may have bonded together to create channels transmitting to the surface of the 3D-printed parts [18]. A steady and slow release of the drugs was thus observed after the burst release. The experimental results demonstrate that the drug-loaded PCL/nHA screws can offer extended elution of high concentrations of vancomycin/ceftazidime (superior to the minimum inhibitory concentrations) for more than 14 days. All of these findings demonstrate the great potential of solution-extrusion 3D printing for the manufacture of drug-loaded implants.

5. Conclusions

This study explored the solution-extrusion 3D printing of vancomycin- and ceftazidime-loaded PCL/nHA materials utilizing a lab-made printer. The influence of distinct parameters on the printed part quality was examined. The empirical outcomes illustrate that the tensile property of printed nanocomposites increases with the fill density yet diminishes with a decrease in the ratio of PCL/nHA to DCM and print speed. Nanocomposite parts printed with a 90° orientation demonstrated the most superior mechanical properties. In addition, the drug-loaded PCL/nHA screws can provide extended elution of high levels of vancomycin/ceftazidime over a 14-day period. Eventually solution-extrusion 3D printing technology may be used to print drug-loaded implants for various medical applications.

Author Contributions: Conceptualization, P.-Y.C.; methodology, Y.-H.L. and Y.-T.L.; formal analysis, Y.-H.L. and Y.-T.L.; investigation, Y.-H.L., Y.-C.C. and S.-J.L.; data curation, C.-J.L. and P.-Y.C.; writing—original draft preparation, S.-J.L.; writing—review and editing; funding acquisition, Y.-C.C., S.-J.L. All authors have read and agreed to the published version of the manuscript.

Funding: This work was sponsored by the Ministry of Science and Technology, Taiwan (Contract No. 109-2221-E-182-058-MY2) and Chang Gung Memorial Hospital (Contract No. CRRPD2K0011 and CRRPG3K0091).

Conflicts of Interest: The authors declare no conflict of interest.

References

1. Venkatesan, J.; Kim, S.K. Nano-hydroxyapatite composite biomaterials for bone tissue engineering–A review. *J. Biomed. Nanotechnol.* **2014**, *10*, 3124–3140. [CrossRef] [PubMed]
2. Ramesh, N.; Moratti, S.C.; Dias, G.J. Hydroxyapatite-polymer biocomposites for bone regeneration: A review of current trends. *J. Biomed. Mater. Res. B* **2018**, *106*, 2046–2057. [CrossRef] [PubMed]
3. Molino, G.; Palmieri, M.C.; Montalbano, G.; Fiorilli, S.; Vitale-Brovarone, C. Biomimetic and mesoporous nano-hydroxyapatite for bone tissue application: A short review. *Biomed. Mater.* **2020**, *15*, 022001. [CrossRef] [PubMed]
4. Eshragi, S.; Das, S. Mechanical and microstructural properties of polycaprolactone scaffolds with one-dimensional, two-dimensional, and three-dimensional orthogonally oriented porous architectures produced by selective laser sintering. *Acta Biomater.* **2010**, *6*, 2467–2476. [CrossRef] [PubMed]
5. Dwivedi, R.; Kumar, S.; Pandey, R.; Mahajan, A.; Nandana, D.; Katti, D.S.; Mehrotra, D. Polycaprolactone as biomaterial for bone scaffolds: Review of literature. *J. Oral Biol. Craniofac. Res.* **2020**, *10*, 381–388. [CrossRef]
6. Ngo, T.D.; Kashani, A.; Imbalzano, G.; Nguyen, K.T.Q.; Hui, D. Additive manufacturing (3D printing): A review of materials, methods, applications and challenges. *Compos. B* **2018**, *143*, 172–196. [CrossRef]
7. Bekas, D.G.; Hou, Y.; Liu, Y.; Panesar, A. 3D printing to enable multifunctionality in polymer-based composites: A review. *Compos. B* **2019**, *179*, 107540. [CrossRef]
8. Alafaghani, A.; Qattawi, A.; Alrawi, B.; Guzman, A. Experimental optimization of Fused Deposition Modelling processing parameters: A design-for-manufacturing approach. *Procedia Manuf.* **2017**, *10*, 791–803. [CrossRef]
9. Woodruff, M.A.; Hutmacher, D.W. The return of a forgotten polymer–Polycaprolactone in the 21st century. *Prog. Polym. Sci.* **2010**, *35*, 1217–1256. [CrossRef]
10. Labeta, M.; Thielemans, W. Synthesis of polycaprolactone: A review. *Chem. Soc. Rev.* **2009**, *38*, 3484–3504. [CrossRef]
11. Beheshtizadeh, N.; Lotfibakhshaiesh, N.; Pazhouhnia, Z.; Hoseinpour, M.; Nafari, M. A review of 3D bio-printing for bone and skin tissue engineering: A commercial approach. *J. Mater. Sci.* **2020**, *55*, 3729–3749. [CrossRef]
12. Egan, P.F. Integrated design approaches for 3D printed tissue scaffolds: Review and outlook. *Materials* **2019**, *12*, 2355. [CrossRef] [PubMed]
13. Guerra, A.J.; Ciurana, J. 3D-printed bioabsordable polycaprolactone stent: The effect of process parameters on its physical features. *Mater. Design.* **2018**, *137*, 430–437. [CrossRef]
14. Jiao, Z.; Luo, B.; Xiang, S.; Ma, H.; Yu, Y.; Yang, W. 3D printing of HA/PCL composite tissue engineering scaffolds. *Adv. Ind. Eng. Polym. Res.* **2019**, *2*, 196–202. [CrossRef]
15. Holländer, J.; Genina, N.; Jukarainen, H.; Khajeheian, M.; Rosling, A.; Makila, E.; Sandler, N. Three-dimensional printed PCL-based implantable prototypes of medical devices for controlled drug delivery. *J. Pharm. Sci.* **2016**, *105*, 2665–2676. [CrossRef] [PubMed]
16. Visscher, L.E.; Dang, H.P.; Knackstedt, M.A.; Hutmacher, D.W.; Tran, P.A. 3D printed Polycaprolactone scaffolds with dual macro-microporosity for applications in local delivery of antibiotics. *Mater. Sci. Eng. C* **2018**, *87*, 78–89. [CrossRef]
17. Yi, H.G.; Choi, Y.J.; Kang, K.S.; Hong, J.M.; Pati, R.G.; Park, M.N.; Shim, I.K.; Lee, C.M.; Kim, S.C.; Cho, D.W. A 3D-printed local drug delivery patch for pancreatic cancer growth suppression. *J. Control. Release* **2016**, *238*, 231–241. [CrossRef]
18. Chen, J.M.; Lee, D.; Yang, J.W.; Lin, S.H.; Lin, Y.T.; Liu, S.J. Solution-extrusion additive manufacturing of biodegradable polycaprolactone. *Appl. Sci.* **2020**, *10*, 3189. [CrossRef]
19. Hsu, Y.H.; Chen, D.W.; LI, M.J.; Yu, Y.H.; Chou, Y.C.; Liu, S.J. Sustained delivery of analgesic and antimicrobial agents to knee joint by direct injections of electrosprayed multipharmaceutical-loaded nano-microparticles. *Polymers* **2018**, *10*, 890. [CrossRef]
20. Gheisari, H.; Karamian, E.; Abdellahi, M. A novel hydroxyapatite–hardystonite nanocomposite ceramic. *Ceram. Int.* **2015**, *41*, 5967–5975. [CrossRef]
21. Murei, A.; Ayinde, W.B.; Gitari, M.W.; Samie, A. Functionalization and antimicrobial evaluation of ampicillin, penicillin and vancomycin with Pyrenacantha grandifora Baill and silver nanoparticles. *Sci. Rep.* **2020**, *10*, 11596. [CrossRef] [PubMed]
22. Huang, Y.; Zhang, Y.; Yan, Z.; Liao, S. Assay of ceftazidime and cefepime based on fluorescence quenching of carbon quantum dots. *Luminescence* **2015**, *30*, 1133–1138. [CrossRef] [PubMed]
23. Mondal, D.; Griffith, M.; Venkatraman, S.S. Polycaprolactone-based biomaterials for tissue engineering and drug delivery: Current scenario and challenges. *Int. J. Polym. Mat. Polym. Biomat.* **2016**, *65*, 255–265. [CrossRef]
24. Anil, S.; Chalisserry, E.P.; Man, S.Y.; Venkatesan, J. Biomaterials for craniofacial tissue engineering and regenerative dentistry. In *Advanced Dental Biomaterials*; Woodhead Publishing: New York, NY, USA, 2019; pp. 643–674.
25. Yokomizo, K.; Banno, Y.; Yoshikawa, T.; Kotaki, M. Effect of molecular weight and molecular weight distribution on weld-line interface in injection-molded polypropylene. *Polym. Eng. Sci.* **2013**, *53*, 2336–2344. [CrossRef]
26. Yizong, T.; Ariff, Z.M.; Liang, K.G. Evaluation of weld line strength in low density polyethylene specimens by optical microscopy and simulation. *J. Eng. Sci.* **2017**, *13*, 53–62.
27. Costa, L.M.M.; Bretas, R.E.S.; Gergorio, R., Jr. Effect of solution concentration on the electrospray/electrospinning transition and on the crystalline phase of PVDF. *Mater. Sci. Appl.* **2010**, *1*, 247–252. [CrossRef]

Article

Substantial Effect of Water on Radical Melt Crosslinking and Rheological Properties of Poly(ε-Caprolactone)

Angelica Avella [1], Rosica Mincheva [2], Jean-Marie Raquez [2] and Giada Lo Re [1,*]

[1] Department of Industrial and Materials Science, Division of Engineering Materials, Chalmers University of Technology, SE-412 96 Gothenburg, Sweden; avella@chalmers.se
[2] Laboratory of Polymeric and Composite Materials (LPCM), Center of Innovation and Research in Materials and Polymers (CIRMAP), University of Mons (UMONS), B-7000 Mons, Belgium; rosica.mincheva@umons.ac.be (R.M.); jean-marie.raquez@umons.ac.be (J.-M.R.)
* Correspondence: giadal@chalmers.se

Abstract: One-step reactive melt processing (REx) via radical reactions was evaluated with the aim of improving the rheological properties of poly(ε-caprolactone) (PCL). In particular, a water-assisted REx was designed under the hypothesis of increasing crosslinking efficiency with water as a low viscous medium in comparison with a slower PCL macroradicals diffusion in the melt state. To assess the effect of dry vs. water-assisted REx on PCL, its structural, thermo-mechanical and rheological properties were investigated. Water-assisted REx resulted in increased PCL gel fraction compared to dry REx (from 1–34%), proving the rationale under the formulated hypothesis. From dynamic mechanical analysis and tensile tests, the crosslink did not significantly affect the PCL mechanical performance. Dynamic rheological measurements showed that higher PCL viscosity was reached with increasing branching/crosslinking and the typical PCL Newtonian behavior was shifting towards a progressively more pronounced shear thinning. A complete transition from viscous- to solid-like PCL melt behavior was recorded, demonstrating that higher melt elasticity can be obtained as a function of gel content by controlled REx. Improvement in rheological properties offers the possibility of broadening PCL melt processability without hindering its recycling by melt processing.

Keywords: reactive melt processing; water-assisted; radical crosslinking; peroxide initiators; biopolymers; poly(ε-caprolactone); rheology

Citation: Avella, A.; Mincheva, R.; Raquez, J.-M.; Lo Re, G. Substantial Effect of Water on Radical Melt Crosslinking and Rheological Properties of Poly(ε-Caprolactone). *Polymers* **2021**, *13*, 491. https://doi.org/10.3390/polym13040491

Academic Editor: Khalid Lamnawar
Received: 21 January 2021
Accepted: 1 February 2021
Published: 4 February 2021

Publisher's Note: MDPI stays neutral with regard to jurisdictional claims in published maps and institutional affiliations.

Copyright: © 2021 by the authors. Licensee MDPI, Basel, Switzerland. This article is an open access article distributed under the terms and conditions of the Creative Commons Attribution (CC BY) license (https://creativecommons.org/licenses/by/4.0/).

1. Introduction

Increased environmental concerns regarding the accumulation of plastic waste in landfills and the marine environment are pushing industry and academia to consider relevant eco-friendly alternatives to conventional plastics [1,2]. Biodegradable thermoplastic polyesters represent a possible solution but generally their processability needs to be enhanced to enable their commercial scaling-up, beyond their economic barriers [3]. Among the commercially available biopolyesters, poly(ε-caprolactone) (PCL) is biodegradable with high ductility and toughness, as well as other mechanical properties comparable with low density polyethylene [4,5]. These characteristics make PCL desirable for various commercial applications like packaging and single-use items, in addition to biomedical applications [6].

However, because of its high linearity, PCL possesses low melt viscosity and strength [7] which limit its processability for example in film blowing, film extrusion or foaming, as seen for other aliphatic biopolyesters [8,9]. Moreover, PCL is characterized by a relatively low crystallization rate [10], reducing the effectiveness of its quenching and injection molding. Several strategies have been proposed to overcome the PCL drawbacks and improve its processability such as blending with other polymer matrices [11,12] and its chemical modification.

In the latter case, branching and/or crosslinking via ionizing radiation [7,13] or with the use of organic peroxides [13] is a widely explored route to tailor thermo-mechanical and rheological properties of aliphatic biodegradable polyesters such as polylactide and polyhydroxyalkanoates. Moreover, crosslinking PCL has been proven to maintain [14,15] or even favor [16] its biodegradability. In this context, reactive melt processing (REx) is the most employed technique for these chemical modifications in both solvent-free and continuous manners, it does not require any post-purification and eases an industrial uptake [3].

The use of peroxides as radical initiators has been previously reported for PCL backbone modification [17,18] or PCL blends compatibilization [19–21]. However, to our knowledge only few works have explored the effect of the only peroxide on PCL branching/crosslinking in solvent casting [15,22] or during melt processing [9,23–25]. Gandhi et al. [23] studied the PCL crosslinking with dicumyl peroxide during melt processing, in comparison with radiation crosslinking. The study indicates a more efficient gelation and improvement of PCL rheological properties using peroxide via REx, due to the lower chain scission induced by peroxide compared to radiation. The same system was also presented by Di Maio et al. [9] with the aim of improving PCL viscoelastic properties for foaming. Dynamic rheology of the modified PCL shows an increase in viscosity and melt elasticity with higher amounts of dicumyl peroxide. Structural PCL modification via REx comparing two types of peroxide was investigated by Przybysz et al. [24]. The results point out that different degrees of branching/crosslinking can be achieved as a function of the peroxide structure and amount. Han et al. [25] reported the use of benzoyl peroxide (BPO) to crosslink PCL in a two-step method: first blending at low temperature, followed by crosslinking during compression molding. A maximum in tensile strength was registered at 1 wt.% of BPO, while higher amount of peroxide increased the gel fraction but lowered the mechanical properties. It is worth to note that after a critical gel fraction ($\approx$ 40–50%) the thermoplasticity of the polymer is hampered, together with an increase in its elastic melt behavior, hindering its melt processing and possibility to be melt-recycled.

Looking into the mechanism of peroxide-induced gelation, peroxide decomposes at high temperatures into free radicals, which then abstract hydrogens directly from the polymer backbone and consequently generate covalent bonds between chains macroradical recombination, yielding branching/crosslinking [26]. It has been demonstrated that radical reaction kinetics can be improved using low viscous media in polyphasic solvent or bulk systems [27]. In the context of REx, the diffusivity of macroradicals can be hindered due to their high viscosity, thus limiting the radical propagation. Therefore, water molecules can be considered as a potential aid in melt radical reactions because of their high mobility. A beneficial effect of water in polymer grafting during free radical graft copolymerization in water suspension has been proven by Kaur et al. [28]. Water-assisted radical reactive melt processing has been successfully exploited for the preparation of cellulose-based PCL nanocomposites [4]. Moreover, water has demonstrated to act as: (i) true catalyst through its ability to form complexes via hydrogen bonds or radical conjugation (catalysis) [29]; (ii) enhancer of the recombination processes (radical complex) [30]; (iii) highly efficient collision partner stabilizing intermediates in radical reactions (energy-transfer) [30].

Following these findings, the objectives of our work were the design and the control of PCL crosslinking with low BPO amounts via one-step water-assisted REx to broaden the PCL processability by improving its rheological properties. Low levels in BPO (up to 1 wt.%) were chosen to limit the gelation up to 40%, preserving PCL thermoplasticity and possible melt-recyclability. Our REx design considered the use of water as a temporary low viscosity medium in the melt, to increase the rate of radical propagation, hence the reaction efficiency. The rationale behind the design was to exploit an expected larger diffusivity of water-borne hydroxyl radical compared to PCL macroradicals, before water evaporation. Therefore, a water-assisted processing was carried out at 120 °C in comparison with traditional dry REx. BPO was chosen as peroxide initiator due to its low decomposition temperature [31], consistent with the selected processing temperature. Structural, thermo-mechanical and rheological properties of neat and reacted PCL were analyzed to assess the effectiveness of this strategy. The results showed the formation of higher gel fraction in water-assisted REx, confirming the validity of the formulated hypotheses. PCL melt viscosity increased with the extent of crosslinking (up to 34%), achieving a predominant elasticity of the PCL melt, synonym of improved melt strength and processability [7,32], at the expense of mechanical properties slightly limited by PCL chain scission.

This work provides relevant insights for future controlled water-assisted reactive melt processing of PCL and its composites with hydrophilic reinforcements. Our results deeper explain successful water-assisted REx of cellulose/PCL biocomposites [4], paving the route for the new generation of biodegradable composites with polysaccharides, with improved mechanical and rheological properties.

2. Materials and Methods

2.1. Materials

Poly(ε-caprolactone) (PCL) Capa6506 was purchased as powder form from Ingevity, (Warrington, UK). According to the supplier, the PCL grade has a mean molecular weight of 50,000 g·mol^{-1}, melting temperature (T_m) of 60 °C and a melt flow rate of 7.9–5.9 g/10 min at 190 °C/2.16 kg. Benzoyl peroxide (BPO), under the trade name Luperox A75 (75%, left water) was purchased from Sigma-Aldrich AB (Stockholm, Sweden) and was used without further purification. Dichloromethane (DCM) was purchased from VWR International AB (Stockholm, Sweden) with purity higher than 99.5%.

2.2. Reactive Melt Processing

For dry reactive melt processing (REx), PCL powder was manually premixed with different BPO amounts (0–0.1–0.25–0.5 and 1 wt.%) and processed in an internal mixer AEV 330 (50 cm^3) (Brabender® GmbH & Co., Duisburg, Germany) with counter-rotating screws W50 (feeding for ≈ 5 min at 30 rpm, then 10 min at 60 rpm). The selected processing temperature was 120 °C to enable water evaporation during REx. and at the same time overcome the onset and melting temperatures of BPO (≈98 and 103 °C, respectively [33]). It is worth to note that we assumed the overall 15 min of REx to be sufficient for BPO decomposition, due to its half lifetime at 120 °C of ≈ 3 min [34]. The reacted PCL was coded as PCL-xL, with x indicating the wt.% of BPO present during the reaction. Water-assisted REx was also carried out by premixing PCL with 50 wt.% of deionized water and 0.5 or 1 wt.% BPO. The obtained paste was melt processed under the same conditions of dry REx. It has been assumed a complete water evaporation within the processing time [4]. These two materials were denoted as PCL-0.5Lw and PCL-1Lw, respectively. For further analyses all the reacted PCL samples have been shaped in squared films of 1 mm thickness by compression molding (Buscher-Guyer KHL 100, Zurich, Switzerland) at 120 °C, at 40 bar for 3 min and 500 bar for 1 min.

2.3. Characterization Methods

To verify their solubility, neat and reacted PCL samples with 0.5 and 1 wt.% of peroxide during dry and water-assisted REx were dispersed in dichloromethane (0.6 g in 30 mL) and the mixture was magnetically stirred overnight with Teflon coated stir rods.

To separate the soluble and insoluble fractions from boiling dichloromethane (500 mL), a Soxhlet extraction was carried out for 72 h on 5 g samples. The insoluble fraction was filtered off in glass fiber thimbles (Whatman 603G, VWR), dried at room temperature for 48 h and then weighted to measure the gel content of reacted PCL. Finally, the gel fraction was calculated according to Equation (1):

$$Gel\ fraction\ [\%] = \frac{w_i}{w_0} \times 100 \qquad (1)$$

where w_i indicates the weight of the residual insoluble fraction and w_0 indicates the initial weight of the sample.

Size-exclusion chromatography (SEC) was performed in chloroform (CHCl$_3$) at 30 °C using an Agilent (Diegem, Belgium) liquid chromatograph equipped with an Agilent degasser, an isocratic HPLC pump (flow rate = 1 mL·min^{-1}), an Agilent autosampler (loop volume = 100 µL; solution concentration = 2 mg·mL^{-1}), an Agilent-DRI refractive index detector, and three columns: a PL gel 5 µm guard column and two PL gel Mixed-B 5 µm columns (linear columns for separation of molecular weight (PS) ranging from 200 to 4 × 10^5 g·mol^{-1}). Polystyrene standards were used for calibration.

The thermal stability was studied by thermogravimetric analysis (TGA) with a TGA/DSC 3 + Star system (Mettler Toledo, Greifensee, Switzerland). Approximately 5 mg of each sample were preheated from room temperature to 70 °C, where an isothermal segment was maintained for 15 min to remove residual moisture. Then the samples were heated to 550 °C at a heating rate of 5 °C·min^{-1}, under N$_2$ constant flow of 50 mL·min^{-1}. Temperature of the onset of degradation ($T_{5\%}$), was identified as the temperature at which the weight loss was 5%. Temperature of degradation (T_d) was extrapolated as the temperature of the peak of the first derivative (DTG). Char residue was estimated as the final weight % at 550 °C.

Differential scanning calorimetry (DSC) was performed on a Mettler Toledo DSC 2 calorimeter equipped with a HSS7 sensor and a TC-125MT intercooler. The endotherms were recorded following a heating/cooling/heating temperature profile from −80 °C to 200 °C, at a heating rate of 10 °C·min^{-1}, under N$_2$ constant flow of 50 mL·min^{-1}. The melting temperature (T_m) was detected as the temperature of the maximum of the melting transition peak in the second heating scan, while the glass transition temperature (T_g) at the inflection point of the transition step. Crystallization temperature (T_c) was evaluated as the temperature of the crystallization peak minimum in the cooling scan. The degree of crystallinity (χ_{DSC}) was calculated according to Equation (2):

$$\chi_{DSC}[\%] = \frac{\Delta H_M}{\Delta H_0} \times 100 \quad (2)$$

where ΔH_M is the specific melting enthalpy and ΔH_0 is the melting enthalpy of 100% crystalline PCL (136 J·g^{-1} [4]).

X-Ray diffraction (XRD) spectra were recorded by a D8 Advance Diffractometer (Bruker AXS, Karlsruhe, Germany) with Cr Kα radiation (35 kV, 50 mA) in a 2θ range between 5 and 100°, at a speed of 0.6 deg·min^{-1}. The crystallinity of the samples was calculated according to Equation (3):

$$\chi_{XRD}[\%] = \frac{A_c}{A_{tot}} \times 100 \quad (3)$$

where A_c is the area under the crystalline peaks of the spectra, while A_{tot} is the total area under the spectra between 2θ = 5 and 100°. The Scherrer equation (Equation (4)) was used to calculate the crystallite size in the direction normal to the 110 lattice planes (D_{110}):

$$D_{110} = \frac{0.9\lambda}{B_{110}\cos\theta} \quad (4)$$

where λ is the radiation wavelength (2.29 Å), B_{110} is the full width at half-maximum X-ray diffraction line in radians and θ is the Bragg angle. The 0.9 constant was previously used in literature for similar systems [35–37].

Dynamic mechanical properties were evaluated by dynamic mechanical thermal analysis (DMTA) with a DMA Q800 (TA Instruments, New Castle, DE, USA) apparatus in tension-film mode on rectangular bars (25 × 5 × 1 mm^3). The bars were cut from compression molded films and conditioned for at least 48 h at 23 °C and 53% relative humidity. Temperature ramps were performed from −80 °C to 45 °C at a heating rate of 2 °C·min^{-1}, at a frequency of 1 Hz and strain amplitude of 1%, selected in the linear viscoelastic region from strain amplitude sweeps. The glass (T_g) and alpha (T_α) transition temperatures were determined as the temperatures of the maxima of the loss moduli and tanδ, respectively. The damping factor (DF) values were recorded as the maxima of tanδ.

Tensile properties (Young's modulus, yield stress, ultimate tensile strength, and elongation at break) were carried out according to the standard ASTM D638-14. Dumbbell specimens, with 25 mm gauge length and 1 mm thickness, were cut from compression molded films and conditioned for at least 48 h at 23 °C and 53% relative humidity, prior to testing. At least five specimens were tested for each material at a crosshead speed of 6 mm·min^{-1} with a Zwick/Z2.5 tensile tester (ZwickRoell Ltd., Leominster, UK) equipped with a load cell of 2 kN. Tests were performed and evaluated.

Dynamic rheological measurements were carried out using an Anton Paar MCR 702 rheometer (Graz, Austria) in single-drive mode with a parallel plate geometry (15 mm ø). Disks (20 mm ø) were cut from compression molded films and were conditioned for at least 48 h at 23 °C and 53% relative humidity. The disks were tested at 120 °C and gap of 1 mm, after removal of melt material exceeding the selected geometry. First, oscillatory strain sweep tests were performed in a shear strain range 0.01–100% to determine the linear viscoelastic region. Frequency sweep tests were performed in an angular frequency range from 200 to 0.08 rad·s^{-1} at an applied strain of 1%, within the linear region. Storage modulus (G'), loss modulus (G''), complex modulus (G*), complex viscosity (η*) and phase angle (δ) were recorded.

3. Results and Discussion

3.1. Reactive Melt Processing

Chain extension of PCL with a benzoyl peroxide, was successfully carried out via reactive melt processing (REx) in an internal mixer with peroxide content up to 1 wt.% (Figure 1a). During REx no relevant torque increase was detected, indicating that the selected amounts of peroxide did not lead to high crosslinking level, so that the processability of PCL was not hindered. The system was indeed designed to induce a chain extension, i.e., partially crosslinked structure, preserving PCL thermoplasticity. Herein, the purpose of water-assisted REx was to use water as a relatively low viscous phase to favor the PCL radical diffusion [27] and evaluate its catalytic function in radical reactions [29], consequently improving their efficiency.

At relatively high temperature the peroxide decomposes into benzoate (PO*) or phenyl (P*) radicals initiating a free radical mechanism (Figure 1b) [38]. In the water-assisted REx, the interaction of water with the PO* or P* should induce the formation of hydroxyl radical (HO*), as previously reported [29,30,39]. The radicals are then expected to propagate by proton abstraction from the α-carbon relative to the carbonyl group of PCL [17]. Finally, PCL macroradicals can recombine, resulting in branching/crosslinking, or undergo β-scission (Figure 1c) [24,25]. Water can also stabilize the radical structures, suppressing the scission reactions and enhance recombination thus increasing the overall reaction rate [29,30].

To find clear evidence on branching/crosslinking formed during REx, the produced materials have been structurally analyzed.

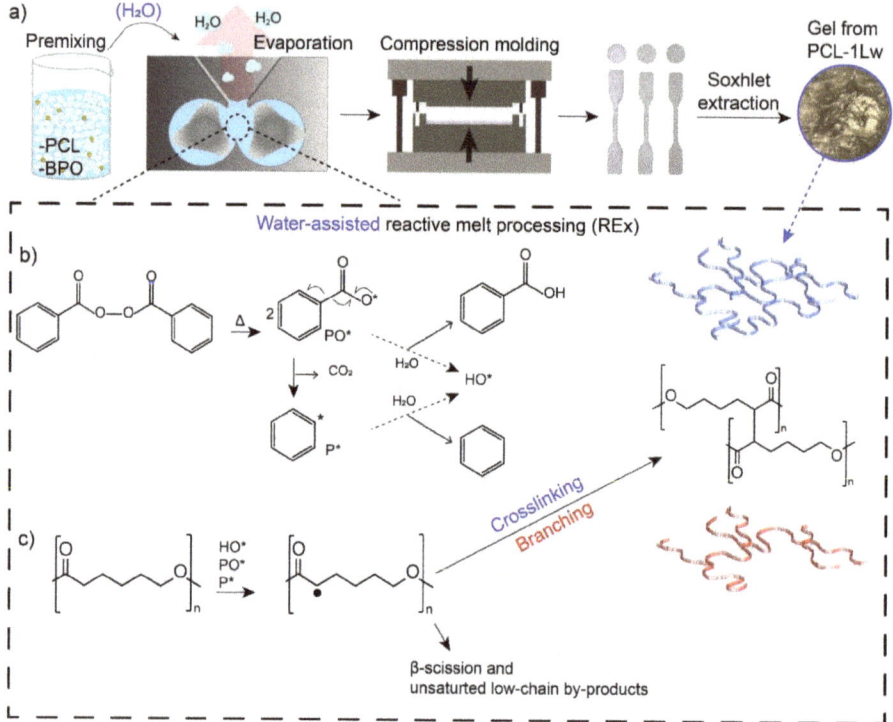

Figure 1. (a) Scheme of sequential steps: REx with or without water, compression molding of test specimens and Soxhlet extraction to recover gel fractions. The photo refers to insoluble fraction of PCL reacted with 1 wt.% peroxide during water-assisted REx. Schematics of (b) peroxide activation and reactions with water and (c) reaction products.

3.2. Structural Analysis

During REx, linear PCL gets first increasingly branched and finally converted into an insoluble gel network [23]. Therefore, its solubility can give an indication about the structure of the reacted PCL and to more extent about the presence of crosslinked chains. In this case, solubility tests have been performed in dichloromethane (DCM), a good solvent for the linear PCL [24]. At a first glance only in the PCL-1Lw dispersion a visible gel was floating, indicating that the radical reaction has taken place during the REx, while all the other samples appeared soluble (Figure S1).

In order to quantify the gel content of the materials, the insoluble fractions were Soxhlet extracted from DCM. It is assumed that the gel contains only the PCL over a critical molecular weight, i.e., 3D crosslinked network and highly branched chains. PCL processed with 1 wt.% peroxide revealed only 1% gel content, in contrast to what found previously by Han et al. [25] that obtained around 40% gel fraction using the same peroxide type and content. However, it is worth to note that they used a higher PCL molecular weight (80,000 g·mol^{-1}, from supplier's datasheet) compared to our study. In this regard, lower molecular weight chains, even if crosslinked, might still be more soluble or have less statistical chance of crosslinking [26], thus leading to lower gel content. This aspect was also evaluated in the work of Navarro et al. [40], illustrating how the gel content of PCL crosslinked by irradiation was strongly dependent on its initial molecular weight.

Instead, in the presence of water at the same peroxide content, a comparable gel fraction was measured (34%). This result confirmed a more efficient crosslinking of PCL structure during water-assisted REx. Similar results on the effect of water were reported for graft free radical copolymerization in a different polymer system [28].

The designed chain extension can be verified by size exclusion chromatography on the soluble fractions. The SEC curves show a monomodal distribution (Figure 2) and are broadened after REx. The reaction of polyesters with peroxide can lead to two reaction products: branched/crosslinked chains and β-scission [24,25]. REx with 0.5 wt.% of peroxide resulted in higher weight-average molecular weight ($\overline{M_w}$) and dispersity (Đ), compared to PCL (Table 1), highlighting a successful branching/crosslinking reaction. However, from the curve an increased population of low molecular weight (M_w) chains is visible, resulting from the β-scission, and contributing to the increased dispersity. The presence of water results in a further increased $\overline{M_w}$ and dispersity, but slightly lower amount of low M_w chains. Water strongly improves the reaction efficiency, leading to higher branching/crosslinking, compared to the competing β-scission. This is consistent with the gel content results (at 1 wt.% peroxide). It is worth to note that large presence of water (>50 wt.%) in PCL melt processing does not significantly decrease its molecular weight due to water-induced hydrolysis [41].

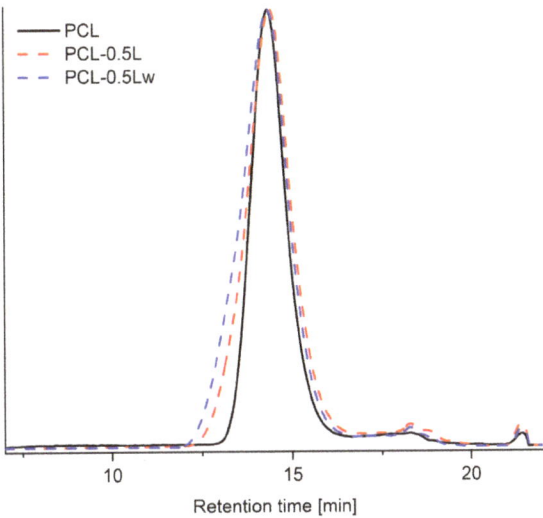

Figure 2. SEC chromatograms of fractions completely soluble in chloroform of PCL and PCL reacted with 0.5 wt.% peroxide during dry (PCL-0.5L) or water-assisted (PCL-0.5Lw) REx.

Table 1. Number ($\overline{M_n}$) and weight ($\overline{M_w}$) average molecular weights rounded off and dispersity (Đ) detected from SEC of fractions completely soluble in chloroform of PCL and PCL reacted with 0.5 wt.% peroxide during dry (PCL-0.5L) or water-assisted (PCL-0.5Lw) REx.

Material	$\overline{M_n}$ (g·mol^{-1})	$\overline{M_w}$ (g·mol^{-1})	Đ
PCL	77,000	150,000	2
PCL-0.5L	74,000	220,000	2.92
PCL-0.5Lw	86,000	290,000	3.37

3.3. Thermal Properties

Reduced thermal stability and a decreased onset of degradation are generally indicative of any degradative phenomena or chain scission and can be assessed by thermogravimetric analysis [42].

The thermograms show a reduced degradation onset of PCL, that is particularly evident at low amount of peroxide (<0.5 wt.%) (Table 2 and Table S1). This result suggests that PCL chain scission is prevalent at lower amount of peroxide (<0.5 wt.%), while branching/crosslinking increases at larger amounts (0.5 and 1 wt.%) [42]. However, the degradation behavior and temperature of PCL have no significant variation after REx with peroxide (Figure S2 and Figure 3).

Structural changes can be reflected on the PCL transition temperatures and crystallinity, so the materials were analyzed by DSC and XRD.

Table 2. Thermal properties of neat and reacted PCL detected by TGA, DSC and XRD analyses. Onset temperatures of degradation ($T_{5\%}$) evaluated at 5% weight losses in TGA. Crystallization temperatures (T_c) detected from the DSC cooling scans. Crystallinity (χ_{XRD}) and crystal size in the direction perpendicular to (110) lattice plane (D_{110}) calculated from XRD diffractograms.

Material	$T_{5\%}$ (°C)	T_c (°C)	χ_{XRD} (%)	D_{110} (Å)
PCL	363	35.1	38.2	223
PCL-0.5L	331	36.7	-	-
PCL-0.5Lw	325	36.4	-	-
PCL-1L	345	35.5–40.2	41.4	260
PCL-1Lw	347	41.2	42.8	291

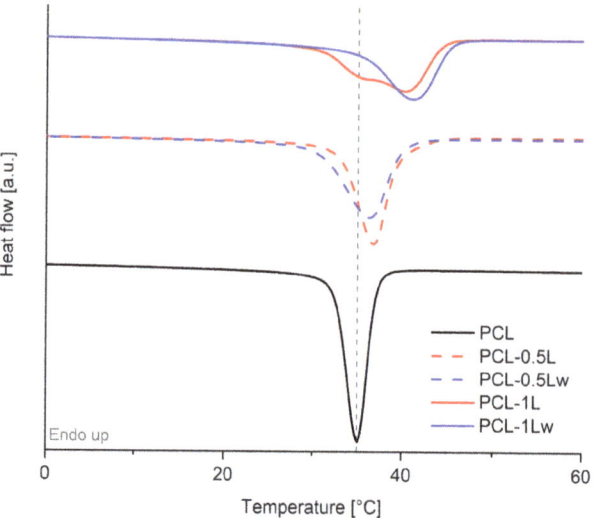

Figure 3. DSC cooling scan of neat and reacted PCL.

The T_m of the materials is not affected by REx (Figure S4) [43] while noticeable differences are observed in the DSC cooling scans in which a shift of the T_c to higher values is recorded with increasing the peroxide amount (1 wt.%) (Figure 3 and Table 2) [24,25,44]. This effect can be ascribed to low M_w fractions that facilitate the crystallization process due to their higher mobility and easier folding [43,45]. However, it is worth to note that our results can also be explained by a heterogeneous nucleation effect of the branching/crosslinking points, already observed in similar systems [46,47]. Moreover, the crystallization peaks are broadened with increasing amount of peroxide. According to SEC analysis, PCL dispersity increased after REx due to the formation of fractions both at higher and lower M_w. The increased dispersity can result in more heterogeneous spherulite species, leading to broader crystallization [45,48]. In presence of water the crystallization process is completed at higher temperatures and the increase of the T_c is more significant, in agreement with its higher gel content measured.

For a better understanding of PCL crystalline features after REx, XRD spectra have been collected on representative samples at higher gel content (1 wt.% peroxide). The diffraction spectra show two main crystalline peaks around 32.1° and 35.6°, previously assigned to (110) and (200) crystallographic lattices [36], which are not shifted in the reacted PCL (Figure S5). The ratio between the area of crystalline peaks and the total area of the spectrum provides indications on the crystallinity degree. Compared to neat PCL, the calculated crystallinity degree slightly increases by ≈ 10% after reaction with 1 wt.% peroxide (Table 2). Increase in crystallinity has been similarly reported in other crosslinked systems [40,47,49], explained by a nucleating effect of the crosslinked network. Additionally, to verify if REx led to different crystalline structures, Scherrer equation was used to calculate the crystallite size. The results indicate an increased crystallite size in the direction normal to the 110 lattice after REx, conceivably due to steric effects induced by PCL branching points. Both crystallinity degree and crystallite size are further enhanced by the presence of water, in line with the larger branching/crosslinking achieved. Mishra et al. [45,50] also recorded similar increases in PCL crystallinity and crystallite size in crosslinked PCL blends by XRD analysis. Nevertheless, hereafter reported bulk properties of modified PCLs cannot be inferred by the small changes observed in their crystallinity. To correlate PCL structural and thermal properties to its performance, dynamic mechanical and tensile behaviors were investigated. Moreover, DMTA can provide more accurate information than DSC on the transition temperatures, *e.g.*, being the glass transition a first order transition in the loss moduli.

3.4. Mechanical Properties

The curves (Figure 4) from DMTA in tensile mode indicate that only one transition occurs for neat and reacted PCL. Both the glass (T_g) and alpha (T_α) transition temperatures (Table 3) show a slight shift towards higher temperatures with increasing amount of peroxide.

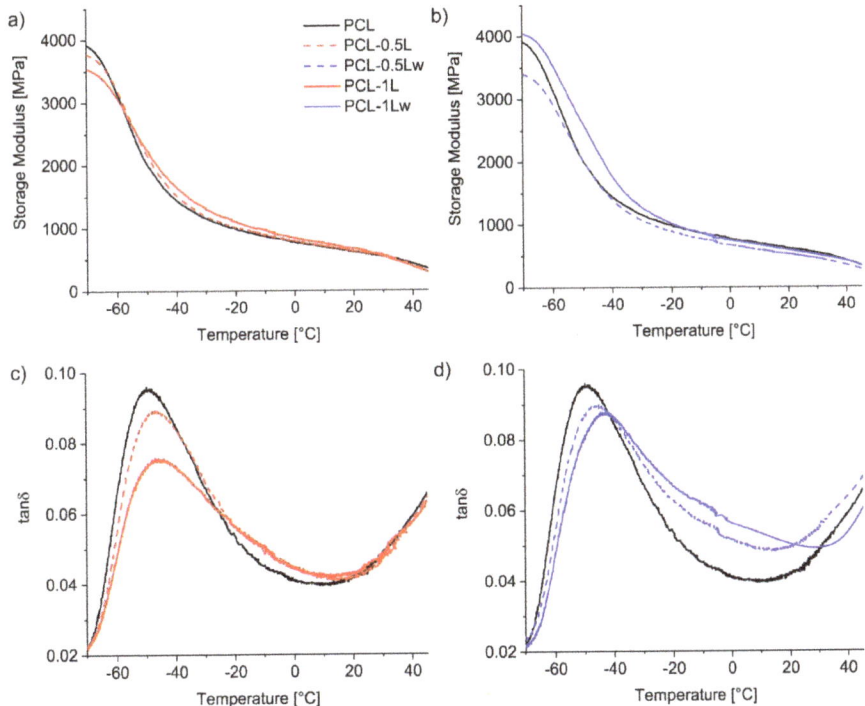

Figure 4. Representative curves of storage moduli (**a**), (**b**) and tanδ (**c**),(**d**) recorded from DMTA temperature sweep of PCL and reacted PCL during dry (**a**),(**c**) and water-assisted (**b**), (**d**) REx.

Table 3. DMTA main results of neat and reacted PCL recorded by DMTA Glass transition temperature (T_g) as temperature of loss modulus peak maximum. Alpha transition temperature (T_α) as temperature of tanδ peak maximum. Damping factor (DF) as the maximum of tanδ. Storage modulus (E') values at −70 and 20 °C. Scattering of the data below 3%.

Material	T_g (°C)	T_α (°C)	DF	$E'_{-70\,°C}$ (MPa)	$E'_{20\,°C}$ (MPa)
PCL	−56.2	−49.4	0.095	3920	613
PCL-0.5L	−54.8	−46.6	0.089	3770	621
PCL-0.5Lw	−54.4	−45.7	0.090	3405	528
PCL-1L	−54.0	−45.7	0.075	3544	669
PCL-1Lw	−52.1	−42.9	0.088	4051	591

The delay in these transitions further demonstrates the formation of branching/crosslinking during REx, which hinder the macromolecular mobility [51]. Indeed, the largest shift in transition temperatures is reported for PCL-1Lw, according to its higher gel content achieved. Only for this material, the storage modulus improves for a large range of temperatures (from −70 to −20 °C) (Figure 4). For all the other materials, no reinforcement is shown due to the low gel content and the plasticizing effect of the low M_w chains.

It is worth noting that limited reinforcement of crosslinks is expected after the glass transition, as previously observed in similar systems [24]. However, PCL damping factor decreases with increasing peroxide amount, indicating a more elastic behavior of the polymer after REx. The limited reinforcement above the T_g observed from DMTA is also reflected in the tensile properties of the materials recorded at room temperature (Figure 5 and Table 4). The Young's modulus and yield stress of PCL are slightly lowered after REx and further decreased when the reaction was carried out in the presence of water, while PCL elongation and strength are preserved up to 0.5 wt.% in peroxide (Figure S6 and Table S2). Water-assisted REx with 1 wt.% led to a ≈ 15% increase of ultimate tensile strength compared to unmodified PCL, resulting from increased strain hardening, and a slight improvement of elongation compared to PCL-1L. Similar reduction of PCL tensile properties after crosslinking has been previously reported [25], with only exception of a small increase in PCL strength, as also achieved in this work.

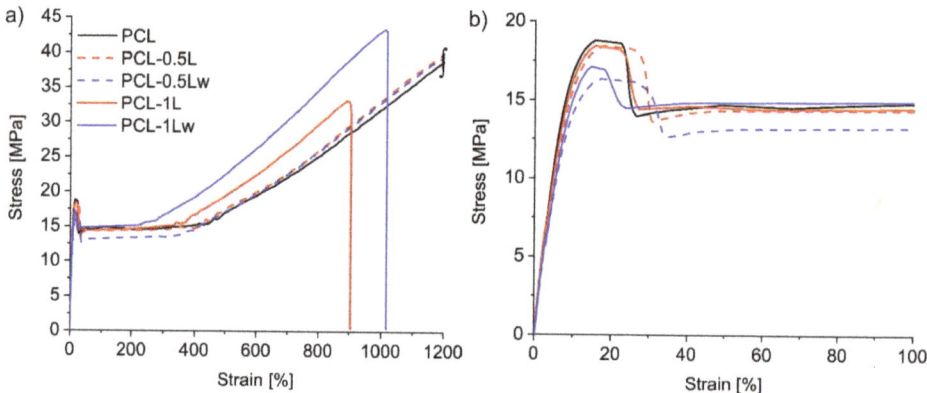

Figure 5. (a) Stress-strain curves from tensile tests at room temperature of neat and reacted PCL and (b) their magnification on the yield region.

Table 4. Mechanical properties assessed from tensile tests at room temperature. Each value represents the average of 5 measurements with the standard deviation. * Data extracted at the upper limit of the instrument; break did not occur.

Material	Young's Modulus (MPa)	Yield Stress (MPa)	Ultimate Tensile Strength (MPa)	Elongation at Break (%)
PCL	284 ± 5	19.2 ± 0.5	> 38.7 *	> 1200 *
PCL-0.5L	248 ± 6	18.4 ± 0.1	> 38.8 *	> 1175 *
PCL-0.5Lw	218 ± 8	16.0 ± 0.4	> 39.8 *	> 1200 *
PCL-1L	242 ± 13	17.3 ± 1	33.6 ± 1.3	941 ± 57
PCL-1Lw	225 ± 14	17.1 ± 0.5	44 ± 1.2	1035 ± 48

After analyzing the thermomechanical properties below the glass transition and at room temperature, parallel plate rheology was used to explore the behavior of the material in the melt state, crucial for investigating changes in PCL melt processability. Moreover, rheological analysis allows further understanding of the macromolecular structure.

3.5. Dynamic Rheology

From the isothermal frequency sweeps at 120 °C (Figure 6a), a substantial improvement of the shear storage modulus of PCL up to four orders of magnitude is observed with increasing crosslinking level and progressively reduced slopes correspond to higher moduli [45]. Even if less pronounced, the loss modulus also increased as a consequence of the increased $\overline{M_w}$ [23]. At higher level of peroxide, PCL melt presented a transition from viscous to elastic behavior. A crossover point around 6 rad·s^{-1} of frequency is observed for PCL-1L, while a complete inversion is recorded for PCL-1Lw. The prevalent elastic melt behavior in the entire frequencies range is inferred by increased chain entanglement for the high molecular weight crosslinked PCL [44].

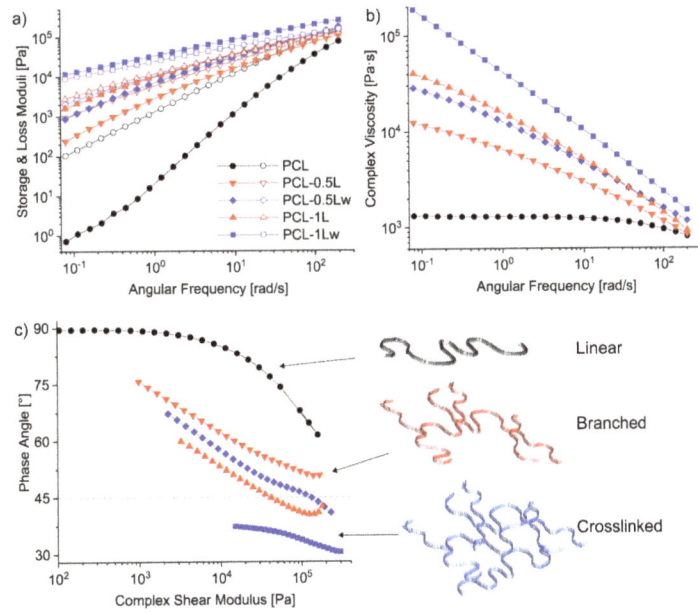

Figure 6. Dynamic rheological behavior in frequency sweeps at 120 °C of neat and reacted PCL: (a) storage (full symbols) and loss (empty symbols) moduli; (b) complex viscosity; (c) Van Gurp-Palmen plot showing phase angles as a function of complex modulus, phase angles below 45° indicate predominant elasticity.

Viscous to elastic transition has been highlighted in the Van Gurp-Palmen plot (Figure 6c), where the recorded below 45° phase angles point out a predominant elasticity of the melt [5]. The Van Gurp-Palmen plot of PCL-1Lw shows the lowest phase angles (<45°), confirming the predominant elastic behavior of the higher crosslinked PCL melt, as illustrated in the figure. Instead, the phase angles recorded for the unmodified PCL are close to 90°, indicating a typical flow behavior of viscoelastic fluid with highly linear structure and low dispersity [47,52]. Prevalent branched structure of PCL is assumed for intermediate gel content and phase angles between 70 and 45° (Figure 6c). The commented rheological transition liquid to solid-like is also reflected in a dramatic increase in viscosity. The complex viscosity of PCL shows a typical Newtonian plateau followed by a minimal shear thinning region at higher frequencies (Figure 6b). The complex viscosity of the modified PCL shows a non-Newtonian behavior, with pronounced shear thinning that increases with the crosslinking level [8]. Accordingly, the shear thinning for water-assisted REx materials it is the more pronounced. Up to two orders of magnitude increase is measured for complex viscosity of the modified PCL at low frequencies, in agreement with the increased $\overline{M_w}$ and higher entanglement of the branched structure [8,49]. At higher frequencies the impact of peroxide-initiated reactions is lower as the viscosities converge (at frequencies around 200 rad·s^{-1}). It is worth to note the substantial effect of water can be already observed at lower level of peroxide. PCL-0.5Lw showed higher resistance against high shear forces than the neat and modified PCL via dry REx.

Accordingly, increase in melt viscosity and shear moduli suggests that the melt strength of PCL can be controlled by the only reaction with peroxide, thus improving its processability [44].

4. Conclusions

Despite the extensive use of peroxide as crosslinking agent, this study showed how water-assisted REx can be employed with low levels of peroxide to control PCL crosslinking and improve its melt elasticity while preserving its thermoplasticity. Water was chosen as catalyst for the designed Rex under the hypothesis that a low viscosity medium can boost the radical reaction by increasing the radical diffusion. The results showed that the presence of water increased PCL molecular weight and gel content compared to the dry process, from 1% to 34% with 1 wt.% peroxide, confirming a more efficient radical propagation in water-assisted REx. Differential scanning calorimetry showed increased crystallization temperatures and easier crystallization process of reacted PCL, compared to neat PCL. From the dynamic mechanical analysis, higher branching/crosslinking slightly increased the transition temperatures and led to a reinforcement effect in the temperature range below the glass transition. After the glass transition, the mechanical reinforcement was limited, both in dynamic mechanical and tensile properties, reflecting the plasticizing effect of low molecular weight chains formed by PCL β-scission during peroxide-initiated reactions. Instead, in the melt state the effect of branching/crosslinking was more evident as shown by melt rheology. The rheological behavior of crosslinked PCL showed a transition from the typical viscous-like to solid-like. The shear storage moduli were increased by the reactive melt processing, confirming the desired improvement of PCL rheological properties. Increasing the gel content, the viscosity of reacted PCL increased at low frequencies, while higher shear thinning behavior was shown in the entire frequency range. At higher frequencies, approaching typical melt processing conditions, viscosity of crosslinked PCL converged to the neat biopolyesters values, underlying that a controlled water-assisted REx can improve PCL rheological properties without affecting its processability. This work represents a relevant reference for future controlled water-assisted reactive melt processing of polymers and composites, in which water could have the role of feeding medium, e.g., for polysaccharides suspension, but also useful to boost peroxide-initiated reactions.

Supplementary Materials: The following are available online at https://www.mdpi.com/2073-4360/13/4/491/s1, Figure S1: Photo of test tubes containing neat poly(ε-caprolactone) (PCL) and PCL reacted with 1 wt.% peroxide in the presence of water (PCL-1Lw), after 24 h in solution with dichloromethane, Table S1: Thermal properties of neat and reacted PCL detected by Thermogravimetrical analysis (TGA) and Differential Scanning Calorimetry (DSC). TGA: Onset temperatures of degradation ($T_{5\%}$) evaluated at 5% weight loss; degradation temperature (T_d) evaluated as the peak temperature of DTG; char residue at 550 °C. DSC: Glass (T_g) and melting (T_m) temperatures detected from the second heating scan; melting enthalpy (ΔH_M) and crystallinity (χ_{DSC}) detected from the melting peak in the second heating scan (from 15 to 65 °C); crystallization temperature (T_c) detected from the cooling scans, Figure S2: Curves from thermogravimetric analysis (TGA) of neat and reacted PCL with different wt.% of peroxide during dry (PCL-xL) or water-assisted (PCL-xLw) melt processing, Figure S3: Derivative thermogravimetric (DTG) curves of neat and reacted PCL, Figure S4: Thermograms from second heating scan of Differential Scanning Calorimetry (DSC) of neat and reacted PCL, Figure S5: X-Ray diffractograms of PCL and PCL reacted with 1 wt.% peroxide during dry (PCL-1L) or water-assisted (PCL-1Lw) melt processing, Figure S6: Stress-strain curves from tensile tests at room temperature neat and reacted PCL, Table S2: Mechanical properties assessed from tensile tests at room temperature. Each value represents the average of 5 measurements with the standard deviation. *Data extracted at the upper limit of the instrument.

Author Contributions: All authors have equally contributed, read, and agreed to the published version of the manuscript.

Funding: This research was funded by Knut och Alice Wallenbergs Stiftelse Biokompositforskning, grant number Dnr. KAW V-2019-0041 and Genie from Chalmers Tekniska Högskola.

Institutional Review Board Statement: Not applicable.

Informed Consent Statement: Not applicable.

Data Availability Statement: The data presented in this study are available on request from the corresponding author.

Acknowledgments: G.L.R. acknowledges the Chalmers Areas of Advance.

Conflicts of Interest: The authors declare no conflict of interest. The funders had no role in the design of the study; in the collection, analyses, or interpretation of data; in the writing of the manuscript, or in the decision to publish the results.

References

1. Geyer, R.; Jambeck, J.R.; Law, K.L. Production, use, and fate of all plastics ever made. *Sci. Adv.* **2017**, *3*, e1700782. [CrossRef]
2. Rydz, J.; Sikorska, W.; Kyulavska, M.; Christova, D. Polyester-based (bio)degradable polymers as environmentally friendly materials for sustainable development. *Int. J. Mol. Sci.* **2015**, *16*, 564–596. [CrossRef]
3. Raquez, J.M.; Narayan, R.; Dubois, P. Recent advances in reactive extrusion processing of biodegradable polymer-based compositions. *Macromol. Mater. Eng.* **2008**, *293*, 447–470. [CrossRef]
4. Kaldéus, T.; Träger, A.; Berglund, L.A.; Malmström, E.; Re, G.L. Molecular engineering of the cellulose-poly(caprolactone) bio-nanocomposite interface by reactive amphiphilic copolymer nanoparticles. *ACS Nano* **2019**, *13*, 6409–6420. [CrossRef] [PubMed]
5. Lo Re, G.; Engström, J.; Wu, Q.; Malmström, E.; Gedde, U.W.; Olsson, R.T.; Berglund, L. Improved cellulose nanofibril dispersion in melt-processed polycaprolactone nanocomposites by a latex-mediated interphase and wet feeding as LDPE alternative. *ACS Appl. Nano Mater.* **2018**, *1*, 2669–2677. [CrossRef]
6. Woodruff, M.A.; Hutmacher, D.W. The return of a forgotten polymer—Polycaprolactone in the 21st century. *Prog. Polym. Sci.* **2010**, *35*, 1217–1256. [CrossRef]
7. Darwis, D.; Nishimura, K.; Mitomo, H.; Yoshii, F. Improvement of processability of Poly(ε-caprolactone) by radiation techniques. *J. Appl. Polym. Sci.* **1999**, *74*, 1815–1820. [CrossRef]
8. Wei, L.; McDonald, A.G. Peroxide induced cross-linking by reactive melt processing of two biopolyesters: Poly(3-hydroxybutyrate) and poly(l-lactic acid) to improve their melting processability. *J. Appl. Polym. Sci.* **2015**, *132*, 1–15. [CrossRef]
9. Di Maio, E.; Iannace, S.; Marrazzo, C.; Narkis, M.; Nicolais, L. Effect of molecular modification on PCL foam formation and morphology of PCL. *Macromol. Symp.* **2005**, *228*, 219–228. [CrossRef]
10. Niaounakis, M. Introduction. In *Biopolymers: Processing and Products*; William Andrew: Kidlington, UK, 2015; pp. 1–77, ISBN 9780323266987.
11. Fortelny, I.; Ujcic, A.; Fambri, L.; Slouf, M. Phase structure, compatibility, and toughness of PLA/PCL blends: A review. *Front. Mater.* **2019**, *6*, 206. [CrossRef]
12. Muthuraj, R.; Misra, M.; Mohanty, A.K. Biodegradable compatibilized polymer blends for packaging applications: A literature review. *J. Appl. Polym. Sci.* **2018**, *135*. [CrossRef]

13. Przybysz-Romatowska, M.; Haponiuk, J.; Formela, K. Reactive extrusion of biodegradable aliphatic polyesters in the presence of free-radical-initiators: A review. *Polym. Degrad. Stab.* **2020**, *182*, 109383. [CrossRef]
14. Koenig, M.F.; Huang, S.J. Evaluation of crosslinked poly(caprolactone) as a biodegradable, hydrophobic coating. *Polym. Degrad. Stab.* **1994**, *45*, 139–144. [CrossRef]
15. Jarrett, P.; Benedict, C.V.; Bell, J.P.; Cameron, J.A.; Huang, S.J. Mechanism of the biodegradation of polycaprolactone. In *Polymers as Biomaterials*; Shalaby, S.W., Hoffman, A.S., Ratner, B.D., Horbett, T.A., Eds.; Springer: Boston, MA, USA, 1984; pp. 181–192, ISBN 978-1-4613-2433-1.
16. Yoshii, F.; Darwis, D.; Mitomo, H.; Makuuchi, K. Crosslinking of poly(ε-caprolactone) by radiation technique and its biodegradability. *Radiat. Phys. Chem.* **2000**, *57*, 417–420. [CrossRef]
17. Kim, C.H.; Cho, K.Y.; Park, J.K. Grafting of glycidyl methacrylate onto polycaprolactone: Preparation and characterization. *Polymer* **2001**, *42*, 5135–5142. [CrossRef]
18. John, J.; Tang, J.; Yang, Z.; Bhattacharya, M. Synthesis and characterization of anhydride-functional polycaprolactone. *J. Polym. Sci. Part. A Polym. Chem.* **1997**, *35*, 1139–1148. [CrossRef]
19. Semba, T.; Kitagawa, K.; Ishiaku, U.S.; Hamada, H. The effect of crosslinking on the mechanical properties of polylactic acid/polycaprolactone blends. *J. Appl. Polym. Sci.* **2006**, *101*, 1816–1825. [CrossRef]
20. Przybysz, M.; Marć, M.; Klein, M.; Saeb, M.R.; Formela, K. Structural, mechanical and thermal behavior assessments of PCL/PHB blends reactively compatibilized with organic peroxides. *Polym. Test.* **2018**, *67*, 513–521. [CrossRef]
21. Garcia-Garcia, D.; Rayón, E.; Carbonell-Verdu, A.; Lopez-Martinez, J.; Balart, R. Improvement of the compatibility between poly(3-hydroxybutyrate) and poly(ε-caprolactone) by reactive extrusion with dicumyl peroxide. *Eur. Polym. J.* **2017**, *86*, 41–57. [CrossRef]
22. Sedov, I.; Magsumov, T.; Abdullin, A.; Yarko, E.; Mukhametzyanov, T.; Klimovitsky, A.; Schick, C. Influence of the cross-link density on the rate of crystallization of poly(ε-Caprolactone). *Polymers* **2018**, *10*, 902. [CrossRef] [PubMed]
23. Gandhi, K.; Kriz, D.; Salovey, R.; Narkis, M.; Wallerstein, R. Crosslinking of polycaprolactone in the pre-gelation region. *Polym. Eng. Sci.* **1988**, *28*, 1484–1490. [CrossRef]
24. Przybysz, M.; Hejna, A.; Haponiuk, J.; Formela, K. Structural and thermo-mechanical properties of poly(ε-caprolactone) modified by various peroxide initiators. *Polymers* **2019**, *11*, 1101. [CrossRef]
25. Han, C.; Ran, X.; Su, X.; Zhang, K.; Liu, N. Effect of peroxide crosslinking on thermal and mechanical properties of poly(ε-caprolactone). *Polym. Int.* **2007**, *600*, 593–600. [CrossRef]
26. Tolinski, M. Crosslinking. In *Additives for Polyolefins*; William Andrew: Kidlington, UK, 2009; pp. 215–220.
27. Denisov, E.T.; Denisov, E.T. Effect of solvent on free radical reactions. In *Liquid-Phase Reaction Rate Constants*; Springer: Boston, MA, USA, 1974; pp. 427–441.
28. Kaur, I.; Gautam, N.; Khanna, N.D. Modification of polypropylene through intercrosslinking graft copolymerization of poly(vinyl alcohol): Synthesis and characterization. *J. Appl. Polym. Sci.* **2008**, *107*, 2238–2245. [CrossRef]
29. Vöhringer-Martinez, E.; Hansmann, B.; Hernandez, H.; Francisco, J.S.; Troe, J.; Abel, B. Water catalysis of a radical-molecule gas-phase reaction. *Science* **2007**, *315*, 497–501. [CrossRef] [PubMed]
30. Troe, J. The Polanyi lecture. The colourful world of complex-forming bimolecular reactions. *J. Chem. Soc. Faraday Trans.* **1994**, *90*, 2303–2317. [CrossRef]
31. D'Haene, P.; Remsen, E.E.; Asrar, J. Preparation and characterization of a branched bacterial polyester. *Macromolecules* **1999**, *32*, 5229–5235. [CrossRef]
32. Dean, K.M.; Petinakis, E.; Meure, S.; Yu, L.; Chryss, A. Melt strength and rheological properties of biodegradable poly(lactic aacid) modified via alkyl radical-based reactive extrusion processes. *J. Polym. Environ.* **2012**, *20*, 741–747. [CrossRef]
33. Shen, Y.; Zhu, W.; Papadaki, M.; Mannan, M.S.; Mashuga, C.V.; Cheng, Z. Thermal decomposition of solid benzoyl peroxide using advanced reactive system screening tool: Effect of concentration, confinement and selected acids and bases. *J. Loss Prev. Process. Ind.* **2019**, *60*, 28–34. [CrossRef]
34. Cicogna, F.; Coiai, S.; Rizzarelli, P.; Carroccio, S.; Gambarotti, C.; Domenichelli, I.; Yang, C.; Dintcheva, N.T.; Filippone, G.; Pinzino, C.; et al. Functionalization of aliphatic polyesters by nitroxide radical coupling. *Polym. Chem.* **2014**, *5*, 5656–5667. [CrossRef]
35. Oliveira, J.E.; Mattoso, L.H.C.; Orts, W.J.; Medeiros, E.S. Structural and morphological characterization of micro and nanofibers produced by electrospinning and solution blow spinning: A comparative study. *Adv. Mater. Sci. Eng.* **2013**, *2013*. [CrossRef]
36. Castilla-Cortázar, I.; Vidaurre, A.; Marí, B.; Campillo-Fernández, A.J. Morphology, crystallinity, and molecular weight of poly(ε-caprolactone)/graphene oxide hybrids. *Polymers* **2019**, *11*, 1099. [CrossRef]
37. Magnani, C.; Idström, A.; Nordstierna, L.; Müller, A.J.; Dubois, P.; Raquez, J.M.; Lo Re, G. Interphase design of cellulose nanocrystals/poly(hydroxybutyrate- ran-valerate) bionanocomposites for mechanical and thermal properties tuning. *Biomacromolecules* **2020**, *21*, 1892–1901. [CrossRef] [PubMed]
38. Chellquist, E.M.; Gorman, W.G. Benzoyl peroxide solubility and stability in hydric solvents. *Pharm. Res. Off. J. Am. Assoc. Pharm. Sci.* **1992**, *9*, 1341–1346. [CrossRef]
39. Mardyukov, A.; Sanchez-Garcia, E.; Crespo-Otero, R.; Sander, W. Interaction and reaction of the phenyl radical with water: A source of OH radicals. *Angew. Chem. Int. Ed.* **2009**, *48*, 4804–4807. [CrossRef]
40. Navarro, R.; Burillo, G.; Adem, E.; Marcos-Fernández, A. Effect of ionizing radiation on the chemical structure and the physical properties of polycaprolactones of different molecular weight. *Polymers* **2018**, *10*, 397. [CrossRef] [PubMed]

41. Lo Re, G.; Spinella, S.; Boujemaoui, A.; Vilaseca, F.; Larsson, P.T.; Adås, F.; Berglund, L.A.; Larsson, P.T.; Ada, F.; Berglund, L.A. Poly(ε-caprolactone) biocomposites based on acetylated cellulose fibers and wet compounding for improved mechanical performance. *ACS Sustain. Chem. Eng.* **2018**, *6*, 6753–6760. [CrossRef]
42. Carlson, D.; Dubois, P.; Nie, L.; Narayan, R. Free radical branching of polylactide by reactive extrusion. *Polym. Eng. Sci.* **1998**, *38*, 311–321. [CrossRef]
43. Skoglund, P.; Fransson, A. Continuous cooling and isothermal crystallization of polycaprolactone. *J. Appl. Polym. Sci.* **1996**, *61*, 2455–2465. [CrossRef]
44. Kim, D.J.; Kim, W.S.; Lee, D.H.; Min, K.E.; Park, L.S.; Kang, I.K.; Jeon, I.R.; Seo, K.H. Modification of poly(butylene succinate) with peroxide: Crosslinking, physical and thermal properties, and biodegradation. *J. Appl. Polym. Sci.* **2001**, *81*, 1115–1124. [CrossRef]
45. Mishra, J.K.; Chang, Y.W.; Kim, W. The effect of peroxide crosslinking on thermal, mechanical, and rheological properties of polycaprolactone/epoxidized natural rubber blends. *Polym. Bull.* **2011**, *66*, 673–681. [CrossRef]
46. Takamura, M.; Nakamura, T.; Kawaguchi, S.; Takahashi, T.; Koyama, K. Molecular characterization and crystallization behavior of peroxide-induced slightly crosslinked poly(L-lactide) during extrusion. *Polym. J.* **2010**, *42*, 600–608. [CrossRef]
47. Kolahchi, A.R.; Kontopoulou, M. Chain extended poly(3-hydroxybutyrate) with improved rheological properties and thermal stability, through reactive modification in the melt state. *Polym. Degrad. Stab.* **2015**, *121*, 222–229. [CrossRef]
48. Signori, F.; Boggioni, A.; Righetti, M.C.; Rondán, C.E.; Bronco, S.; Ciardelli, F. Evidences of transesterification, chain branching and cross-linking in a biopolyester commercial blend upon reaction with dicumyl peroxide in the melt. *Macromol. Mater. Eng.* **2015**, *300*, 153–160. [CrossRef]
49. Ai, X.; Wang, D.; Li, X.; Pan, H.; Kong, J.; Yang, H.; Zhang, H.; Dong, L. The properties of chemical cross-linked poly(lactic acid) by bis(tert-butyl dioxy isopropyl) benzene. *Polym. Bull.* **2019**, *76*, 575–594. [CrossRef]
50. Mishra, J.K.; Chang, Y.W.; Kim, D.K. Green thermoplastic elastomer based on polycaprolactone/epoxidized natural rubber blend as a heat shrinkable material. *Mater. Lett.* **2007**, *61*, 3551–3554. [CrossRef]
51. Suhartini, M.; Mitomo, H.; Nagasawa, N.; Yoshii, F.; Kume, T. Radiation crosslinking of poly(butylene succinate) in the presence of low concentrations of trimethallyl isocyanurate and its properties. *J. Appl. Polym. Sci.* **2003**, *88*, 2238–2246. [CrossRef]
52. Trinkle, S.; Freidrich, C. Van Gurp-Palmen-plot: A way to characterize polydispersity of linear polymers. *Rheol. Acta* **2001**, *40*, 322–328. [CrossRef]

Article

Shear and Extensional Rheology of Linear and Branched Polybutylene Succinate Blends

Violette Bourg [1,2,3,4,*], **Rudy Valette** [5], **Nicolas Le Moigne** [1], **Patrick Ienny** [2], **Valérie Guillard** [3] **and Anne Bergeret** [1]

1. Polymers Composites and Hybrids (PCH)-IMT Mines Ales, 30100 Ales, France; nicolas.le-moigne@mines-ales.fr (N.L.M.); anne.bergeret@mines-ales.fr (A.B.)
2. LMGC, IMT Mines Ales, University of Montpellier, CNRS, 30100 Ales, France; patrick.ienny@mines-ales.fr
3. Ingénierie des Agropolymères et Technologie Emergentes—IATE, Univ Montpellier, INRAE, Institut Agro, 34000 Montpellier, France; valerie.guillard@umontpellier.fr
4. Nestlé Research SA, Nestlé Institute of Packaging Sciences, Route du Jorat, 57, 1000 Lausanne, Switzerland
5. MINES ParisTech, CEMEF-Centre de Mise en Forme des Matériaux, CNRS UMR 7635, PSL Research University, CS 10207 rue Claude Daunesse, CEDEX, 06904 Sophia Antipolis, France; rudy.valette@mines-paristech.fr
* Correspondence: violette.bourg@rd.nestle.com

Abstract: The molecular architecture and rheological behavior of linear and branched polybutylene succinate blends have been investigated using size-exclusion chromatography, small-amplitude oscillatory shear and extensional rheometry, in view of their processing using cast and blown extrusion. Dynamic viscoelastic properties indicate that a higher branched polybutylene succinate amount in the blend increases the relaxation time due to an increased long-chain branching degree. Branched polybutylene succinate exhibits pronounced strain hardening under uniaxial elongation, which is known to improve processability. Under extensional flow, the 50/50 wt % blend exhibits the same behavior as linear polybutylene succinate.

Keywords: molecular architecture; long-chain branching; polybutylene succinate; biodegradable; size-exclusion chromatography; shear rheology; extensional rheology

1. Introduction

Nowadays, polymeric packaging materials are mostly based on non-renewable resources. However, due to growth of environmental issues, it becomes more and more obvious that alternatives must be considered to replace these conventional petroleum-based polymers by either biodegradable (or compostable) and/or bio-based polymers. For decades, the mostly used polymer matrices for packaging purposes have belonged to the polyolefin family (polyethylene (PE), polypropylene (PP)). The reasons for these polyolefins to hold such a monopoly are quite clear: they exhibit a very good processability due to their large operating window and thermal stability, they are relatively cheap (around 1–2 euros per kilogram depending on the properties), and a tremendous choice of tailored properties is available (e.g., improved thermal stability, various molecular features with controlled rheological properties . . .).

The blow and cast film extrusion processes used to produce thin films for packaging purposes require high melt strength/elasticity and high viscosity polymer in order to avoid processing instabilities such as bubble instabilities in the case of blown-film extrusion [1,2]. Long-chain branched (LCB) structures have already demonstrated their efficiency in improving the melt strength compared to the linear structures. Indeed, LCB are known to exhibit strain hardening under elongational flow, which is required to obtain a homogeneous deformation during the stretching stage in the blown-film extrusion [3,4]. Beyond their impact on the rheological performances, LCB are also known to affect the crystallization and phase behavior in the solid state, leading to modified mechanical properties compared to their linear homolog.

Many studies on the influence of long-chain branching on the rheological properties of conventional polyolefins such as PE [1,3,5–11] or PP [4,12–14] have been carried out. In general, the authors agree on a definition of LCB (from a rheological point of view) as being branches whose weight is higher than the critical mass for entanglements, M_c. The studies reveal that LCB with branches longer than twice M_c have a significant impact on the processing properties [15] and have a strong effect on the viscosity, the elasticity, and the activation energy under small amplitude oscillatory shear flow. In addition, the effect of polydispersity on these parameters has been reported to be very similar to the one expected from LCB, leading to difficulties in separating the effects of different molecular features. Even if such relationships between molecular structures and resulting rheological properties have been extensively investigated, the lack of controlled molecular architectures in terms of degree of branching, branch length, and structure have seriously limited the type of polymers studied, being mostly restricted to some model polyolefins. Nonetheless, some studies have been performed on LCB biodegradable or bio-based polymer matrices. Very recently, Nouri et al. studied the impact of different types of branching on the rheological behavior under the extensional flow of poly (lactic acid) (PLA) [16,17]. They showed that branched PLAs exhibit large strain hardening, but they did not conclude on the effect of the type of branching on the results. Although the low melt strength and viscosity of PLAs usually prevent them from being used in a wide range of thermoplastic processes such as blown extrusion [18–21], they demonstrated that the addition of some branches can greatly improve the processability of PLA. Meanwhile, another biopolyester has attracted interest in these past 15 years, namely polybutylene succinate (PBS) thanks to its biodegradability and thermomechanical properties close to PE and PP [22]. PBS exhibits balanced performances in terms of thermal and mechanical properties as well as a good thermoplastic processability. PBS is a highly crystalline (around 50%) aliphatic polyester with a melting point of 115 °C (similar to PE) and a tensile yield strength (in the case of unoriented samples) that can reach up to 30–35 MPa (comparable to PP) [23]. PBS is usually synthetized via polycondensation of succinic acid and 1,4 butanediol (BDO), and its monomer can also be derived from renewable resources [23].

Linear grades of PBS have been developed and commercialized over the years, but to our knowledge, only the company Showa Denko High Polymer (SDK) was commercializing one grade of LCB PBS. Unfortunately, SDK abandoned the production of PBS in 2016.

LCB polymers in general, and in particular newly developed (bio)polymers, are more expensive than linear polymers due to the more complex synthesis or additional reactive extrusion step to obtain them. As for PP, blending linear and branched PBS is therefore interesting to improve the processability of linear polymer while decreasing the price of the polymer.

The very few studies that report on the rheological properties of PBS concern linear or branched PBS and especially never point out any relationships between their molecular structure and rheological behavior. To our knowledge, there are no published works dealing with the influence of blending linear and branched PBS on the rheological behavior of PBS. In addition, only a few studies concerned blends of polyolefins especially in extensional flow [3]. This study aims to provide a deep structural and rheological characterization of the long-chain branching (so-called BPBS) and linear (so-called LPBS) PBS and to understand the influence of their blending on linear and nonlinear viscoelastic properties. First, an analysis of the macromolecular structure of both BPBS and LPBS was carried out using size-exclusion chromatography; then, the impact of their blending on linear and nonlinear viscoelastic properties was analyzed. Finally, the relationship between their macromolecular structure and rheological properties has been discussed.

2. Experimental

2.1. Materials and Processing

A linear and a branched polybutylene succinate respectively referred as LPBS and BPBS produced by Showa Denko High Polymer (Tokyo, Japan) and commercialized under the trade names Bionolle 1001MD™ and Bionolle 1903MD™ (sourced through Sojitz Europe Plc, Paris branch, Paris, France), respectively, were chosen in this study. Since LPBS is obtained by the polycondensation of bifunctionnal monomers, the structure produced no longer contains any LCB neither SCB. On the other hand, BPBS results from the copolymerization of butylene succinate monomer with an additional monomer including multifunctional groups, leading to branched structures [24]. BPBS and LPBS properties (density ρ, Melt Flow Rate MFR, melting temperature T_m, and architecture) according to the supplier datasheet are given in Table 1.

Table 1. Bionolle™ branching polybutylene succinate (BPBS) and linear polybutylene succinate (LPBS) grades properties.

	BPBS	LPBS
ρ (g/cm^3)	1.24	1.24
MFR (2.16 kg/190 °C)	4.5	1.5
T_m (°C)	114–115	114–115
Topology	LCB	Linear

Blending and sample preparation were performed with a laboratory-scale extruder Polylab system (Thermo Fisher Scientific, Waltham, MA, USA) composed of a HAAKE RheoDrive4 motor unit/torque rheometer (Thermo Fisher Scientific, Waltham, MA, USA) with three thermoregulated zones coupled with a HAAKE Rheomex 19/25 OS single screw extruder (Thermo Fisher Scientific, Waltham, MA, USA). The extruder was equipped with a fish-tail designed die (270 mm width, 0.45 mm die gap) in order to obtain film samples. The extruder temperature zones were fixed at 240–235–230 °C and the die was fixed at 180 °C. The screw speed was kept constant at 80 rpm, 4.08 $\dot{m}$ (kg·h^{-1}).

In addition to the two pure polymers, three BPBS/LPBS blends (20/80 wt %, 50/50 wt %, and 80/20 wt % BPBS/LPBS) were dry-mixed and melt-compounded in the extruder. PBS is known to be sensitive to water or residual carboxylic acids terminals [23], which can lead to degradation issues (i.e., hydrolysis) during processing; all materials were dried at 80 °C during 4 h in a vacuum oven prior to extrusion.

2.2. Characterizations

2.2.1. Molecular Architecture Analysis

Triple detection size exclusion chromatography (SEC) was used to determine both the absolute molecular weight and level of branching of the blends. In the case of the use of a Low-Angle Laser Light Scattering detector (LALLS) or a Right-Angle Laser Light Scattering detector (RALLS) with a viscometer detector, both the molecular weight and intrinsic viscosity (and their distribution) of each fraction of the eluted polymer can be determined.

In this study, SEC measurements were carried out on a chromatograph composed of a Waters 515 HPLC pump with PLgel Mixed-C 5 µm 600 mm column set (AGILENT TECHNOLOGIES Inc., Santa Clara, CA, USA), a Waters 410 differential refractometer (DRI) sourced from AGILENT TECHNOLOGIES Inc. (Santa Clara, CA, USA), a Viscotek T60A dual detector (Right-Angle Laser Light Scattering detector at λ = 670 nm, RALLS, and a differential viscometer, IV-DP), sourced from Malvern Panalytical Ltd. (Malvern, UK).

Pellets were dissolved in chloroform to obtain a sample concentration of 20 mg/mL and solutions were filtered by a polytetrafluoroethylene (PTFE) membrane filter with 0.2 µm pores prior to SEC measurements. Chloroform was chosen as an eluent because of its good solubility and differential refractive index for polyesters (around 0.06 for PBS, calculated value). The mobile phase was CHCl$_3$ at a flow rate 1.0 mL/min. Five times the injection volume of 20 µL was used in order to avoid a dead volume effect. The detector temperature was 35 °C and the column temperature was 20 °C. Universal calibration was performed using a linear polystyrene with a narrow polydispersity and an average M_w of

19.760 g/mol as a standard. All analyses were carried out on Viscotek software OmniSEC® (v4.6), sourced from Malvern Panalytical Ltd. (Malvern, UK).

2.2.2. Small Amplitude Oscillatory Shear Rheometry/Linear Viscoelasticity

Dynamical rheological measurements of LPBS, BPBS, and their blends were performed using a rheometer ARES (TA Instruments, New Castle, DE, USA) equipped with a 25 mm parallel plates geometry in oscillatory shear and steady mode at 150, 170, and 190 °C. Stacked films were used to obtain a 1.5 mm thick sample and then cut into a 25 mm diameter disk. Preliminary time sweeps were conducted at $\gamma = 5\%$ to check if any degradation occurred during the test. A slight decrease of complex modulus was observed for the LPBS at 190 °C, which was probably due to thermal degradation, after a few minutes. Then, measurements were conducted with three points per decade in order to shorten the test and prevent samples from degradation during the test. The strain value, for all experiments, was set to 5% (small amplitude), which was checked to keep all measurements in the linear regime. Then, frequency sweep measurements were performed, within the frequency range from 100 to 0.01 rad. s^{-1}. During loading, the gap was progressively set to 0.8 mm, and a varying delay was applied to the polymer to reach equilibrium (zero normal force) before starting each test. All measurements were reproduced twice and were well reproducible. The viscoelastic parameters, namely, storage modulus (G'), loss modulus (G''), complex viscosity (η^*), and loss angle (δ) were calculated by using TA Orchestrator® software, sourced from TA Instruments (New Castle, DE, USA).

2.2.3. Steady Shear Rheometry

Additional steady shear viscosity measurements were performed at 150, 170, and 190 °C, within the range of shear rate from 0.01 to 10 s^{-1}, in order to check consistency with oscillatory measurements and access to lower shear rates.

2.2.4. Elongational Viscosity Measurements

Elongational viscosity was measured using the TA Extensional Viscosity Fixture (EVF) of the ARES rotational rheometer (both sourced from TA Instruments, New Castle, DE, USA). Films of 10 mm width and 18 mm length were directly cut into the center of the width of the extruded film (to avoid edge bead effects and hence an inhomogeneous thickness of the films across their width) and stacked to obtain the 0.7 mm targeted thickness required to reach a sufficient level of force. Careful attention was paid to avoid slippage within the stack when clamped on each rotating drum. The tests were performed at constant strain rates of 0.01, 0.1, 1, and 10 s^{-1} and at 150 °C to prevent the samples from sagging. At least two trials were performed at each extension rate for each sample to ensure reproducibility of the results.

3. Results and Discussion

3.1. Analysis of the Macromolecular Architecture of Linear and Branched PBS

The values of weight- and number-average molecular weights (M_w and M_n respectively), polydispersity index (PI as the M_w/M_n ratio), intrinsic viscosity [η] and Mark Houwink Sakurada (MHS) parameters related to each polymer are listed in Table 2. The mean value of the radius of gyration (r_g) was directly calculated by the software from the measured value of the hydrodynamic radius r_h thanks to the Fox–Flory relationship [25].

Table 2. Molecular characteristics of linear and branched PBS.

	M_w (g/mol)	M_n (g/mol)	PI	$[\eta]$ (dL/g)	r_g (nm)	MHS Parameters $K \times 10^5$ (dL/g)	a
LPBS	123,500	37,800	3.3	1.37	14.7	42.4	0.725
BPBS	122,900	14,700	8.4	1.21	14.7	325	0.528

Molecular weight distributions of LPBS and BPBS samples are illustrated in Figure 1 and show the relatively narrow distribution of the linear PBS (LPBS) compared to the broader distribution of the branched PBS (BPBS) that is in agreement with the polydispersity values. The high polydispersity index and broad molecular weight distribution of BPBS is a first evidence of the presence of LCB in this PBS grade [26]. Both samples exhibit a log-Gaussian molar mass monomodal distribution. The intrinsic viscosities of LPBS and BPBS, denoted $[\eta_{LPBS}]$ and $[\eta_{BPBS}]$, respectively, were directly measured from the viscosity detector for each eluted fraction of polymer. In a semi-logarithmic graph, MHS parameters were determined from the slope and intercept of the intrinsic viscosity as a function of molecular weight:

$$[\eta] = K \times M_v^a \quad (1)$$

where K and a are dependent on the considered polymer–solvent system and on the measurement temperature.

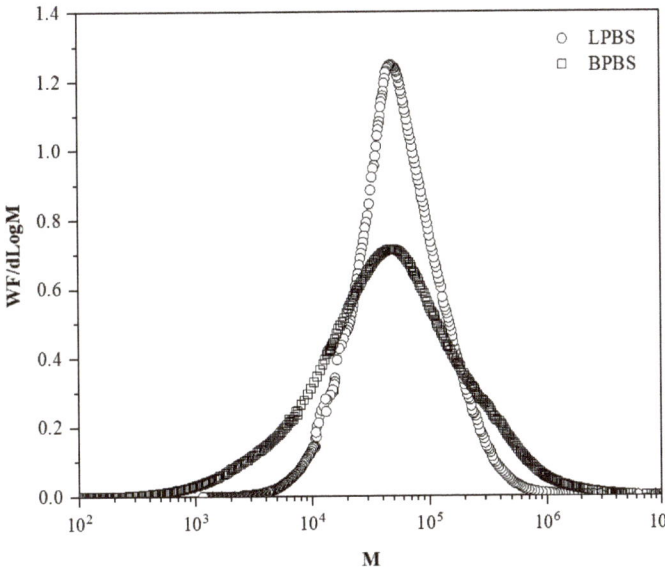

Figure 1. Molecular weight distributions of BPBS (□) and LPBS (○).

The exponent a gives information on the conformation of the polymer chains in the solvent. A value between 0 and 0.5 is generally obtained for branched chains, and a value comprised between 0.5 and 0.8 for is obtained for flexible chains [25]. The MHS parameters of PBS were reported only once by Garin et al. on an in-house synthetized linear PBS characterized by SEC coupled with multi-angle laser light scattering (SEC-MALLS). They found an average value of $a = 0.71$ (+/−0.1) and $K = 39.94 \times 10^{-5}$ (+/−6.31 × 10^{-5}) dL·g^{-1} [27], which are in total agreement with our values for LPBS. However, to the best of our knowledge, no study has previously reported the MHS parameters for long-chain branched PBS. For BPBS, we found a much higher value of $K = 325 \times 10^{-5}$ dL·g^{-1} and a lower value of $a = 0.528$, indicating a lower flexibility of polymer chains related to branching. In their study on the role of the architecture on the conformation in dilute solution of in-house synthetized polystyrenes (PS), Kharchenko et al. proposed a deep characterization of linear, star, and hyperbranched PS. For the same branching density, they found that a star PS exhibited a lower value ($a = 0.68$) and even much lower value for hyperbranched PS ($a = 0.39$) compared to the linear counterpar of equivalent molecular weight.

Figure 2 shows the hydrodynamic radius of LPBS and BPBS as a function of molecular weight M. At low molecular weight, BPBS and LPBS exhibit the same value of hydrodynamic radius, meaning that the macromolecules below 40,000 g/mol are mostly linear. Above 40,000 g/mol, the radius of BPBS fall below that of LPBS, as expected by the lower volume occupied by the coil of a branched chain which is more compact than a linear one of the same molar mass [4]. Thus, the continuous decrease of the radius depicts the increase of LCB as the molecular weight increases.

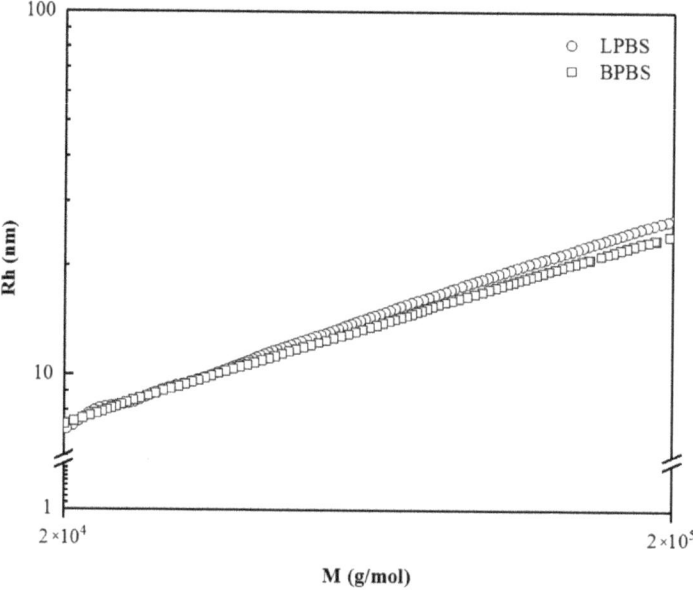

Figure 2. Log-log plot of hydrodynamic radius of BPBS (□) and LPBS (○).

Consequently, to the reduction of the hydrodynamic radius in this range of molecular weight (i.e., >40,000 g/mol), the intrinsic viscosity of BPBS is lower than that of LPBS, as shown in Figure 3. At the same time, the viscosity branching index (g') (given by Equation (2)) decreases almost monotonically from 0.9 (g' would be equal to 1 for linear chains) to 0.65 on the range of molecular weights accordingly with the BPBS decreasing viscosity.

$$g' = \frac{[\eta]_b}{[\eta]_l}\bigg|_{=M} \qquad (2)$$

where $[\eta]$ is the intrinsic viscosity and the subscripts b and l correspond to the branched and linear counterpart, respectively, which are considered at the same molecular weight.

The decrease of viscosity branching index indicates that the chains are mostly linear below 40,000 g/mol and that higher molecular weight molecules have a larger branching degree.

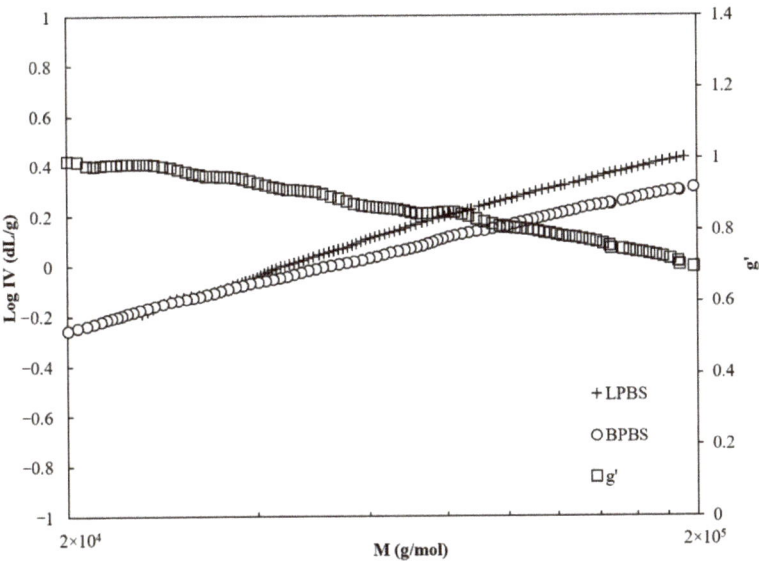

Figure 3. Intrinsic viscosity of LPBS (□) and BPBS (◇) and g' the viscosity branching index (○), dotted line representing Mark Houwink Sakurada (MHS) semi-logarithmic line.

Assuming that the BPBS is randomly branched, which is the case for most branched polyesters in nature [28], a Zimm–Stockmayer model for randomly distributed branch points per molecule with a random distribution of branch length (polydispersed) tri-functional polymers [29] (Equation (3)) was used to determine the branching characteristics of the polymer.

$$g = \frac{6}{B_w}\left[\frac{1}{2}\left(\frac{2+B_w}{B_w}\right)^{\frac{1}{2}} ln\left(\frac{(2+B_w)^{\frac{1}{2}}+B_w^{\frac{1}{2}}}{(2+B_w)^{\frac{1}{2}}-B_w^{\frac{1}{2}}}\right)-1\right] \quad (3)$$

The former relationship links the branching index (g) to the weight-average number of branch points per molecule (B_w) and the LCB frequency (λ) (average number of branch points per 1000 monomers) given by:

$$\lambda = \frac{1000 B_w M_0}{M_n} \quad (4)$$

where M_0 is the molecular weight of a monomer and M_n is the number-average molecular weight. For PBS, $M_0 = 172$ g/mol.

In the present study, only the viscosity branching index g' was determined. The viscosity branching index and the branching index g are related through ε known as the shielding factor (sometimes called the drainage factor) equal to 3/2 in the non-draining case. This factor is known to be dependent on the branching structure and has shown to be equal to 0.5 for star polymers while ε is closer to 1.5 for comb-shaped structures [30]:

$$g' = g^\varepsilon = g^{\frac{3}{2}} \quad (5)$$

Figure 4 shows the average number of branch points per molecule B_w and the LCB frequency λ plotted versus the molecular weight. The number of branches is ranged from 0 and 0.5 between 20,000 and 40,000 g/mol and starts to grow sharply above 40,000 g/mol from 0.5 to 3.7 at 200,000 g/mol.

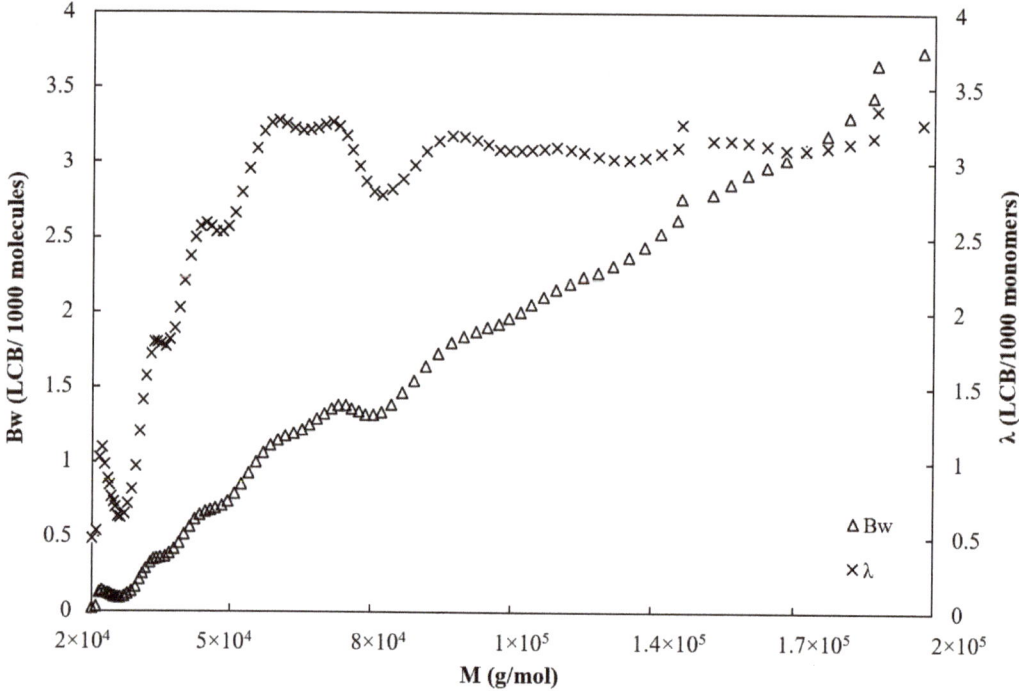

Figure 4. Average number of branch points per molecule B_w (△) and long-chain branched (LCB) frequency per 1000 monomer units λ (×) for BPBS.

Meanwhile, the branching frequency increases sharply from 0.5 LCB per 1000 units to approximately 3 LCB per 1000 units in the range of molecular weight comprised between 20,000 and 60,000 g/mol and then remains constant at 3 LCB/1000 units on the range of 60,000 and 200,000 g/mol.

Values of the average viscosity branching index, average number of LCB per molecule, and LCB frequency calculated by OmniSEC software are given in Table 3.

Table 3. Long-chain branching characteristics. Viscosity branching index, average LCB per molecule and average branching frequency per 1000 units for BPBS between 20,000 and 300,000 g/mol.

	g'	B_w (LCB/Molecule)	λ (LCB/1000 Monomers)
BPBS	0.82	2.03	2.8

Only very scarce studies were carried out on the characterization of long-chain branched polybutylene succinate. Although Wang et al. [31] studied the synthesis, characterization, and properties of long-chain branched PBS, they did not evaluate the branching characteristics and only determined the conventional molecular features (M_w, M_n, PI) by the use of conventional calibration (leading to possible misinterpretation of M_w). Kim et al. [32] used SEC-MALLS to determine the molecular characteristics of in-house synthetized branched PBS in chloroform. Unfortunately, they did not report any branching characteristics nor MHS coefficients. Two decades ago, Yoshikawa et al. performed SEC coupled with multi-angle laser light scattering (MALLS) detector on the same polymer grades used in this study [24]. The authors did not report the distribution of the number of LCB per molecule as a function of molecular weight and did not report the branching frequency.

They only give a mean value of the branching index as a function of the molecular weight calculated from pretty scattered data. They found an average number of branch points close to 2 LCB per molecule, which is consistent with our results and could confirm the shielding factor hypothesis we made. Finally, with a mean value of 2 LCB per molecule, Yoshikawa et al. suggested that the branching architecture could be an H-shaped molecule.

Focusing on the studies reporting the branching structure of polyolefins, the rheological behavior of the blends of linear and branched polymers depended heavily on the amount of LCB, on the molecular weight, on the type of branching (or topology), and on the relative composition of the mixture.

Stange et al. studied the impact of blending linear and long-chain branched PP on the shear and elongational flow behavior [4]. They showed that increasing the amount of LCB in the LCB-PP/linear PP blend resulted in a more pronounced shear thinning behavior, which is beneficial for processing. This effect was attributed to both the higher polydispersity and the increasing amount of LCB with the addition of LCB-PP in the blend. The type of architecture has also been proven to greatly influence the rheological behavior. For instance, the Zero-Shear rate Viscosity (ZSV) of long-chain branched PE with a moderate degree of star-like molecular structure has been shown to be enhanced over that of its linear counterpart, whereas a highly tree-like long-chain branched one exhibits a lower ZSV than its linear homologue [4].

Despite the difficulty of understanding the relationships between the molecular architecture and the resulting rheological properties, especially regarding the commercial nature and the impact of blending architectures, we show, in the following, a relationship with the rheological characterization of these architectures and blends.

3.2. Impact of Blending Linear and Branched PBS on Rheological Behavior

3.2.1. Rheometry under Oscillatory Shear Flow

Storage (G') and loss moduli (G'') of LPBS, BPBS, and their blends at 150 °C are illustrated in Figures 5 and 6, respectively. The loss modulus is found to dominate the flow of every composition on the entire range of frequency tested (i.e., no crossover is highlighted). At low frequencies, an enhancement of storage modulus is found for BPBS compared to LPBS and the blends lie in between the pure components, while the opposite trend is observed at high frequencies. No major difference of G'' according to the composition is highlighted at low shear rates, while at high shear rates, the LPBS shows the highest loss modulus and BPBS shows the lowest.

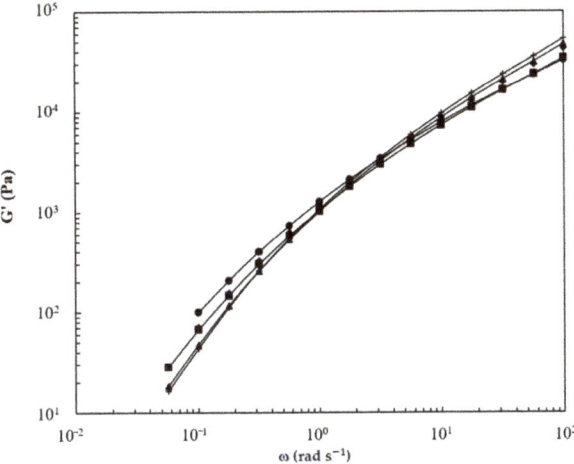

Figure 5. Storage modulus of BPBS (●), LPBS (+), and BPBS/LPBS blends (■) 80/20%, (♦) 50/50%, (▲) 20/80% at 150 °C.

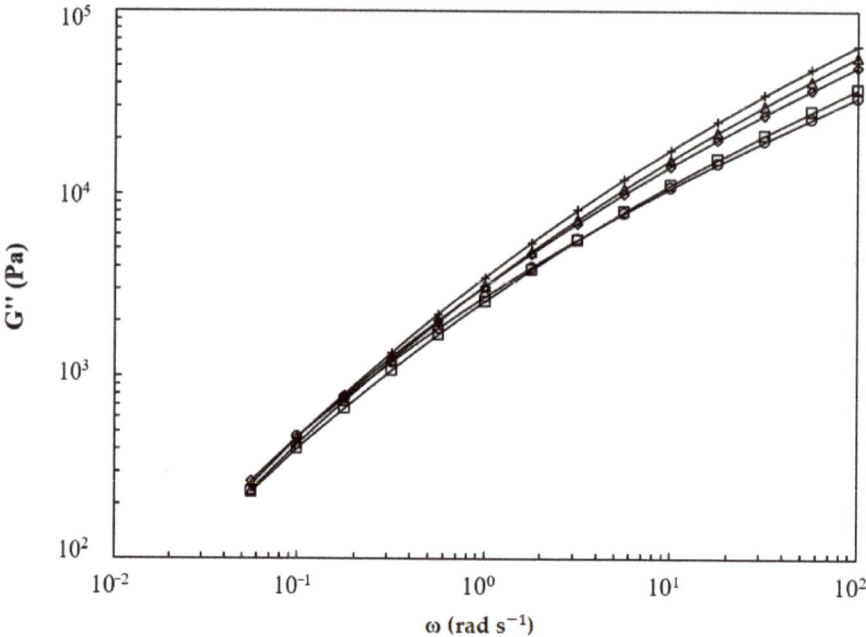

Figure 6. Loss modulus of BPBS (○), LPBS (+) and BPBS/LPBS blends (□) 80/20%, (◇) 50/50%, (Δ) 20/80% at 150 °C.

The storage and loss moduli slopes at low frequency for LPBS, BPBS, and their blends at 150 °C are listed in Table 4. As expected, the terminal zone was almost reached for LPBS, especially for the G'' slope. The value of 1.67 for the G' slope (power law exponent) indicates that there are long (>10 s) relaxation processes, which is probably due to the high molecular mass tail of the distribution (see Figure 1). When the amount of BPBS increases, the slope of G' decreases from 1.67 for the LPBS to 1.21 for the BPBS and the slope of G'' decreases from 0.99 for the LPBS to 0.82 for the BPBS. Clearly, a lower frequency (longer time) is needed for the LCB polymer to reach the terminal zone, and increasing the BPBS weight fraction results in a terminal zone reached at lower frequency. Wang et al. [33] reported the same trend when blending linear and branched PLA. Increasing the PLA branched weight fraction resulted in a higher deviation from the linear terminal relaxation zone. They argued that at low frequency, where only the high relaxation times contribute to the viscoelastic behavior, the behavior exhibited by the blends is ascribed to the presence of LCB. Since BPBS and LPBS have a similar weight-average molecular weight, this result confirms the effect of LCB on the increase in relaxation times. Indeed, LCB favor relaxation by arm fluctuation, which is associated with much larger relaxation times than the relaxation by reptation encountered in linear chains [34]. Moreover, as BPBS is more polydispersed than LPBS, the broadening of relaxation times shifts the terminal zone down to lower frequencies. Furthermore, it is also noted that the rubbery region has not been reached, whatever the composition of the mixture considered. As a matter of fact, it is also well known that LCB and polydispersity lead both to similar effects in broadening the transition region between the terminal zone and the plateau modulus. Therefore, in the frequency range tested, only the behavior in the transition zone can be noticed.

Table 4. Slopes (power law exponents) of storage and loss moduli for BPBS, LPBS and their blends at 150 °C.

BPBS/LPBS (wt %)	0/100	20/80	50/50	80/20	100/0
G' slope	1.67	1.60	1.45	1.43	1.21
G' slope	0.99	0.97	0.93	0.91	0.82

At high frequency ($\omega = 100$ rad·s^{-1}), G' and G'' decrease when the amount of BPBS increases. Conversely, at low frequency ($\omega = 0.05$ s^{-1}), the incorporation of BPBS leads to an enhancement of G' while G'' remains almost constant. The low-frequency region is known to be the most sensitive to molecular features [15,33]. Therefore, the enhancement of G' in the low-frequency region is attributed to the longer relaxation time ascribed to the LCB. Therefore, the more LCB, the higher the G' enhancement, and thus, the higher elasticity and melt strength (i.e., a more notable solid-like behavior).

The ZSV of LPBS, BPBS, and their blends was estimated from steady-state experiments carried out on the shear rate range 0.01 to 1 s^{-1} at 150 °C. At low shear rates, the complex viscosity $|\eta^*|$ (not shown) exhibits a constant value for all blends and pure polymers. The BPBS exhibits a slightly higher ZSV than LPBS (4569 Pa.s for LPBS vs. 5555 Pa s for BPBS). Given that the weight-average molecular weight is considered similar between the two neat components and that the ZSV is known to be insensitive to the molecular weight distribution, it is suggested that this improvement in elasticity is due to LCB.

3.2.2. Cox–Merz Rule and Carreau–Yasuda Model

The steady-state η viscosity and complex viscosity $|\eta^*|$ curves were superposed following the Cox–Merz rule [35], and the Carreau–Yasuda [36] model (Equation (6)) was fitted on each curve (as illustrated in Figure 7):

$$|\eta^*(\omega)| = \eta_0 \left[1 + (\lambda \omega)^\alpha\right]^{\frac{(n-1)}{\alpha}} \quad (7)$$

where η_0 is the ZSV, λ is the characteristic relaxation time, ω is the angular frequency, n is the flow index reflecting shear-thinning, and α is the parameter describing the width of the transition between the terminal and the shear thinning regimes of slope ($n - 1$).

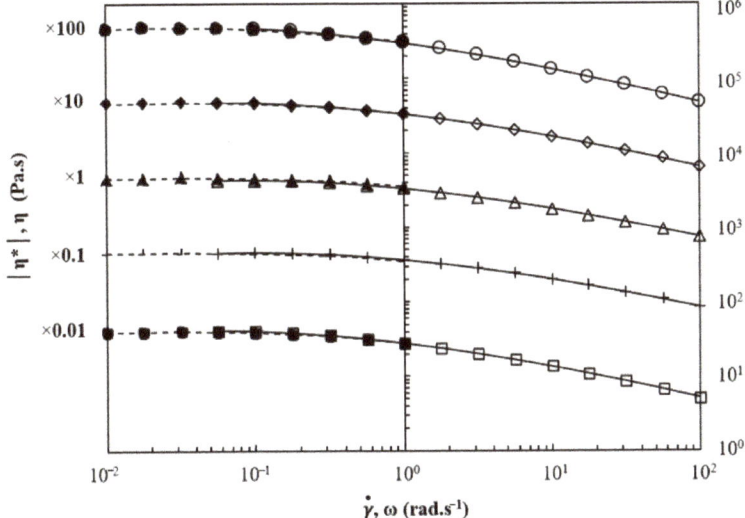

Figure 7. Cox–Merz rule and Carreau–Yasuda model applied to BPBS (○), LPBS (+), and BPBS/LPBS blends (□) 80/20% (◇) 50/50%, and (△) 20/80% at 150 °C (steady shear full symbols, dashed line; oscillatory shear open symbols, Carreau–Yasuda fit solid line).

For sake of clarity, every piece of experimental data was shifted by one or two decades higher or lower according to its rank when not shifted. The superposition seems to hold for every blend. No major deviation is clearly seen from the Cox–Merz rule, but the log-log plot often hides small deviations. As a matter of fact, a slight deviation is observed for the BPBS. In their study on the consequence of blending PLA of different chain architectures, Lehermeier and Dorgan [37] reported the same lack of superposition between steady shear viscosity and dynamic oscillatory (small amplitude) shear viscosity. Better agreement with the Cox–Merz rule is found for linear architecture than for long-chain branched polymers [35]. Utracki and Gendron [38] in their study on the behavior of LDPE (i.e., LCB Low-Density Polyethylene) during extrusion reported a similar breakdown that they suggested to be related to the existence of strain-hardening in elongational flow. This latter suggestion leads to think that the LCB is the cause of the lack of superposition.

The resulting parameters of the Carreau–Yasuda fit are summarized in Table 5. The fitted data of zero-shear rate viscosity show a good agreement with the experimental ones for all blends. The same trend as that reported above can be depicted from the fitted data concerning the behavior of ZSV as a function of the PBS weight fraction.

Table 5. Carreau–Yasuda parameters for LPBS, BPBS, and their blends at 150 °C.

BPBS/LPBS (wt %)	0/100	20/80	50/50	80/20	100/0
η_0 [Pa s]	4569	4616	4855	4144	5555
λ [s]	0.69	1.04	1.24	1.35	2.21
n	0.61	0.61	0.59	0.58	0.55
a	0.90	0.89	0.88	0.92	0.89

It can be seen from Table 5 that increasing BPBS in LPBS results in an increase of the characteristic relaxation time and a progressive decrease of the shear thinning index n. Nevertheless, the 20/80% BPBS/LPBS blend exhibits the same shear thinning than the LPBS and a close ZSV, which seems to indicate that adding 20% of BPBS in LPBS has no strong effect on the rheological behavior under shear flow (except from the slight increase of relaxation time).

The increase in relaxation time and shear thinning of branched polymers compared to their linear counterparts has already been reported in the case of PE [39], PP [4], and PLA [33]. The progressive increase of the relaxation time below 80% of BPBS in the blend suggests a good correlation between the relaxation time and the amount of branching (since the type and length should be similar in all blends) and supports the theory of miscibility at those BPBS weight fractions. Similar findings were also reported by Liu et al. [10] for LDPE/LLDPE blends. In addition, the improvement of shear thinning index can also be attributed to the large polydispersity of BPBS. In fact, the polydispersity affects the shear thinning index in the same way that LCB does, and separating the coupled effects appears as a complicated task. Hence, the shear thinning index includes both the effect of LCB and of the molecular weight distribution.

The relaxation spectrum gives valuable information on the distribution of relaxation times and therefore on the structure of the blend. The spectrum in its discrete form is constructed from the Generalized Maxwell Model (GMM) with seven elements, according to Laun's method [40].

The spectrum of the miscible blends is expected to result in a single peak with a smooth transition from one pure component to another [8]. The sum of the resulting relaxation strengths G_i is assimilated to the plateau modulus G_N^0 [15] as:

$$G_N^0 = \sum G_i \tag{7}$$

In addition, the discrete constants given by the general Maxwell model allows the estimation of the zero-shear rate viscosity η_0 and the mean relaxation time $\overline{\lambda}$ according to Equations (8) and (9).

$$\eta_0 = \sum_i^N G_i \lambda_i \tag{8}$$

$$\overline{\lambda} = \frac{\sum \lambda_i^2 G_i}{\sum \lambda_i G_i} \tag{9}$$

The results of relaxation strength normalized by the instantaneous modulus (weighed relaxation strength ω_i) are summarized in Table 6.

Table 6. Weighted relaxation strengths from the Generalized Maxwell Model (GMM) for BPBS, LPBS, and their blends at 150 °C.

BPBS/LPBS (wt %)	0/100	20/80	50/50	80/20	100/0
$\lambda_1 = 10^{-5}$ s	6.45×10^{-10}	7.38×10^{-10}	8.25×10^{-10}	1.09×10^{-9}	1.20×10^{-9}
$\lambda_2 = 10^{-4}$ s	1.54×10^{-9}	1.76×10^{-9}	1.97×10^{-9}	2.61×10^{-9}	2.87×10^{-9}
$\lambda_3 = 10^{-3}$ s	7.46×10^{-1}	7.48×10^{-1}	7.48×10^{-1}	7.37×10^{-1}	7.28×10^{-1}
$\lambda_4 = 10^{-2}$ s	2.06×10^{-1}	2.03×10^{-1}	1.98×10^{-1}	2.03×10^{-1}	2.02×10^{-1}
$\lambda_5 = 10^{-1}$ s	4.17×10^{-2}	4.27×10^{-2}	4.68×10^{-2}	$5 \times 10 \times 10^{-2}$	5.79×10^{-2}
$\lambda_6 = 1$ s	5.45×10^{-3}	5.90×10^{-3}	7.39×10^{-3}	8.82×10^{-3}	1.15×10^{-2}
$\lambda_7 = 10$ s	1.25×10^{-4}	1.78×10^{-4}	3.58×10^{-4}	4.68×10^{-4}	8.74×10^{-4}

On the frequency range tested (i.e., where the plateau and terminal zones are missing), we can observe a clear increase of the weighted relaxation strengths with the BPBS weight fraction at higher relaxation times. Given that the weight-average molecular weights of the LPBS and BPBS are similar, it can be suggested that the increase of the BPBS amount in the blend is responsible for the weighted strengths moduli to shift to longer relaxation times. Furthermore, the broadening of the relaxation spectrum together with the apparition of a shoulder for higher BPBS-rich blends for the higher relaxation times leads to the same conclusion that the weighted relaxation strengths are more significant at longer times because of the increased number of LCB in the blend [12].

Then, the relaxation modulus and relaxation time constants $\{G_i, \lambda_i\}$ were used as fitting parameters (in a nonlinear mean square minimization procedure) so that G_N^0 and η_0 were estimated more accurately (Equations (7) and (8)). The results of the second fitting procedure in term of G_N^0, $\overline{\lambda}$, and η_0 are listed in Table 7.

Table 7. Plateau modulus G_N^0, mean relaxation time $\overline{\lambda}$, and the zero-shear rate viscosity η_0 of BPBS, LPBS, and their blends at 150 °C.

BPBS/LPBS (wt %)	0/100	20/80	50/50	80/20	100/0
G_N^0 (Pa)	1.99	1.75	1.57	1.20	1.15
$\overline{\lambda}$ (s)	1.10	1.36	2.10	2.37	2.61
η_0 (Pa s)	4865	4580	5109	4481	5331

The estimated plateau modulus is found to be in the same range as other polyesters such as Polylactic acid (PLA) ($G_N^0 = 0.90 \times 10^5$ Pa for L-PLA and 2×10^5 Pa for D,L-PLA [41]) or Polyhydroxybutyrate-co-3-hydroxyvalerate (PHBV) ($G_N^0 = 1.53\text{–}1.97 \times 10^5$ Pa) [41]) and even Polyethylene terephthalate (PET) ($G_N^0 = 0.72\text{–}1.06 \times 10^5$ Pa [42]). Moreover, the plateau modulus decreased with increasing BPBS weight fraction in the blend due to LCB. To our knowledge, the PBS plateau modulus value was reported only once before by Garin et al. [27]

(mean value of $G_N^0 = 1.5 \times 10^5$ Pa) and is in total agreement with the value reported in this paper.

3.2.3. Extensional Viscosity

Up to now, all investigated viscoelastic properties were measured under oscillatory shear flow and especially at low deformation ($\gamma = 5\%$). However, these small deformations and deformation rates are quite far from those experienced during the cast film extrusion process used in this study. Therefore, elongational experiments may provide relevant information about the structural features of molecules that are not revealed by shear data and that actually influence the final properties of the polymer. The latter being not as effective as extensional flow in generating a significant stretching of the chains.

In the case of a Newtonian flow, Trouton [43] firstly noted that the elongational viscosity η_E equals three times the zero-shear rate viscosity. This quantity is usually referred as Trouton ratio (Equation (10)).

$$\eta_E(\dot{\varepsilon}) = 3\eta_0 \eta_E(\dot{\varepsilon}) = 3\eta_0 \tag{10}$$

It is of universal practice to compare the results of elongational flow experiments (the nonlinear material viscosity function) $\eta_E^+(t, \dot{\varepsilon})$ with the linear response given by:

$$\eta_E^+ = 3\sum_{i=1}^{N} G_i \lambda_i \left(1 - e^{\frac{-t}{\lambda_i}}\right) \eta_E^+ = \eta_E - 3\sum_{i=1}^{N} G_i \lambda_i e^{\left(\frac{-t}{\lambda_i}\right)} \tag{11}$$

where $\{G_i, \lambda_i\}$ are the previously mentioned GMM sets of parameters.

In this context, if the datasets are accurately obtained, then the nonlinear response should agree with the linear one at short times and low strain rates. In addition, the deviation of the nonlinear response from the linear one (also called linear viscoelastic envelope, LVE) is used to qualify the behavior of the material under elongational flow. If the material under elongational flow exhibits a sudden rise deviating above the LVE, the melt is said to be strain hardening, which improves processability, whereas if it drops below the LVE curve, it is said to be strain softening.

Strain hardening has been shown for several LCB polymers such as PP [13], PS [44], and even PLA [45]. Stange et al. [4] showed that even a small amount of LCB-PP (<10 wt %) in blend with linear PP was responsible for strain-hardening behavior at all strain rates tested (from 0.01 to 1 s^{-1}).

The extensional viscosity curves versus time of LPBS, BPBS, and 50/50% BPBS/LPBS at 150 °C and various strain rates ($\dot{\varepsilon} = 0.01$ s^{-1}, 0.1 s^{-1}, 1 s^{-1} and 10 s^{-1}) are presented in Figures 8–10. For LPBS (Figure 8), at $\dot{\varepsilon} = 0.1$ s^{-1}, LPBS acts as a Newtonian fluid falling on the LVE. Above 4 s, the polymer exhibits slight strain hardening. Given that the average relaxation time $\overline{\lambda}$ (given in Table 7) of LPBS is 1.10 s, this strain hardening can be possibly due to a few of the longest chains (i.e., higher relaxation time), which remain stretched before the sample breaks. At $\dot{\varepsilon} = 1$ s^{-1}, the elongational viscosity curve deviates from the LVE exhibiting strain softening above 0.30 s. At this rate, the major part of the chains tends to disorient and relax due to high mobility. The strain-softening behavior of linear chains polymer is explained by a fast retraction dynamic, as occurs for shear thinning. Finally, slight strain hardening is observed for the highest stretching rate of $\dot{\varepsilon} = 10$ s^{-1} from 0.06 s as the inverse Rouse (stretch) timescale of the chains is reached.

BPBS elongational viscosity (Figure 9) exhibits strain softening at $\dot{\varepsilon} = 0.01$ s^{-1} due to the higher time of stretching compared to the average time of relaxation (given in Table 7). However, above these strain rates, the BPBS shows strain hardening with the level of strain hardening going up with the strain rates. In fact, LCB results in a longer relaxation time for BPBS compared to LPBS as mentioned earlier; therefore, when the strain rates increase, there are more and more chains extended accordingly to their high stretch relaxation time [34].

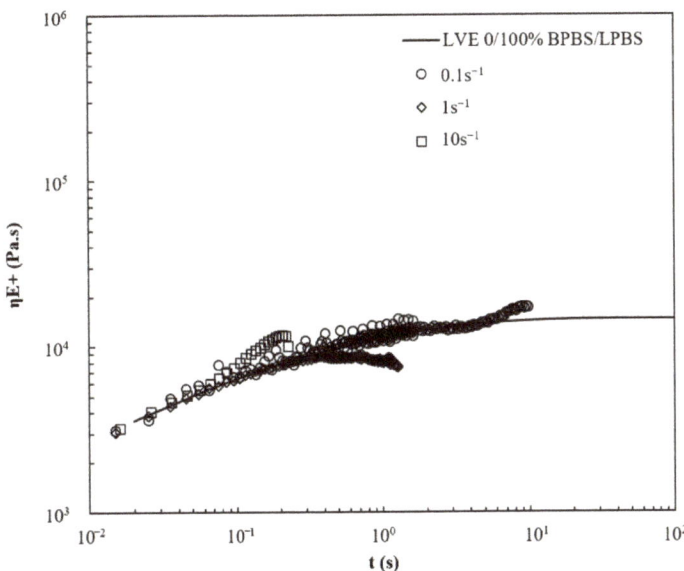

Figure 8. Viscosity curves under extensional flow at various strain rates for LPBS at 150 °C.

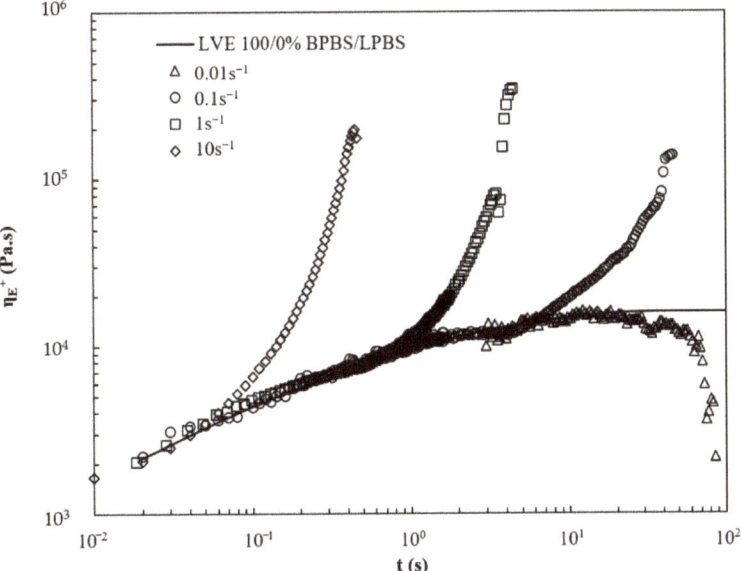

Figure 9. Viscosity curves under extensional flow at various strain rates for BPBS at 150 °C.

The 50/50% BPBS/LPBS blend in Figure 10 shows close behavior to that of LPBS. At the lower rate, it shows a linear response but exhibits strain softening at intermediate rate. At $\dot{\varepsilon} = 10\ \text{s}^{-1}$, a slight strain hardening is observed.

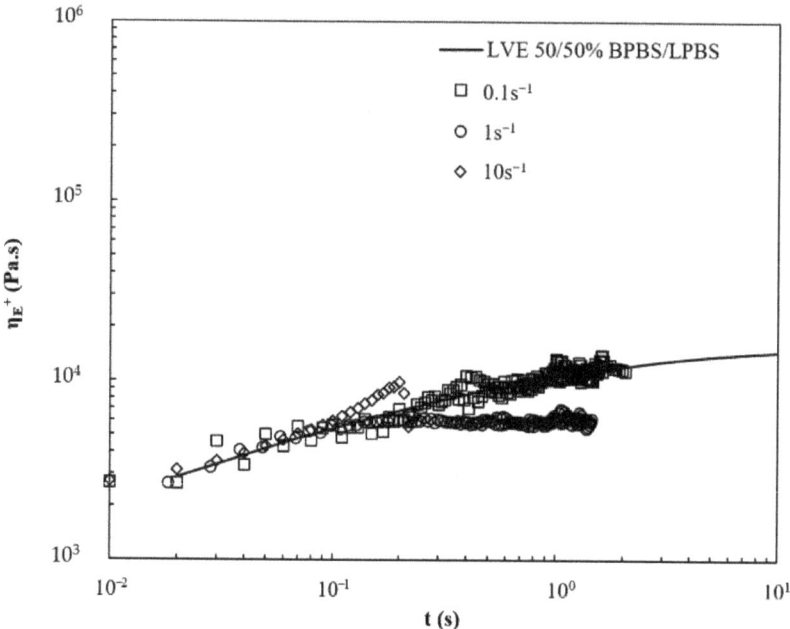

Figure 10. Viscosity curves under extensional flow at various strain rates for 50/50% BPBS/LPBS at 150 °C.

From these results of elongational viscosity, it can be suggested that BPBS strain hardening is related to the presence of LCB. The structure (number and length) of these LCB causes BPBS to exhibit strain hardening. Conversely, high strain rates are required to obtain slight strain hardening in LPBS. The 50/50% BPBS/LPBS blend behaves identical to LPBS. From this last observation, it can be suggested that either the LCB content is too low or the chain length is too small to produce significant strain hardening on the range of strain rates tested. Further characterizations of blends with higher BPBS content might reveal an increased strain hardening on this strain rates range.

4. Concluding Remarks

In this study, the rheological shear and elongational behavior of linear, long-chain branched PBS and their blends was investigated with further effort in understanding the impact of blending linear and LCB structures. An effort was made to deeply characterize the architecture of the commercial LCB PBS grade chosen for the purpose of this study.

Molecular characterization confirmed that LPBS was purely linear and that BPBS exhibits approximately two branches per molecule (supposed H-shape). Although the number-average molecular weights were found to be similar for the two pure BPBS and LPBS, their polydispersity and architecture were found to be extensively different. These dissimilarities prevented us from separating the effect of each molecular feature on the rheological properties of pure components. It appears even more complex if considering their blends.

The discrete relaxation time spectrum based on a seven-mode Generalized Maxwell Model allowed us to focus the interpretation based on the highest relaxation time, for which we observed a clear increase of the weighted relaxation strengths when the BPBS weight fraction increases.

However, if increasing the content of BPBS in the blend has only a marginal effect on the previous results under shear flow, it profoundly influences the transient elongational viscosity.

Under elongational flow, the rheological response of LPBS and BPBS changed from shear thinning to strain hardening with the increase of strain rate. The BPBS showed a marked strain hardening at a lower strain rate compared to LPBS due to the higher relaxation time needed for its arm to disentangle.

An incorporation of 50% of BPBS in LPBS did not significantly change the elongational response compared to LPBS.

Author Contributions: Formal analysis, R.V., V.B.; Funding acquisition, A.B.; Methodology, V.B.; Supervision, A.B., N.L.M., P.I., V.G.; Validation, A.B., N.L.M., P.I., R.V.; Writing—original draft, V.B.; Writing—review & editing, A.B., V.B., N.L.M., P.I., R.V., V.G. All authors have read and agreed to the published version of the manuscript.

Funding: This work was funded by Ceisa Packaging and the French Environment and Energy Management Agency-ADEME (n° TEZ11-26).

Institutional Review Board Statement: Not applicable.

Informed Consent Statement: Not applicable.

Data Availability Statement: Data is contained within the article.

Acknowledgments: The authors gratefully thank Ceisa Packaging and the French Environment and Energy Management Agency-ADEME (n° TEZ11-26) for their technical and financial support, Jean Coudane and Sylvie Hunger (Institut des Biomolécules Max Mousseron, Université de Montpellier, France) as well as Sebastien Rouzeau from Tosoh Bioscience for their help with the SEC analysis.

Conflicts of Interest: The authors declare no conflict of interest.

References

1. Field, G.J.; Micic, P.; Bhattacharya, S.N. Melt strength and film bubble instability of LLDPE/LDPE blends. *Polym. Int.* **1999**, *48*, 461–466. [CrossRef]
2. Ho, K.; Kale, L.; Montgomery, S. Melt strength of linear low-density polyethylene/low-density polyethylene blends. *J. Appl. Polym. Sci.* **2002**, *85*, 1408–1418. [CrossRef]
3. Wagner, M.H.; Kheirandish, S.; Yamaguchi, M. Quantitative analysis of melt elongational behavior of LLDPE/LDPE blends. *Rheol. Acta* **2004**, *44*, 198–218. [CrossRef]
4. Stange, J.; Uhl, C.; Münstedt, H. Rheological behavior of blends from a linear and a long-chain branched polypropylene. *J. Rheol.* **2005**, *49*, 1059. [CrossRef]
5. Yamaguchi, M.; Shigehiko, A. LLDPE/LDPE blends. I. Rheological, thermal, and mechanical properties. *J. Appl. Polym. Sci.* **1999**, *74*, 3153–3159. [CrossRef]
6. Delgadillo-Velázquez, O.; Hatzikiriakos, S.G.; Sentmanat, M. Thermorheological properties of LLDPE/LDPE blends. *Rheol. Acta* **2008**, *47*, 19–31. [CrossRef]
7. Micic, P.; Bhattacharya, S.N.; Field, G. Transient elongational viscosity of LLDPE/LDPE blends and its relevance to bubble stability in the film blowing process. *Polym. Eng. Sci.* **1998**, *38*, 1685–1693. [CrossRef]
8. Dordinejad, A.K.; Jafari, S.H. Miscibility analysis in LLDPE/LDPE blends via thermorheological analysis: Correlation with branching structure. *Polym. Eng. Sci.* **2014**, *54*, 1081–1088. [CrossRef]
9. Zhu, H.; Wang, Y.; Zhang, X.; Su, Y.; Dong, X.; Chen, Q.; Zhao, Y.; Geng, C.; Zhu, S.; Han, C.C.; et al. Influence of molecular architecture and melt rheological characteristic on the optical properties of LDPE blown films. *Polymer* **2007**, *48*, 5098–5106. [CrossRef]
10. Liu, C.; Wang, J.; He, J. Rheological and thermal properties of m-LLDPE blends with m-HDPE and LDPE. *Polymer* **2002**, *43*, 3811–3818. [CrossRef]
11. Ajji, A.; Sammut, P.; Huneault, M.A. Elongational rheology of LLDPE/LDPE blends. *J. Appl. Polym. Sci.* **2003**, *88*, 3070–3077. [CrossRef]
12. McCallum, T.J.; Kontopoulou, M.; Park, C.B.; Muliawan, E.B.; Hatzikiriakos, S.G. The rheological and physical properties of linear and branched polypropylene blends. *Polym. Eng. Sci.* **2007**, *47*, 1133–1140. [CrossRef]
13. Sugimoto, M.; Suzuki, Y.; Hyun, K.; Ahn, K.H.; Ushioda, T.; Nishioka, A.; Taniguchi, T.; Koyama, K. Melt rheology of long-chain-branched polypropylenes. *Rheol. Acta* **2006**, *46*, 33–44. [CrossRef]
14. Suleiman, M.A. Rheological Investigation of the Influence of Short Chain Branching and Mw of LDPE on the Melt Miscibility of LDPE/PP Blends. *Open Macromol. J.* **2011**, *5*, 13–19. [CrossRef]
15. Dealy, J.M.; Larson, R.G. *Structure and Rheology of Molten Polymers*; Hanser: Munich, Germany, 2006.
16. Nouri, S.; Dubois, C.; Lafleur, P.G. Synthesis and characterization of polylactides with different branched architectures. *J. Polym. Sci. Part B Polym. Phys.* **2015**, *53*, 522–531. [CrossRef]

17. Nouri, S.; Dubois, C.; Lafleur, P.G. Effect of chemical and physical branching on rheological behavior of polylactide. *J. Rheol.* **2015**, *59*, 1045–1063. [CrossRef]
18. Al-Itry, R.; Lamnawar, K.; Maazouz, A. Reactive extrusion of PLA, PBAT with a multi-functional epoxide: Physico-chemical and rheological properties. *Eur. Polym. J.* **2014**, *58*, 90–102. [CrossRef]
19. Al-Itry, R.; Lamnawar, K.; Maazouz, A. Improvement of thermal stability, rheological and mechanical properties of PLA, PBAT and their blends by reactive extrusion with functionalized epoxy. *Polym. Degrad. Stab.* **2012**, *97*, 1898–1914. [CrossRef]
20. Khoo, H.H.; Tan, R.B.H. Environmental impacts of conventional plastic and bio-based carrier bags. *Int. J. Life Cycle Assess.* **2010**, *15*, 338–345. [CrossRef]
21. Pivsa-Art, W.; Pavasupree, S.; O-Charoen, N.; Insuan, U.; Jailak, P.; Pivsa-Art, S. Preparation of Polymer Blends Between Poly (L-Lactic Acid), Poly (Butylene Succinate-Co-Adipate) and Poly (Butylene Adipate-Co-Terephthalate) for Blow Film Industrial Application. *Energy Procedia.* **2011**, *9*, 581–588. [CrossRef]
22. Fujimaki, T. Processability and properties of aliphatic polyesters, "BIONOLLE", synthesized by polycondensation reaction. *Polym. Degrad. Stab.* **1998**, *59*, 209–214. [CrossRef]
23. Xu, J.; Guo, B. Poly (butylene succinate) and its copolymers: Research, development and dndustrialization. *Biotechnol. J.* **2010**, *5*, 1149–1163. [CrossRef]
24. Yoshikawa, K.; Ofuji, N.; Imaizumi, M.; Moteki, Y.; Fujimaki, T. Molecular weight distribution and branched structure of biodegradable aliphatic polyesters determined by s.e.c.-MALLS. *Polymer* **1996**, *37*, 1281–1284. [CrossRef]
25. Flory, P.J.; Fox, T.G. Treatment of Intrinsic Viscosities. *J. Am. Chem. Soc.* **1951**, *73*, 1904–1908. [CrossRef]
26. Vega, J.; Aguilar, M.; Peón, J.; Pastor, D.; Martínez-Salazar, J.; Peon, J.; Pastor, D.; Martinez-Salazar, J. Effect of long chain branching on linear-viscoelastic melt properties of polyolefins. *e-Polymers* **2002**, *2*, 1–35. [CrossRef]
27. Garin, M.; Tighzert, L.; Vroman, I.; Marinkovic, S.; Estrine, B. The influence of molar mass on rheological and dilute solution properties of poly(butylene succinate). *J. Appl. Polym. Sci.* **2014**, *131*, 1–7. [CrossRef]
28. McKee, M.G.; Unal, S.; Wilkes, G.L.; Long, T.E. Branched polyesters: Recent advances in synthesis and performance. *Prog. Polym. Sci.* **2005**, *30*, 507–539. [CrossRef]
29. Zimm, B.H.; Stockmayer, W.H. The Dimensions of Chain Molecules Containing Branches and Rings. *J. Chem. Phys.* **1949**, *17*, 1301. [CrossRef]
30. Douglas, J.F.; Roovers, J.; Freed, K.F. Characterization of Branching Architecture through "Universal" Ratios. *Macromolecules* **1990**, *23*, 4168–4180. [CrossRef]
31. Wang, G.; Guo, B.; Li, R. Synthesis, characterization, and properties of long-chain branched poly(butylene succinate). *J. Appl. Polym. Sci.* **2012**, *124*, 1271–1280. [CrossRef]
32. Kim, E.U.N.K.; Bae, J.S.; Im, S.S.; Kim, B.C.; Han, Y.K. Preparation and Properties of Branched Polybutylenesuccinate. *J. Appl. Polym. Sci.* **2001**, *80*, 1388–1394. [CrossRef]
33. Wang, L.; Jing, X.; Cheng, H.; Hu, X.; Yang, L.; Huang, Y. Blends of linear and long-chain branched poly(l-lactide)s with high melt strength and fast crystallization rate. *Ind. Eng. Chem. Res.* **2012**, *51*, 10088–10099. [CrossRef]
34. McLeish, T.C.B. Tube theory of entangled polymer dynamics. *Adv. Phys.* **2002**, *51*, 1379–1527. [CrossRef]
35. Cox, W.P.; Merz, E.H. Correlation of Dynamic and Steady Flow Viscosities. *J. Polym. Sci.* **1958**, *28*, 619–622. [CrossRef]
36. Carreau, P.J. Rheological Equations from Molecular Network Theories. *J. Rheol.* **1972**, *16*, 99. [CrossRef]
37. Lehermeier, H.J.; Dorgan, J.R. Melt rheology of poly(lactic acid): Consequences of blending chain architectures. *Polym. Eng. Sci.* **2001**, *41*, 2172–2184. [CrossRef]
38. Utracki, L.A.; Gendron, R. Pressure Oscillation during Extrusion of Polyethylenes. II. *J. Rheol.* **1984**, *28*, 601. [CrossRef]
39. Yan, D.; Wang, W.J.; Zhu, S. Effect of long chain branching on rheological properties of metallocene polyethylene. *Polymer* **1999**, *40*, 1737–1744. [CrossRef]
40. Laun, H.M. Description of the non-linear shear behaviour of a low density polyethylene melt by means of an experimentally determined strain dependent memory function. *Rheol. Acta* **1978**, *17*, 1–15. [CrossRef]
41. Ramkumar, D.H.S.; Bhattacharya, M. Steady shear and dynamic properties of biodegradable polyesters. *Polym. Eng. Sci.* **1998**, *38*, 1426–1435. [CrossRef]
42. Yilmazer, U.; Xanthos, M.; Bayram, G.; Tan, V. Viscoelastic characteristics of chain extended/branched and linear polyethylene terephthalate resins. *J. Appl. Polym. Sci.* **2000**, *75*, 1371–1377. [CrossRef]
43. Trouton, F.T. On the Coefficient of Viscous Traction and Its Relation to that of Viscosity. *Proc. R. Soc. A Math. Phys. Eng. Sci.* **1906**, *77*, 426–440. [CrossRef]
44. Hine, P.J.; Duckett, A.; Read, D.J. Influence of Molecular Orientation and Melt Relaxation Processes on Glassy Stress-Strain Behavior in Polystyrene. *Macromolecules* **2007**, *40*, 2782–2790. [CrossRef]
45. Liu, J.; Lou, L.; Yu, W.; Liao, R.; Li, R.; Zhou, C. Long chain branching polylactide: Structures and properties. *Polymer* **2010**, *51*, 5186–5197. [CrossRef]

Article

Applicability of the Cox-Merz Rule to High-Density Polyethylene Materials with Various Molecular Masses

Raffael Rathner [1,*], Wolfgang Roland [1], Hanny Albrecht [1], Franz Ruemer [2] and Jürgen Miethlinger [1]

1. Institute of Polymer Extrusion and Compounding, Johannes Kepler University Linz, Altenberger Str. 69, 4040 Linz, Austria; wolfgang.roland@jku.at (W.R.); hanny.albrecht@Pro2Future.at (H.A.); juergen.miethlinger@gmail.com (J.M.)
2. Borealis Polyolefine GmbH, Sankt-Peter-Straße 25, 4021 Linz, Austria; franz.ruemer@borealisgroup.com
* Correspondence: raffael.rathner@jku.at

Citation: Rathner, R.; Roland, W.; Albrecht, H.; Ruemer, F.; Miethlinger, J. Applicability of the Cox-Merz Rule to High-Density Polyethylene Materials with Various Molecular Masses. *Polymers* **2021**, *13*, 1218. https://doi.org/10.3390/polym13081218

Academic Editor: Khalid Lamnawar

Received: 23 March 2021
Accepted: 8 April 2021
Published: 9 April 2021

Publisher's Note: MDPI stays neutral with regard to jurisdictional claims in published maps and institutional affiliations.

Copyright: © 2021 by the authors. Licensee MDPI, Basel, Switzerland. This article is an open access article distributed under the terms and conditions of the Creative Commons Attribution (CC BY) license (https://creativecommons.org/licenses/by/4.0/).

Abstract: The Cox-Merz rule is an empirical relationship that is commonly used in science and industry to determine shear viscosity on the basis of an oscillatory rheometry test. However, it does not apply to all polymer melts. Rheological data are of major importance in the design and dimensioning of polymer-processing equipment. In this work, we investigated whether the Cox-Merz rule is suitable for determining the shear-rate-dependent viscosity of several commercially available high-density polyethylene (HDPE) pipe grades with various molecular masses. We compared the results of parallel-plate oscillatory shear rheometry using the Cox-Merz empirical relation with those of high-pressure capillary and extrusion rheometry. To assess the validity of these techniques, we used the shear viscosities obtained by these methods to numerically simulate the pressure drop of a pipe head and compared the results to experimental measurements. We found that, for the HDPE grades tested, the viscosity data based on capillary pressure flow of the high molecular weight HDPE describes the pressure drop inside the pipe head significantly better than do data based on parallel-plate rheometry applying the Cox-Merz rule. For the lower molecular weight HDPE, both measurement techniques are in good accordance. Hence, we conclude that, while the Cox-Merz relationship is applicable to lower-molecular HDPE grades, it does not apply to certain HDPE grades with high molecular weight.

Keywords: Cox-Merz rule; high-viscosity HDPE materials; extrusion; modelling and simulation; rheology

1. Introduction

The rheological behaviour of polymer melts is of major significance in polymer processing as it describes the deformation and flow behaviour of the material. A suitable choice of rheological model is essential for predicting the behaviour of a polymer during processing. Despite the wealth of publications in this context, the rheological behaviour of polymer melts remains a subject of scientific and technological interest [1–3] because it can be used to optimize a range of processing parameters and extrusion equipment. As it is influenced by a large number of parameters (e.g., concentration of the fluid, morphology, chemical structure), the rheological behaviour of polymer melts is very complex and sometimes difficult to relate to various physical properties of fluid polymer blends and alloys [4–6]. When determining the rheological behaviour of polymeric fluids, for which a variety of methods exist, the non-linear viscoelastic properties in particular give rise to high complexity. One approach—which is used, for example, in high-pressure capillary and extrusion rheometry—is to determine the rheological properties based on the pressure drop in a known geometry. Other methods are based on oscillatory measurements using, for instance, plate-plate and cone-plate rheometers. The big advantages of oscillatory over pressure-driven approaches are that they are fast, cheap, and easy to use [7]. Additionally, a relatively small amount of material is needed, and low shear rates can be measured. In comparison to purely rotational experiments, in oscillatory mode the imposed shear is relatively low, which avoids shear-induced heating of the test specimen [8]. It would thus be very useful to determine the rheological properties of polymers in oscillatory measurements. Cox and Merz [9] postulated an empirical relation, which states that the frequency dependence of the complex viscosity $\eta^*(\omega)$ of polystyrene melts with a range of molecular weights is equivalent to the shear rate dependence of the steady shear viscosity $\eta(\dot{\gamma})$. Since then, the Cox-Merz rule has been investigated in the context of other polymers and was found to apply to various linear and branched polymers [10–12]. However, several research groups have reported that the Cox-Merz relation does not hold for some polymers (e.g., concentrated suspensions compounds, highly branched polymers, polymer blends, thermoplastic elastomers, functionalized polymers and in some cases high-molecular-weight polymers) [13–18]. For example, Snijkers and Vlassopoulos [19] found that the Cox-Merz relation did not apply to certain well-defined branched polymers they studied. Robertson and Roland [20] demonstrated the non-validity of the Cox-Merz rule for a variety of branched polyisobutylenes. Järvela [21] reported that the Cox-Merz rule does not apply to blends of polypropylene and maleated polypropylene. All three groups found several materials for which the Cox-Merz rule does not hold. Further reports [22,23] showed the non-validity of the Cox-Merz rule for filled polymers and materials that are able to form hydrogen bonds or exhibit other complex intermolecular binding phenomena (e.g., polyacrylamide and polyvinylchloride). Venkatraman and Okano [24] examined the applicability of the Cox-Merz rule to a variety of polyethylene types and found that it depends strongly on chemical structure, molecular weight, and entanglement. Additionally it has been shown that the Cox-Merz rule cannot be applied to certain low-density polyethylens (LDPE) and for random branched polystyrenes [25–27]. Another research group [28] stated that the Cox-Merz rule can be applied only in the low shear-rate region and that viscosity will be overestimated when the rule does not apply. Since the Cox-Merz rule cannot be applied in several cases, new postulations of the Cox-Merz rule have been made [29–31]. Overestimation or incorrect measurement of the viscosity has extreme consequences for the layout of processing tools, resulting, for instance, in incorrect pressure drops, residence time, and shear-rate distributions in the flow geometry, which can ultimately render the tools unusable.

Although there are no clear guidelines for when it can be applied, the Cox-Merz rule is widely used in the context of polyolefins in industrial practice because of the advantages of oscillatory measurements mentioned above. Since various studies have shown the non-validity of the Cox-Merz relation for particular kinds of polymers, we investigated whether it applies to commercially available—and in industry highly relevant—linear thermoplastic high-density polyethylene (HDPE) pipe grades. HDPE exhibits branching only to a very limited degree and consists of long chains that ensure its excellent mechanical properties [32]. To assess the suitability of the Cox-Merz rule for determining the viscosity of highly viscous HDPE pipe materials, we measured three commercially available polyethylene materials from Borealis using three well-known methods: (i) parallel-plate rheometry in oscillatory mode, (ii) high-pressure capillary rheometry, and (iii) extrusion rheometry. Based on the viscosity curves obtained, we then simulated the pressure drop along a pipe head and validated the results against experimental data.

2. Materials and Methods

2.1. Materials

In this study we used three commercially available HDPE grades:

- Material 1 was a high viscosity hexene copolymer polyethylene compound (HDPE) for pipe applications (PE 100) with high density and an outstanding resistance to slow crack growth.
- Material 2 was a high-density polyethylene for injection and compression moulding.
- Material 3 was another high-density polyethylene for injection and compression moulding.

Table 1 summarizes the melt flow rate (MFR) according to ISO 1133 (5.0 kg at 190 °C), mass average molecular weight Mw, and z-average molar mass Mz of the three different materials. Mw and Mz were measured with gel permeation chromatography (GPC).

Table 1. Melt flow rate (MFR) and molecular weight distributions of the high-density polyethylene (HDPE) materials.

	MFR (g/10 min)	M_W (g/mol)	M_Z (g/mol)
Material 1	0.25	230,000	1,190,000
Material 2	1.5	110,000	550,000
Material 3	4.0	85,500	387,000

2.2. Parallel-Plate Rheometer

The complex dynamic shear viscosity was measured by means of a combined motor-transducer (CMT) MCR302 rheometer from Anton Paar. The experiments were carried out at 200 °C under nitrogen atmosphere to prevent degradation, using a parallel-plate geometry with a diameter of 25 mm and a thickness of 0.8 mm. The frequency-sweep method with an amplitude of $\gamma_0 = 0.03$ was chosen to obtain frequency-dependent storage (G') and loss module (G'') for angular frequencies ω in the range between 0.01 rad/s and 628 rad/s. These modules were used to determine the shear-rate-dependent complex viscosity $\eta^*(\omega)$. The Cox-Merz relation was used to convert the oscillatory shear data into shear-rate-dependent viscosity $\eta(\dot{\gamma})$:

$$\eta(\dot{\gamma}) = \eta^*(\omega) \tag{1}$$

The complex viscosity η^* was calculated by [33]:

$$|\eta^*| = \frac{|G^*|}{\omega} \cdot \sqrt{1 + \frac{G'^2}{G''}} \tag{2}$$

where G'' is the loss modulus, G' the storage modulus, and ω the angular frequency. To determine the ratio of G' and G'', we calculated the phase shift (δ) by

$$\delta = \tan^{-1}\left(\frac{G''}{G'}\right) \tag{3}$$

where a value of 45° indicates that G' and G'' have the same value, and values above or below 45° indicate that the elastic part or the viscous part dominates, respectively.

2.3. High-Pressure Capillary Rheometer (HPCR)

A Rheograph 25 (Göttfert Werkstoff-Prüfmaschinen GmbH) high-pressure capillary rheometer with two different dies (diameter = 1 mm, length = 1 mm; diameter = 1 mm; length = 20 mm) was used to measure the polymer at a temperature of 200 °C. The shear viscosity was measured at shear rates between 1 and 5000 s^{-1}. We calculated the wall shear stress τ_w according to [34]

$$\tau_w = \frac{R}{2} \cdot -\frac{\Delta p}{l} \tag{4}$$

where R is the radius of the capillary, Δp the pressure drop, and l the length of the die capillary. The apparent shear rate $\dot{\gamma}_{app}$ was calculated by [34]

$$\dot{\gamma}_{app} = \frac{4 \cdot \dot{V}}{\pi \cdot R^3} \tag{5}$$

where $\dot{V}$ is the volume flow through the capillary die. The Bagley correction [35] was used to correct for entry flow effects of the capillary. To obtain the true shear rate $\dot{\gamma}_{true}$ at the wall, we applied the Weissenberg–Rabinowitsch correction [36]:

$$\dot{\gamma}_{true} = \frac{1}{3} \cdot \left(2 + \frac{dlog(\dot{\gamma}_{app})}{dlog(\tau)}\right) \cdot \dot{\gamma}_{app} \qquad (6)$$

The shear viscosity was then calculated by

$$\eta = \frac{\tau_w}{\dot{\gamma}_{true}} \qquad (7)$$

2.4. Slit-Die Extrusion Rheometer

A Thermo Haake Rheomex system consisting of a single-screw extruder (screw diameter 19 mm, length 33 times the diameter) equipped with a melt pump (2.4 cm³/rev.), a bypass valve, and a slit die with a defined gap height of 0.8 mm and a width of 20 mm was used to determine the shear-rate-dependent viscosity through the slit. The temperature profile was adjusted for the measurement to always reach a melt temperature of 200 °C at the entrance of the die for every measurement point.

We calculated the wall shear stress τ_w relative to the pressure drop Δp and the apparent shear rate $\dot{\gamma}_{app}$ relative to the volume flow $\dot{V}$ in a capillary slit of defined height H, length L, and width W according to [34]:

$$\tau_w = \frac{\Delta p \cdot H}{2L} \qquad (8)$$

$$\dot{\gamma}_{app} = \frac{\Delta p \cdot \dot{V}}{W \cdot H} \qquad (9)$$

To obtain the true shear rate $\dot{\gamma}_{true}$ at the wall, we applied the Weissenberg–Rabinowitsch correction Equation (6) [36], and the shear viscosity was then calculated by Equation (7). The measurement range of the shear rate of the extrusion rheometer was between 25 and 300 s⁻¹.

3. Simulation

3.1. Fitting of Experimental Data

The data obtained from oscillatory and capillary rheometry were fitted independently by using the ANSYS Polymat software module [37] for the modified Cross model [38] given by

$$\eta(\dot{\gamma}) = \frac{\eta_0}{(1 + \lambda \cdot \dot{\gamma})^m} \qquad (10)$$

where $\eta(\dot{\gamma})$ is the shear viscosity, λ is the time constant [s], $\dot{\gamma}$ is the shear rate, η_0 is the zero shear rate viscosity, and m is the Cross-law flow behaviour index.

3.2. Simulation of Extrusion Equipment

Using the commercial finite-volume software ANSYS FLUENT [37], we simulated the HDPE melt flow in a pipe head. We assumed the flow to be (i) steady-state, (ii) creeping, (iii) incompressible, (iv) isothermal, and (v) we ignored gravity. We considered as the computational flow domain the full three-dimensional geometry of the spiral mandrel pipe head (Figure 1). A non-uniform mesh with 2,821,732 cells was generated for the numerical simulation and considered adequate for capturing the flow in the spiral distribution section.

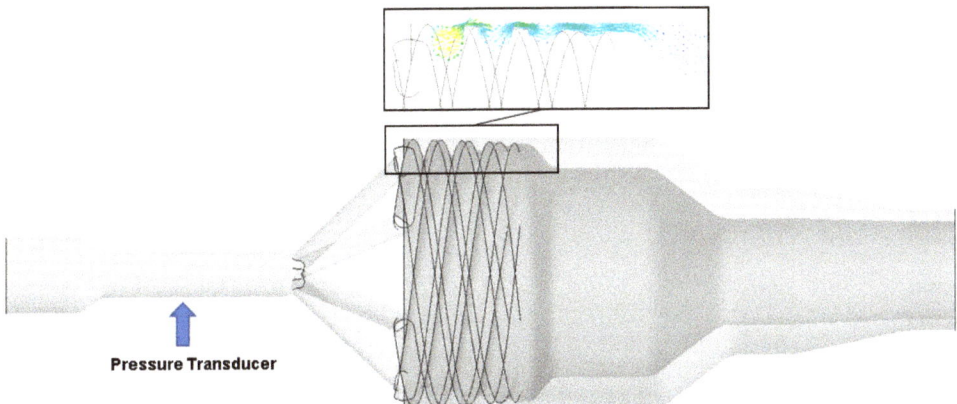

Figure 1. Real geometry of the experimentally validated and simulated pipe head.

At the inlet we specified the mass flow rate, and the pressure outlet was defined to have zero pressure. Further, we assumed that the fluid sticks to the die walls. We simulated five different setups with mass flowrates of 5, 10, 15, 20, and 25 kg/h.

All simulations were performed using a pressure-based coupled solver, and for the gradient computation we used the Green–Gauss node-based solver, which is recommended for an unstructured mesh. The second-order and second-order upwind schemes were employed to solve the pressure and momentum equations. Convergence was achieved when the scaled residual of mass conservation and momentum equations fell below 10^{-5}. Subsequently, we evaluated the pressure drop of the flow geometry.

3.3. Extrusion Experiments with the Real Pipe Head

For the extrusion experiments we used a single screw extruder from ESDE with a diameter of 25 mm and a length of 18 D with a barrier screw and combined it with the pipe head shown in Figure 1. The pressure at the entrance of the pipe head was measured at the position indicated in Figure 1. The temperature profiles of the pipe head and extruder were adjusted to reach a melt temperature of 200 °C at the end of the pipe head. In accordance with the simulations the outputs were adjusted to 5, 10, 15, 20, and 25 kg/h for the experiments.

4. Results and Discussion

4.1. Comparison of Plate-Plate Rheometry (PPR) to High-Pressure Capillary Rheometry (HPCR) and Extrusion Slit Rheometry

High-pressure capillary rheometry (HPCR) and extrusion slit rheometry measurements are both based on calculating the shear-rate-dependent viscosity via a pressure flow through a capillary die. Our comparison shows that the results of the two methods are in good accordance (see Figure 2). In contrast to HPCR and extrusion slit rheometry, which measure a constant pressure flow, plate-plate rheometry is based on oscillatory measurements. As can be seen in Figure 3, the two rheological approaches yield significantly different results for Materials 1 and 2, but are in good accordance for Material 3.

Since extrusion rheology and HPCR are almost identical, we carried out computational fluid dynamics (CFD) simulations only for HPCR and PPR data. The modified cross-law parameters for the three materials (derived from the PPR and HPCR measurements (Tables 2 and 3)) were subsequently used to simulate the two methods. A comparison between the experimental data of PPR and HPCR with the modified cross model can be seen in Figure 3.

A comparison between the experimental viscosity data and the modified cross model is shown in Figure 3.

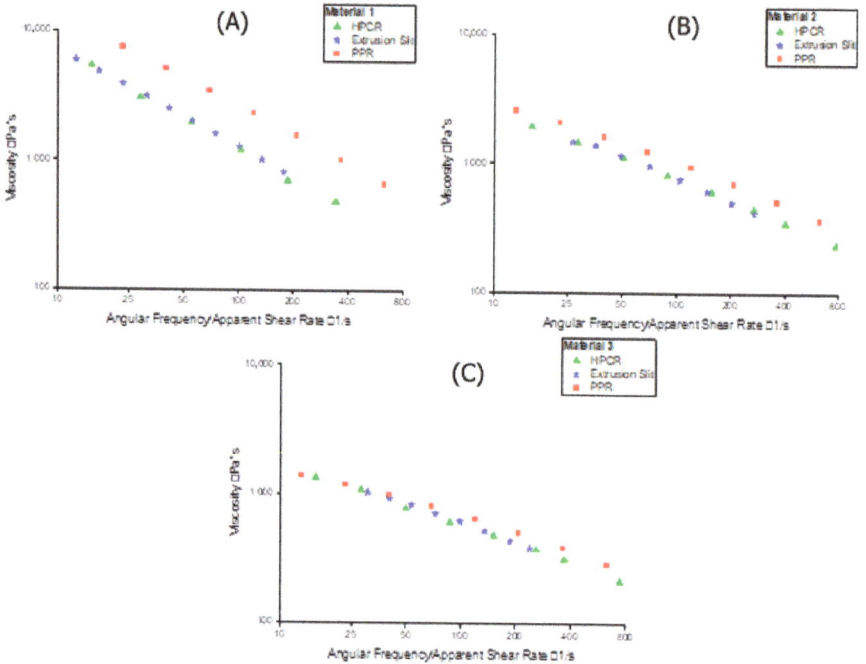

Figure 2. Comparison of pressure-flow-based measurements and oscillatory parallel-plate measurement of HDPEs at 200 °C. (**A**–**C**) show the data for Materials 1, 2, and 3, respectively.

Table 2. Plate-plate rheometry (PPR)-based modified cross-law parameters for HDPE melt at 200 °C.

Parameter	Unit	Material 1	Material 2	Material 3
η_0	Pa.s	43,226	4559	1930
λ	s	0.362	0.125	0.0711
m	-	0.761	0.571	0.491

Table 3. High-pressure capillary rheometry (HPCR)-based modified cross-law parameters for HDPE melt at 200 °C.

Parameter	Unit	Material 1	Material 2	Material 3
η_0	Pa.s	56,796	3715	3175
λ	s	0.323	0.124	0.097
m	-	0.788	0.598	0.494

The cross-law flow behaviour index m is tends to unity for increasingly shear thinning behaviour. Indeed, only Material 3 indicates a Newtonian behaviour plateau with low m value. The zero shear viscosity (η_0) is strongly related to the M_w. As η_0 increases so too does the molecular weight of the polymer. The modified cross-law parameters for an HDPE melt are strongly related to the mass average molecular weight M_w according to Equations 11 and 13 [39]. For the tested materials, the exponents of the equations are listed in Table 4. The values are in good accordance to the literature [39].

$$\eta_0 \sim [M_w]^\alpha \tag{11}$$

$$m \sim [M_w]^\beta \tag{12}$$

$$\lambda \sim [M_w]^\kappa \tag{13}$$

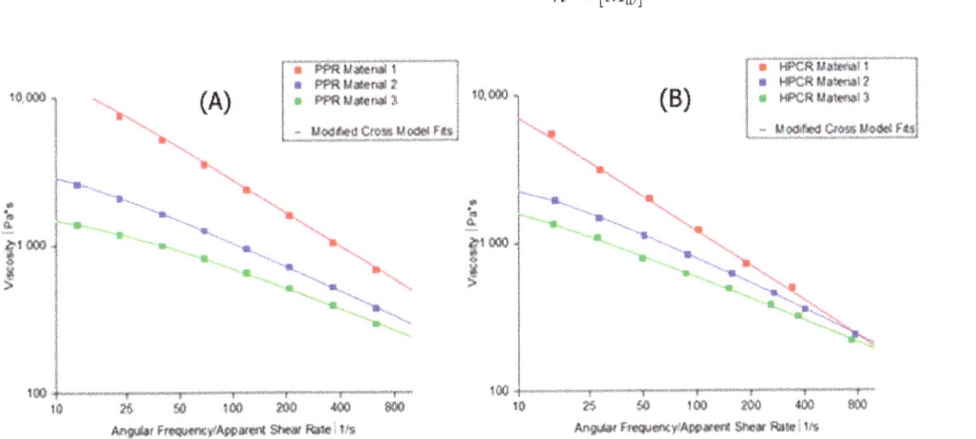

Figure 3. Comparison of the modified cross-law fits of Materials 1,2, and 3 with pressure-flow-based measurements (**B**) and oscillatory parallel-plate measurement (**A**) of HDPEs at 200 °C.

Table 4. Exponents of Equations (11)–(13).

	PPR	HPCR
α	3.09	3.12
β	0.43	0.45
κ	1.59	1.23

To determine the difference in viscosities measured by PPR and HPCR at a range of shear rates, we calculated the viscosity ratios by:

$$\varphi = \frac{\eta_{PPR}}{\eta_{HPCR}} \tag{14}$$

where η_{PPR} and η_{HPCR} are the viscosities measured at a particular shear rate by PPR and by HPCR, respectively.

From Table 5 it can be seen that the difference in rheological data measured by HPCR and PPR is immense for Material 1 and large for Material 2, whereas for Material 3 the two measurement techniques are in good accordance in the lower shear-rate region.

Table 5. Ratios of viscosities measured by PPR and HPCR at various shear rates $\dot{\gamma}$ for Materials 1–3.

	Material 1	Material 2	Material 3
$\dot{\gamma}$	φ	φ	φ
5	2.12	1.27	1.05
50	2.22	1.29	1.13
150	2.33	1.32	1.20
400	2.41	1.36	1.23

4.2. Viscoelasticity of HDPE Materials

The storage and loss modules of the three different materials give insights into their elastic and viscous properties, respectively, and are illustrated in Figure 4.

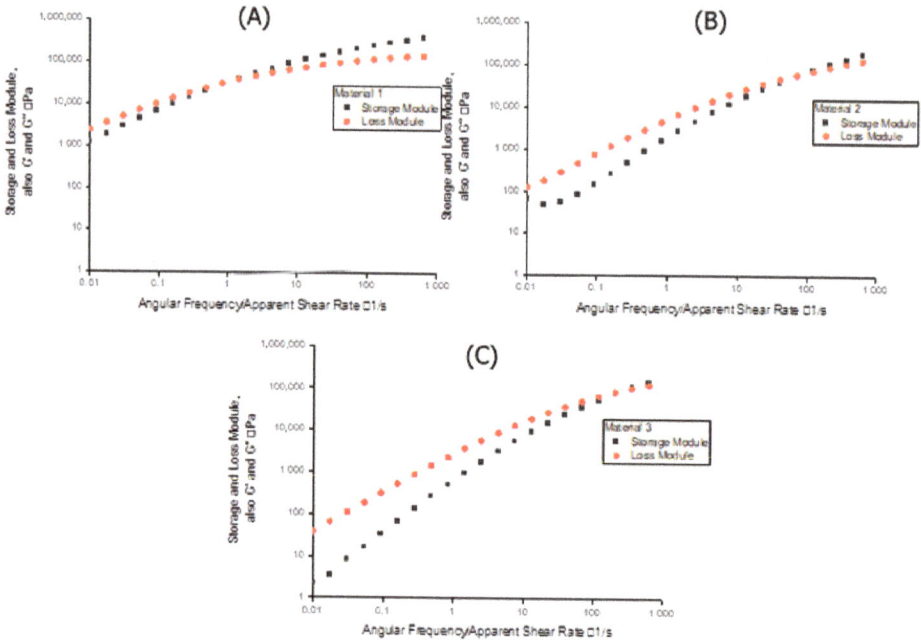

Figure 4. Storage and loss modules of three different HDPE materials. (**A–C**) show, respectively, the data of Materials 1, 2, and 3.

For Material 1, the elastic part is more dominant between 10 s^{-1} and 400 s^{-1}. For Material 2, the viscous part is more dominant up to 100 s^{-1}, beyond which the elastic part becomes more dominant. For Material 3 the viscous part dominates between 10 s^{-1} and 400 s^{-1}.

The cross-over point shifting to the low shear-rate region represents an increase in molecular mass. From Figure 4 it can be concluded that Material 1 has a higher average molecular weight than Material 2, which in turn has a higher average molecular weight than Material 3. Since the cross-over points of the three materials fall within a narrow range according to the storage and loss module, all three can be said to have similar molecular-weight distributions which is confirmed by the polydispersity index PI (Equation (15)). The PI for materials 1, 2 and 3 are 5.2, 5.0 and 4.5, respectively.

When the value of δ is close to 0° the material behaviour is elastic. If the value is close to 90° the material behaviour is viscous. From Figure 5 it can be seen that Material 1 exhibits the highest elastic behaviour and Material 3 the most viscous behaviour. The data in Figure 5 and Table 4 indicate that the Cox-Merz rule is applicable when the value of δ is 60° or higher. Below this value, the PPR and HPCR measurements start to differ significantly, and with decreasing phase shift the difference between PPR and HPCR increases.

$$PI = \frac{M_Z}{M_W} \qquad (15)$$

The cross-over points are in good correlation with the molecular masses from Table 1.

To investigate the influence of the elastic and viscous parts on the applicability of the Cox-Merz rule, we calculated the phase shift δ.

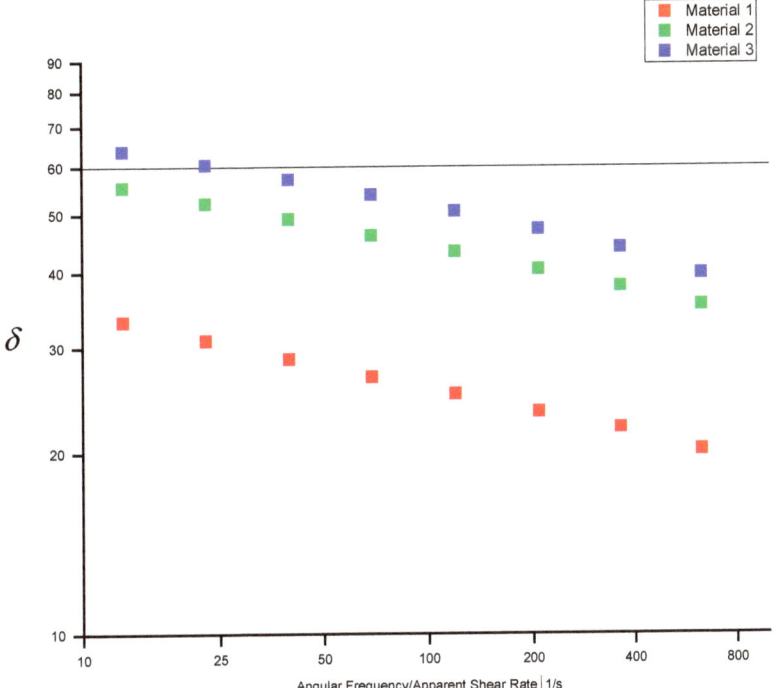

Figure 5. Phase shift of three different HDPE materials.

4.3. Comparison of Pipe-Head Simulations with Measured Rheology Curves

Using the modified cross law to describe the shear-thinning flow behaviour with the parameters given in Table 2, we performed three-dimensional CFD simulations of the pipe head and evaluated the pressure drop along it—calculated as the difference between the area-weighted average pressures between the pressure transducer and the outlet—for the three different HDPE melts. Subsequently, we compared the results to experimental data measured at a range of flow rates (see Figure 6). For Material 1, the pressure drop in the pipe head is shown in Figure 7.

The ratio χ between the pressure drop simulated based on PPR data p_{ppr} and the experimental data p_{exp} was calculated by

$$\chi = \frac{p_{ppr}}{p_{exp}} \tag{16}$$

The ratio ς between the pressure drop simulated based on HPCR data p_{HPCR} and the experimental data was calculated by

$$\varsigma = \frac{p_{HPCR}}{p_{exp}} \tag{17}$$

The results are given in Table 6.

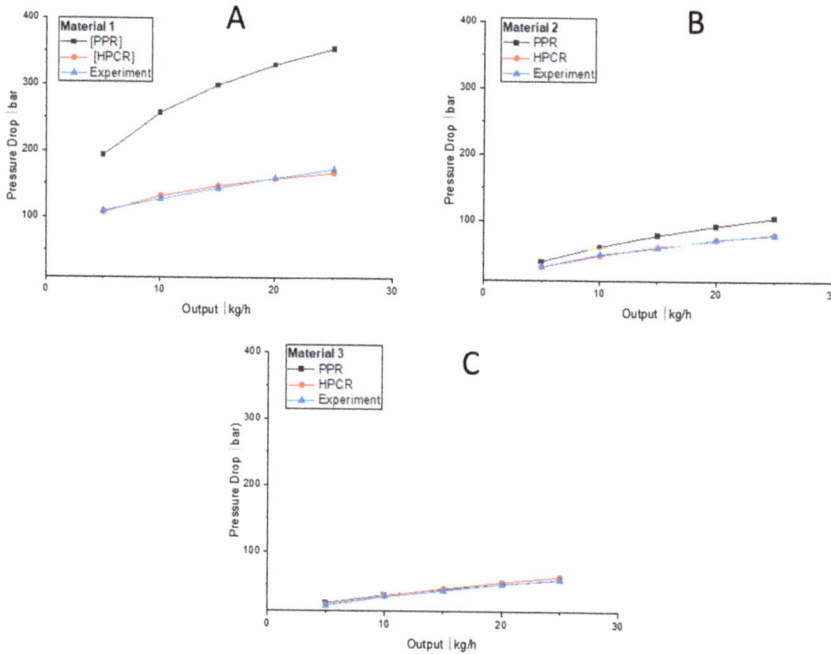

Figure 6. Comparison between the pressure drops according to two rheological models and experimental data at various mass flow rates for HDPE at 200 °C in a 32 mm pipe head. (**A**–**C**) show the simulation results for Materials 1, 2, and 3, respectively.

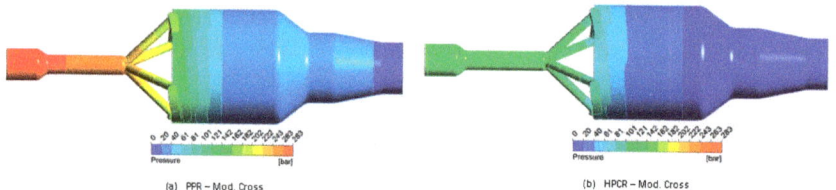

(a) PPR – Mod. Cross   (b) HPCR – Mod. Cross

Figure 7. Pressure distribution in the pipe head for Material 1 at a mass flow rate of 10 kg/h.

Table 6. Comparison of simulated pressure drops with experimental data at various output rates for Materials 1–3.

Output kg/h	Material 1 χ	ς	Material 2 χ	ς	Material 3 χ	ς
5	1.74	0.98	1.23	1	1.16	1
10	2.01	1.03	1.22	0.96	1.03	1
15	2.08	1.03	1.31	1.02	1.07	1.04
20	2.06	0.99	1.27	0.99	1	1.05
25	2.04	0.97	1.31	1.01	1	1.06

The results for Materials 1 and 2 show that the CFD simulations using PPR data overestimate the pressure drop, whereas the simulations based on HPCR data are in good accordance with the experiments. For Material 3, both the PPR-based and the HPTCR-based CFD simulations agree well with the experiments. As can be seen in Figure 6, the choice of measurement technique (PPR or HPCR) is indeed significant, and accurate experimental data is needed to yield good simulation results. Using incorrect rheological data for simulation may result in significant errors and consequently in equipment failure. For Material 1, the estimated pressure drop was twice as high as in the experiments, and for Material 2, the error was also substantial. The simulation results for Material 3, in contrast, show that there is no significant difference between the viscosities obtained.

5. Conclusions

We used three rheological measurement techniques to determine the shear-rate-dependent viscosity of three different HDPE materials: oscillatory parallel-plate, high-pressure capillary, and extrusion slit rheometry. While in parallel-plate rheometry the Cox-Merz relation is used to estimate the shear-dependent viscosity, in high-pressure capillary and extrusion slit rheometry the Weissenberg–Rabinowitsch relation is employed. Our data show that these methods differ significantly in accuracy depending on the material used. For Materials 1 and 2, PPR using the Cox-Merz relation overestimated the shear-rate-dependent viscosity significantly, whereas HPCR using the Bagley correction yielded results that accorded well with the experimental data (Figure 7). The applicability of the Cox-Merz rule to these materials depends heavily on the ratio between storage and loss module: If δ is $60°$ or greater, the Cox-Merz rule can be applied, while it becomes increasingly incorrect for decreasing values of δ. For Material 3, both HPCR- and PPR-based simulation results show good accordance with the rheological data from the experiments.

We conclude that the applicability of the Cox-Merz rule to HDPE materials is strongly dependent on the molecular mass of the material used. The polymers investigated differ significantly in molecular weight, and an increase in molecular weight resulted in considerable divergence from the Cox-Merz rule. The measured data of the very high-molecular-weight polymer Material 1 show a much greater difference between the shear viscosities obtained from PPR and HPCR than that of the data of the lower-molecular-weight Material 2, while for Material 3 no significant difference was observed. For HDPE—a long, straight polymer with limited branching and side chains—we thus conclude that with increasing molecular weight the disparity between the results from HPCR and PPR becomes significant. The Cox-Merz relation applies to HDPE only up to a particular molecular weight, more specifically up to an Mw of 85,000 according to our results.

Since viscosity data are essential to simulating the flow in various polymer-processing equipment, such as pipe heads and plasticizing screws, obtaining accurate viscosity curves is key to producing useful predictions. The consequence of using incorrect viscosity data in designing extrusion equipment is overestimation of the pressure drop in the die, which leads to completely different properties from those expected. Our results demonstrate that choosing the rheometry method according to the properties of the polymer of interest is crucial. In determining the viscosity of high-molecular-weight HDPE melts, capillary pressure flow is more reliable and accurate than is oscillatory measurement applying the Cox-Merz relation. The applicability of the Cox-Merz rule should also be examined critically for other long-chain polymers to ensure reliable rheological simulations when designing polymer-processing equipment.

Author Contributions: Conceptualization, design of experiments, and data treatment, R.R., W.R.; formal analysis, R.R, H.A.; writing—original draft preparation, R.R.; writing—review and editing, W.R., J.M.; funding acquisition, F.R. All authors have read and agreed to the published version of the manuscript.

Funding: This research was funded by Borealis Polyolefine GmbH.

Institutional Review Board Statement: Not applicable.

Informed Consent Statement: Not applicable.

Data Availability Statement: The data presented in this study are available on request from the corresponding author.

Acknowledgments: We would like to thank Borealis Polyolefine GmbH for the financial support.

Conflicts of Interest: There are no conflicts to declare.

References

1. Roller, M.B. Rheology of curing thermosets: A review. *Polym. Eng. Sci.* **1986**, *26*, 432–440. [CrossRef]
2. Rueda, M.M.; Auscher, M.-C.; Fulchiron, R.; Périé, T.; Martin, G.; Sonntag, P.; Cassagnau, P. Rheology and applications of highly filled polymers: A review of current understanding. *Prog. Polym. Sci.* **2017**, *66*, 22–53. [CrossRef]
3. Köpplmayr, T.; Luger, H.-J.; Burzic, I.; Battisti, M.G.; Perko, L.; Friesenbichler, W.; Miethlinger, J. A novel online rheometer for elongational viscosity measurement of polymer melts. *Polym. Test.* **2016**, *50*, 208–215. [CrossRef]
4. Włoch, M.; Datta, J. Rheology of polymer blends. In *Rheology of Polymer Blends and Nanocomposites: Theory, Modelling and Applications*; Thomas, S., Sarathchandran, C., Chandran, N., Eds.; Elsevier: Amsterdam, The Netherlands, 2019; pp. 19–29. ISBN 9780128169575.
5. Kamal, M.R.; Utracki, L.A.; Mirzadeh, A. Rheology of Polymer Alloys and Blends. In *Polymer Blends Handbook*, 2nd ed.; Utracki, L.A., Wilkie, C.A., Eds.; Springer: New York, NY, USA, 2014; pp. 725–873. ISBN 978-94-007-6064-6.
6. Sperling, L.H. Polymer alloys and blends thermodynamics and rheology, by L. A. Utracki, Hanser, Munich, 1989, 356 pp. Price: $90.00. *J. Polym. Sci. C Polym. Lett.* **1990**, *28*, 387. [CrossRef]
7. Schroyen, B.; Vlassopoulos, D.; van Puyvelde, P.; Vermant, J. Bulk rheometry at high frequencies: A review of experimental approaches. *Rheol Acta* **2020**, *59*, 1–22. [CrossRef]
8. Laun, M.; Auhl, D.; Brummer, R.; Dijkstra, D.J.; Gabriel, C.; Mangnus, M.A.; Rüllmann, M.; Zoetelief, W.; Handge, U.A. Guidelines for checking performance and verifying accuracy of rotational rheometers: Viscosity measurements in steady and oscillatory shear (IUPAC Technical Report). *Pure Appl. Chem.* **2014**, *86*, 1945–1968. [CrossRef]
9. Cox, W.P.; Merz, E.H. Correlation of dynamic and steady flow viscosities. *J. Polym. Sci.* **1958**, *28*, 619–622. [CrossRef]
10. Bair, S.; Yamaguchi, T.; Brouwer, L.; Schwarze, H.; Vergne, P.; Poll, G. Oscillatory and steady shear viscosity: The Cox Merz rule, superposition, and application to EHL friction. *Tribol. Int.* **2014**, *79*, 126–131. [CrossRef]
11. Mead, D.W. Analytic derivation of the Cox-Merz rule using the MLD "toy" model for polydisperse linear polymers. *Rheol Acta* **2011**, *50*, 837–866. [CrossRef]
12. Wen, Y.H.; Lin, H.C.; Li, C.H.; Hua, C.C. An experimental appraisal of the Cox-Merz rule and Laun's rule based on bidisperse entangled polystyrene solutions. *Polymer* **2004**, *45*, 8551–8559. [CrossRef]
13. Shan, L.; Tan, Y.; Richard Kim, Y. Applicability of the Cox-Merz relationship for asphalt binder. *Constr. Build. Mater.* **2012**, *37*, 716–722. [CrossRef]
14. Ahmad, N.H.; Ahmed, J.; Hashim, D.M.; Manap, Y.A.; Mustafa, S. Oscillatory and steady shear rheology of gellan/dextran blends. *J. Food Sci. Technol.* **2015**, *52*, 2902–2909. [CrossRef] [PubMed]
15. Stieger, S.; Kerschbaumer, R.C.; Mitsoulis, E.; Fasching, M.; Berger-Weber, G.R.; Friesenbichler, W.; Sunder, J. Contraction and capillary flow of a carbon black filled rubber compound. *Polym. Eng. Sci.* **2020**, *60*, 32–43. [CrossRef]
16. Miao, X.; Guo, Y.; He, L.; Meng, Y.; Li, X. Rheological behaviors of a series of hyperbranched polyethers. *Chin. J. Polym. Sci.* **2015**, *33*, 1574–1585. [CrossRef]
17. Ebagninin, K.W.; Benchabane, A.; Bekkour, K. Rheological characterization of poly(ethylene oxide) solutions of different molecular weights. *J. Colloid Interface Sci.* **2009**, *336*, 360–367. [CrossRef] [PubMed]
18. Yüce, C.; Willenbacher, N. Challenges in Rheological Characterization of Highly Concentrated Suspensions-A Case Study for Screen-printing Silver Pastes. *J. Vis. Exp.* **2017**, *122*, e55377. [CrossRef]
19. Snijkers, F.; Vlassopoulos, D. Appraisal of the Cox-Merz rule for well-characterized entangled linear and branched polymers. *Rheol Acta* **2014**, *53*, 935–946. [CrossRef]
20. Robertson, C.G.; Roland, C.M.; Puskas, J.E. Nonlinear rheology of hyperbranched polyisobutylene. *J. Rheol.* **2002**, *46*, 307–320. [CrossRef]
21. Järvelä, P.; Shucai, L.; Järvelä, P. Dynamic mechanical properties and morphology of polypropylene/maleated polypropylene blends. *J. Appl. Polym. Sci.* **1996**, *62*, 813–826. [CrossRef]
22. Matsumoto, T.; Hitomi, C.; Onogi, S. Rheological Properties of Disperse Systems of Spherical Particles in Polystyrene Solution at Long Time-Scales. *Trans. Soc. Rheol.* **1975**, *19*, 541–555. [CrossRef]
23. Kulicke, W.-M.; Porter, R.S. Relation between steady shear flow and dynamic rheology. *Rheol Acta* **1980**, *19*, 601–605. [CrossRef]
24. Venkatraman, S.; Okano, M. A comparison of torsional and capillary rheometry for polymer melts: The Cox-Merz rule revisited. *Polym. Eng. Sci.* **1990**, *30*, 308–313. [CrossRef]
25. Schulken, R.M.; Cox, R.H.; Minnick, L.A. Dynamic and steady-state rheological measurements on polymer melts. *J. Appl. Polym. Sci.* **1980**, *25*, 1341–1353. [CrossRef]
26. Booij, H.C.; Leblans, P.; Palmen, J.; Tiemersma-Thoone, G. Nonlinear viscoelasticity and the Cox-Merz relations for polymeric fluids. *J. Polym. Sci. Polym. Phys. Ed.* **1983**, *21*, 1703–1711. [CrossRef]
27. Ferri, D.; Lomellini, P. Melt rheology of randomly branched polystyrenes. *J. Rheol.* **1999**, *43*, 1355–1372. [CrossRef]
28. Al-Hadithi, T.S.R.; Barnes, H.A.; Walters, K. The relationship between the linear (oscillatory) and nonlinear (steady-state) flow properties of a series of polymer and colloidal systems. *Colloid Polym. Sci.* **1992**, *270*, 40–46. [CrossRef]

29. Tam, K.C.; Tiu, C. Modified cox-merz rule for charged polymer systems in solution. *J. Macromol. Sci. Part B* **1994**, *33*, 173–184. [CrossRef]
30. Doraiswamy, D.; Mujumdar, A.N.; Tsao, I.; Beris, A.N.; Danforth, S.C.; Metzner, A.B. The Cox-Merz rule extended: A rheological model for concentrated suspensions and other materials with a yield stress. *J. Rheol.* **1991**, *35*, 647–685. [CrossRef]
31. Marrucci, G. Dynamics of entanglements: A nonlinear model consistent with the Cox-Merz rule. *J. Non-Newton. Fluid Mech.* **1996**, *62*, 279–289. [CrossRef]
32. Kissin, Y.V. *Polyethylene: End-Use Properties and Their Physical Meaning*; Kissin, Y.V., Ed.; Hanser Publishers: Munich, Germany, 2013; ISBN 978-1-56990-520-3.
33. Winter, H.H. Three views of viscoelasticity for Cox-Merz materials. *Rheol. Acta* **2009**, *48*, 241–243. [CrossRef]
34. Raj, A.; Rajak, D.K.; Gautam, S.; Guria, C.; Pathak, A.K. Shear Rate Estimation: A Detailed Review. In Proceedings of the Offshore Technology Conference, Houston, TX, USA, 2 May 2016.
35. Bagley, E.B. End Corrections in the Capillary Flow of Polyethylene. *J. Appl. Phys.* **1957**, *28*, 624–627. [CrossRef]
36. Allen, G.M. Rheology of polymeric systems, principles and applications. By P. J. Carreau, D.C.R. De Kee, and R. P. Chhabra, Hanser/Gardner Publications, Cincinnati, OH, 1997, 520 pp., $197.50. *AIChE J.* **1999**, *45*, 1836–1837. [CrossRef]
37. ANSYS POLYFLOW. *User Manual 18.1*; ANSYS: Canonsburg, PA, USA, 2017.
38. Cross, M.M. Rheology of non-Newtonian fluids: A new flow equation for pseudoplastic systems. *J. Colloid Sci.* **1965**, *20*, 417–437. [CrossRef]
39. Wasserman, S.H.; Graessley, W.W. Prediction of linear viscoelastic response for entangled polyolefin melts from molecular weight distribution. *Polym. Eng. Sci.* **1996**, *36*, 852–861. [CrossRef]

Article

Fluorinated Ethylene Propylene Coatings Deposited by a Spray Process: Mechanical Properties, Scratch and Wear Behavior

Najoua Barhoumi [1,2,*], Kaouther Khlifi [1,2], Abderrahim Maazouz [3] and Khalid Lamnawar [3,*]

1. Laboratoire de Mécanique, Matériaux et Procédés, Ecole Nationale Supérieure d'Ingénieurs de Tunis, Université de Tunis, 5, Avenue Taha Husseïn, Montfleury, Tunis 1008, Tunisia; kaouther.khlifi@ipeiem.utm.tn
2. Institut Préparatoire aux Etudes d'Ingénieurs d'El-Manar, Université d'El-Manar, B.P 244, Tunis 2092, Tunisia
3. CNRS, UMR 5223, Ingénierie des Matériaux Polymères, INSA Lyon, Université de Lyon, F-69621 Villeurbanne, France; abderrahim.maazouz@insa-lyon.fr
* Correspondence: najoua.barhoumi@ipeiem.utm.tn (N.B.); khalid.lamnawar@insa-lyon.fr (K.L.)

Abstract: To increase the lifetime of metallic molds and protect their surface from wear, a fluorinated ethylene propylene (FEP) polymer was coated onto a stainless-steel (SS304) substrate, using an air spray process followed by a heat treatment. The microstructural properties of the coating were studied using scanning electron microscopy (SEM) and energy-dispersive X-ray spectroscopy (EDS) as well as X-ray diffraction. The mechanical properties and adhesion behavior were analyzed via a nanoindentation test and progressive scratching. According to the results, the FEP coating had a smooth and dense microstructure. The mechanical properties of the coatings, i.e., the hardness and Young's modulus, were 57 ± 2.35 and 1.56 ± 0.07 GPa, respectively. During scratching, successive delamination stages (initiation, expansion, and propagation) were noticed, and the measured critical loads L_{C1} (3.36 N), L_{C2} (6.2 N), and L_{C3} (7.6 N) indicated a high adhesion of the FEP coating to SS304. The detailed wear behavior and related damage mechanisms of the FEP coating were investigated employing a multi-pass scratch test and SEM in various sliding conditions. It was found that the wear volume increased with an increase in applied load and sliding velocity. Moreover, the FEP coating revealed a low friction coefficient (around 0.13) and a low wear coefficient (3.1×10^{-4} mm^3 N m^{-1}). The investigation of the damage mechanisms of the FEP coating showed a viscoelastic plastic deformation related to FEP ductility. Finally, the coating's resistance to corrosion was examined using electrochemical measurements in a 3.5 wt% NaCl solution. The coating was found to provide satisfactory corrosion protection to the SS304 substrate, as no corrosion was observed after 60 days of immersion.

Keywords: spray process; FEP coating; scratch behavior; friction and wear resistance

1. Introduction

Stainless steel is widely use in the fabrication of molds elements using in the chemical and food processing industry, in metal forming and pharmaceutical molds [1–4]. In sliding mechanical systems during service, wear and friction of these metallic components produce several problems and lead to scratches, severe wear, and plastic deformation [5]. Furthermore, in the strongly aggressive media containing chloride anions, stainless steels lack corrosion resistance [6,7]. These failure mechanisms directly involve the molds' surface, creating a serious problem in the manufacturing of casting, and shortens molds' life [8,9]. To solve this problem, which causes additional costs for mold reconditioning and reduced productivity, various metallic, ceramic, and polymeric coatings have been proposed. Nickel-based metallic and zirconia-based ceramic protective coatings were deposited on steel diecasting molds [8]. According to Óscar Rodríguez-Alabanda [9], fluoropolymer coatings can be applied as anti-adherent coatings on aluminum–magnesium substrates for food containers. In our previous work [10], polymeric perfluoroalkoxy coatings were used to protect agri-food molds to prevent corrosion and wear.

Fluorinated polymeric coatings such as polytetrafluoroethylene (PTFE), perfluoroalkoxy (PFA) and fluorinated ethylene propylene (FEP) are the frequently used fluoroplastics [11–13]. These fluorinated polymers are widely applied on molds elements used in the food sector to prevent adherence to substrates, to achieve low chemical reactivity and good corrosion resistance, to facilitate cleaning, to improve unmolding, and to increase the service life of mold elements by protecting their surface from cavity damage through abrasive wear [14]. In this way, chemical stability at high and low temperatures is obtained, and the coating process is simple, short, and leads to low friction coefficients [6,14–16]. These outstanding properties of anti-adherent fluoropolymers result from the fact that C–C bonds are strengthened by the incorporation of fluorine atoms into organic materials. In addition, the protective properties of fluoropolymer coatings can be related to the chemical (fluorine and chlorine functional groups) and physical (significant ordering of the polymeric chains) characteristics of the resins [17].

Thanks to its self-lubricating properties, FEP is extensively used in polymer/metal tribological systems working in dry, mixing, and wet environments [18,19]; it also induces high corrosion resistance due to its action as a barrier layer between corrosive species and the metal surface [20]. Fluorinated polymers can be produced by two processes. One of them involves the spray of water-based dispersions of fluoropolymer resins, followed by heat treatment at a temperature between 260 °C and 360 °C. The thickness of the smooth film produced by this method can reach 1 mm [21]. The second process involves the electrostatic spraying of polymer powders on a substrate [22]. The first process (spray) is a simple spraying technique with growing scientific interests and industrial applications in the field of industrial molds and components. This method offers corrosion protection and results in increases in mechanical and physical durability and wear resistance.

Recently, efforts have been devoted to exploring the scratch and wear behavior and to enhancing the corrosion and wear resistance of different forms of steel by fluoropolymer coating deposition [8,9,15,16]. The coating adhesion of FEP to SS304 is known to be critical, and many studies have been carried out to investigate the scratch behavior and related damage mechanisms of polymeric materials [23] and the adhesion behavior of coatings [24,25]. Xu et al. [23] studied the scratch behavior of alternating multi-layered PMMA/PC materials and found that the delamination process during scratching could be divided into three stages of delamination (initiation, expansion, and propagation).

The protective properties of an FEP coating in saline aqueous solutions, its adhesion using scratch tests, and its wear behavior using multi-pass scratch tests have not yet been investigated. The present work aimed to develop non-stick coatings to protect the SS304 mold substrate and to ameliorate its surface performance. Stainless-steel specimens (SS304) coated with the fluoropolymer FEP were prepared using the spray coating technique. The process parameters of FEP deposit are illustrated. The structure of the FEP coating and its adhesion were studied. Nanoindentation tests were used to investigate the mechanical properties, and the friction and wear properties were explored using multi-pass scratch testing. The aim of this work was to develop FEP-based coatings, sprayed on SS304 steel, and to propose them as a suitable alternative in the manufacture of wear-resistant, anti-adherent, and corrosion-resistive molds.

2. Materials and Methods

2.1. Material

A stainless steel (SS304) sheet with a chemical composition according to Table 1 was machined to rectangular specimens with dimensions of $100 \times 80 \times 2$ (mm^3). The FEP resin, supplied by Whitford (England) as an aqueous solution with a purity of 99.9%, was used to coat the steel substrates. The coating deposition was carried out by RIET industry (Zaghouan—TUNISIa).

Table 1. Chemical composition of SS304.

Element	Fe	C	Si	Mn	P	S	N	Cr	Mo	Ni	Cu	Co
W_t (%)	70.594	0.049	0.42	1.9	0.024	0.025	0.078	18.1	0.35	8.06	0.35	0.05

2.2. Coating Deposition

Before coating deposition, the steel substrate was cleaned with acetone and polished with a series of silicon carbide sandpapers. Subsequently, it was roughened by sandblasting to a surface finish Ra ≈ 3 µm. Then, it was dried at 60 °C for 24 h. Figure 1 details the deposition of the FEP solution onto the substrate surface using the air spray process. An FEP coating was obtained by repeating the spraying–drying process twice. The spray conditions were maintained at a liquid flow rate of 2.5 mL min^{-1}, an air pressure of 3 bars, and a nozzle-to-substrate distance of 8 cm.

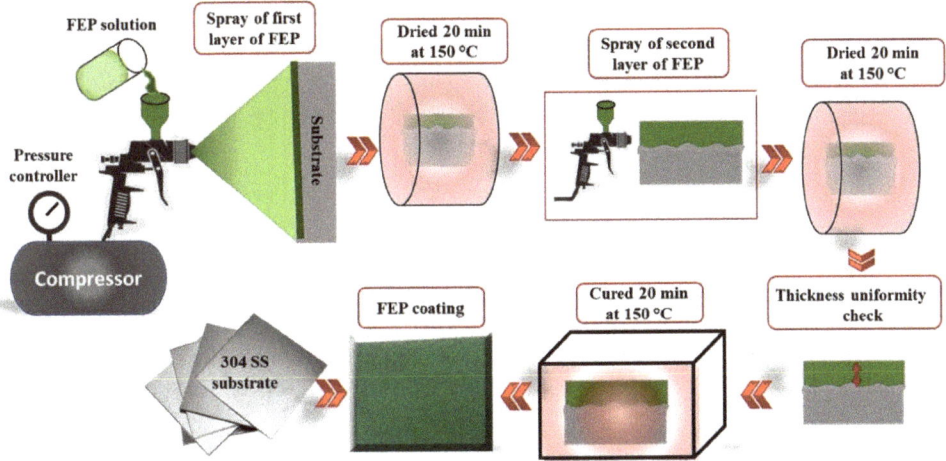

Figure 1. Processing cycle of FEP spraying on stainless steel.

The FEP solution was first applied by air gun spraying to the substrate, after which the materials were dried at 150 °C for 20 min, as shown in Figure 1. This was followed by a second round of the specimens being spray-coated with FEP and dried in an oven at 150 °C for 20 min. Subsequently, they were cured at 380 °C for 25 min for the coating to melt and spread over the substrate. Finally, they were cooled to ambient temperature. The thickness of the coatings was measured with a standard gauge (Ecotest plus device) and was found to be in the range of 44 ± 4 µm. Mechanical and wear characterization tests were conducted and repeated five times on the deposited FEP coatings.

2.3. Structural Characterization

The film composition and the microstructure of FEP coatings were evaluated using Scanning Electron Microscopy (SEM-JEOL JSM 6460LV -Suisse) in combination with energy-dispersive X-ray spectroscopy (EDS) for the identification of the chemical composition. X-ray spectra were obtained at a primary beam energy equal to 20 keV with an acquisition time of 120 s. The study of the phase composition of the samples was performed using X-ray diffraction (Bruker D8 Advance), and X-ray patterns were gathered from 2θ = 10° to 70°.

2.4. Nanoindentation Testing

The mechanical properties of the FEP coatings (hardness and Young's modulus) were studied using the nanoindentation technique developed by CSM Instruments (Anton Paar,

Graz, Austria). Tests were conducted with a nanoindenter equipped with a diamond Berkovich tip with a nominal angle equal to 65.3° and a nominal radius curvature of 20 nm. The maximum applied load was 10 mN, and the penetration depth was set to a value lower than 1/10th of the coating thickness to provide the real film properties and avoid the substrate effect [26]. A minimum of 15 measurements were carried out for each coating to ensure the reproducibility and repeatability of the results. The analysis of load–displacement graphs with the Oliver and Pharr method [27] allowed an estimation of the mechanical properties (Young's modulus, hardness of the film).

2.5. Scratch Testing

Film adhesion was investigated with a micro-scratch tester from CSM Instruments (Anton Paar). The indenter was equipped with a Rockwell diamond tip with a radius of 200 µm. The test was performed in progressive mode in a loading range between 0.3 and 25 N along 3 mm of the scratch length, to study the adhesion behavior and to determine the critical loads of the film. The scratch speed was 10 mm min^{-1}.

The tests were carried out in three consecutive steps: a pre-scan at 300 mN to determine the initial profile of the samples, a progressive scan from 0 to 25 N during which the penetration depth (P_d) was recorded in real time, and a post-scan at 100 mN. SEM was used to study the shape of the residual deformations after the scratch tests. Finally, the indenter conducted a post-scan at a constant load of 300 mN and measured the residual depth (Rd) of the scratch. After the test, the scratch morphology was examined by SEM. At least three progressive load scratch tests were carried out for each sample.

2.6. Tribological Characterization

Tribological tests were performed at a constant normal load using a multi-pass scratch test (CSM Instruments, Anton Paar, Graz, Austria) under a unidirectional loading over a sliding distance (d) of 3 mm. Each test was carried out at a temperature of 22 °C and a relative humidity of 46%. The applied load, F_n (1 and 3 N), the number of passes (100 for each load condition), and the sliding speed (V) were controlled and fixed during the wear testing. Then, wear tracks were examined by SEM.

For each test, the wear volume (V) was calculated by measuring the surface profile on the wear track, and the dissipated energy was determined with the following Formula (1) [28].

$$E = \sum \vec{F_t} \cdot \vec{d} = v \cdot t \sum F_t \qquad (1)$$

where F_t is the tangential load (N), d denotes the sliding distance (m), v is the sliding velocity (m/s), and t is the experiment duration (s). The wear coefficient is defined as the slope of the curve of the wear volume (V) vs. energy (E). A low coefficient wear corresponds to a high wear resistance.

2.7. Corrosion Test

The electrochemical experiments were performed in a conventional three-electrode system in which a saturated calomel electrode (SCE) was used as a reference electrode, a platinum sheet was taken as a counter electrode, and specimens (SS304 and FEP/SS304) with a 1.0 cm^2 area were exposed as the working electrode. Electrochemical measurements were carried out in 3.5% NaCl solution after a stabilization period of 1 h to attain stable E_{corr} values. The experiments were also conducted on the coated samples after continuous immersion in 3.5 wt% NaCl solution for 60 days.

Polarization studies were performed in an Electrochemical VoltaLab PGZ 301 (Lyon, France). Both cathodic and anodic polarization curves were recorded, and Tafel polarization curves were obtained by varying the electrode potential value from −0.6 to 0.4 V at a scan rate of 10 mV S^{-1}. The corrosion parameters included the corrosion potential (E_{corr}) and the corrosion current (I_{corr}), and the inhibition efficiency (I.E) was calculated by using Equation (2):

$$I.E = (1 - \frac{icorr_{coating}}{icorr_{SS304}}) \times 100 \qquad (2)$$

3. Results and Discussion

3.1. Microstructural Properties

The microstructure of the sprayed coatings is presented in Figure 2. From the top-view SEM micrographs (Figure 2a), it can be observed that the FEP coatings had a dense, poreless, and homogeneous structure. According to the EDS analysis (Figure 2b), the elementary composition of the prepared FEP coating consisted of C and F, suggesting that the coating was extraordinarily pure.

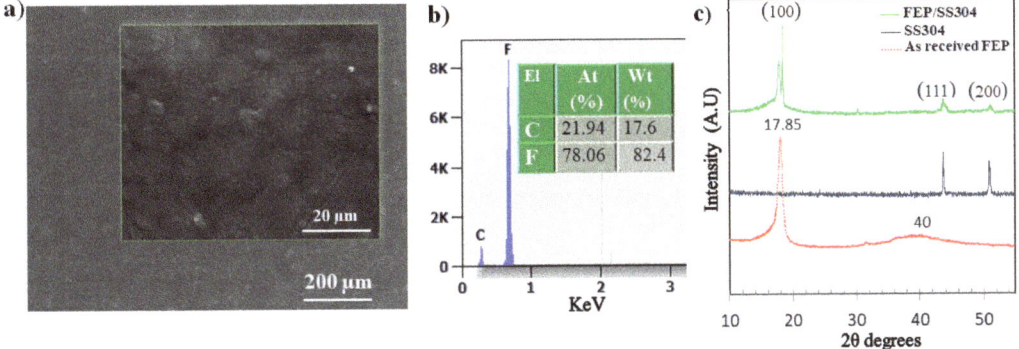

Figure 2. (a) SEM analysis of the FEP coatings, (b) a typical EDS spectrum, and (c) XRD pattern of the coatings.

XRD spectrums of as-received FEP, FEP coating, and SS304 substrate are shown in Figure 2c. The XRD diffractogram of as-received FEP resin was characterized by a sharp peak with the highest intensity and a broad peak. In general, a polymer with a crystalline region presents sharp X-ray diffraction peaks, whereas the X-ray diffraction peaks are broad for an amorphous polymer [29]. The FEP resin had a semicrystalline structure which resulted in a definite crystalline peak at a value of 2θ = 17.85° and a spread amorphous peak at 2θ = 40°. Similar results were obtained by Tcherdyntsev et al. [30]. The XRD spectrum of the FEP coating showed that the crystalline peak in the 2θ range of 17–19° corresponded to the (100) plane of two-dimensional hexagonal packing [31]. According to Wesley et al. [31], the peak with the highest intensity (2θ = 17.85°) is generally associated with the crystalline phase, in which the incident X-ray is diffracted by the plane (100) of the pseudohexagonal lattice structure.

3.2. Mechanical Properties

To investigate the mechanical behavior of the FEP coatings, they were subjected to nanoindentation measurements. During the indentation test, the penetration depth increased as a function of the applied load, and a plastic deformation occurred until the maximum load corresponded to the maximum loading. A typical load–depth curve is plotted in Figure 3, and as can be seen, the FEP coatings demonstrated an elasto-plastic behavior. The analysis of the load–depth curves using the Oliver–Pharr method [27] makes it possible to estimate mechanical properties, i.e., hardness, H, and Young's modulus, E, of films. The elastic modulus and the hardness of the coatings are presented in Figure 3b and were found to be 57 ± 2.35 MPa and 1.56 ± 0.07 GPa, respectively.

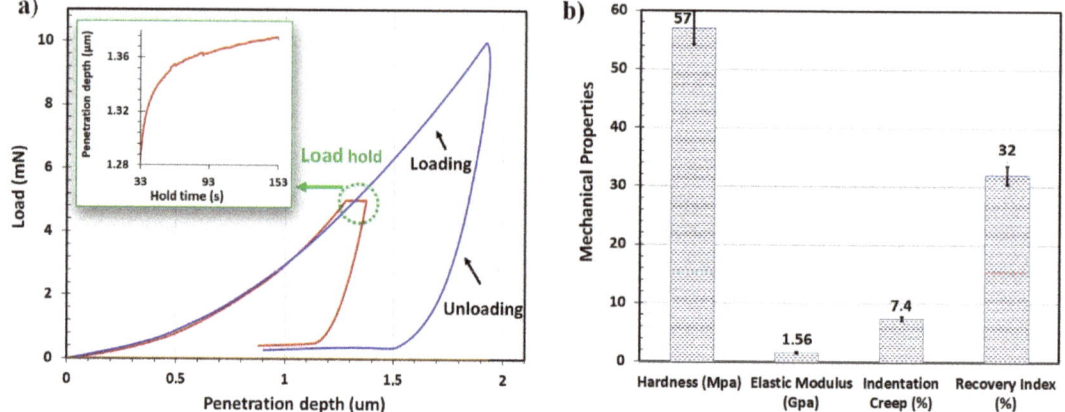

Figure 3. (a) Load–depth curves for FEP coatings with 5 and 10 mN maximum load and (b) measured values from nanoindentation tests on FEP coatings.

The indentation loading/unloading curve of the polymers appeared to be dependent on the holding time, as shown in the inset of Figure 3a. At the applied load of 10 mN, the penetration dept increased with time, indicating creep of FEP. The recovery index and indentation creep were derived from the standard nanoindentation measurements (Figure 3b). The film recovery index W_e/W_t implied an elastic recovery during unloading and a significant plastic deformation (70%) during loading.

3.3. Scratch Behavior

Figure 4a–e shows the results of conventional scratch tests on the FEP coatings. As the normal load increased, the width of the scratch became larger, and partial delamination of the FEP coating occurred at the end of the groove. The scratch progression was accompanied by successive degradations defined by three different critical loads (L_{c1}, L_{c2}, L_{c3}) [25]. These critical loads were determined by combining SEM observations of scratch tracks and the measurements of normal and tangential loads, as well as the penetration and residual depths during the scratch test.

The first critical load L_{c1} corresponding to the appearance of the first crack along the scratch pattern was localized at the edge, thus indicating cohesive failure. The second critical load L_{c2} was the applied normal load at which the extent of fracture events increased both at the bottom and at the edge of the scratch pattern, thereby leading to the observed repeated tensile cracking (adhesive failure). The third critical load L_{c3} was the applied normal load at which the coating exhibited a partial spalling.

Along the scratch path, the photos (Figure 4a–d) revealed that the scratch formed on the FEP coating could be divided into three stages. Stage I (until L_{c1} = 3.36 N), termed as smooth sliding, was the initial deformation stage of the FEP coating. Stage II (between L_{c1} = 3.36 N and L_{c2} = 6.2 N) was the transition region between smooth sliding and material removal featured by periodic micro-cracks. Stage III (from L_{c2} = 6.2 N) was defined as the material removal region where damage occurred and finally led to mass loss of FEP at L_{c3} = 7.6 N.

Figure 4e depicts the trends of the tangential force, penetration, and residual depth vs. normal force during the scratch test. The results were superimposed with the related SEM images of the residual scratch tracks of the investigated coatings. A purely elastic deformation took place at low applied scratch loads, and no scratch marks were detected along the starting segment of the scratch track, as shown in Figure 4a,e. By augmenting the normal load, as demonstrated in Figure 4e, a sudden slope change in the friction load and depth curves related to a first cracking phenomena was detected. This cohesive failure at L_{c1} was confirmed by SEM images (Figure 4b). When the applied load increased, the track demonstrated a propagation of micro-cracks, indicating adhesive failure at L_{c2}. At the highest critical load L_{c3}, a partial delamination was observed, thus indicating that the FEP coating bestowed good scratch resistance on the stainless-steel substrate.

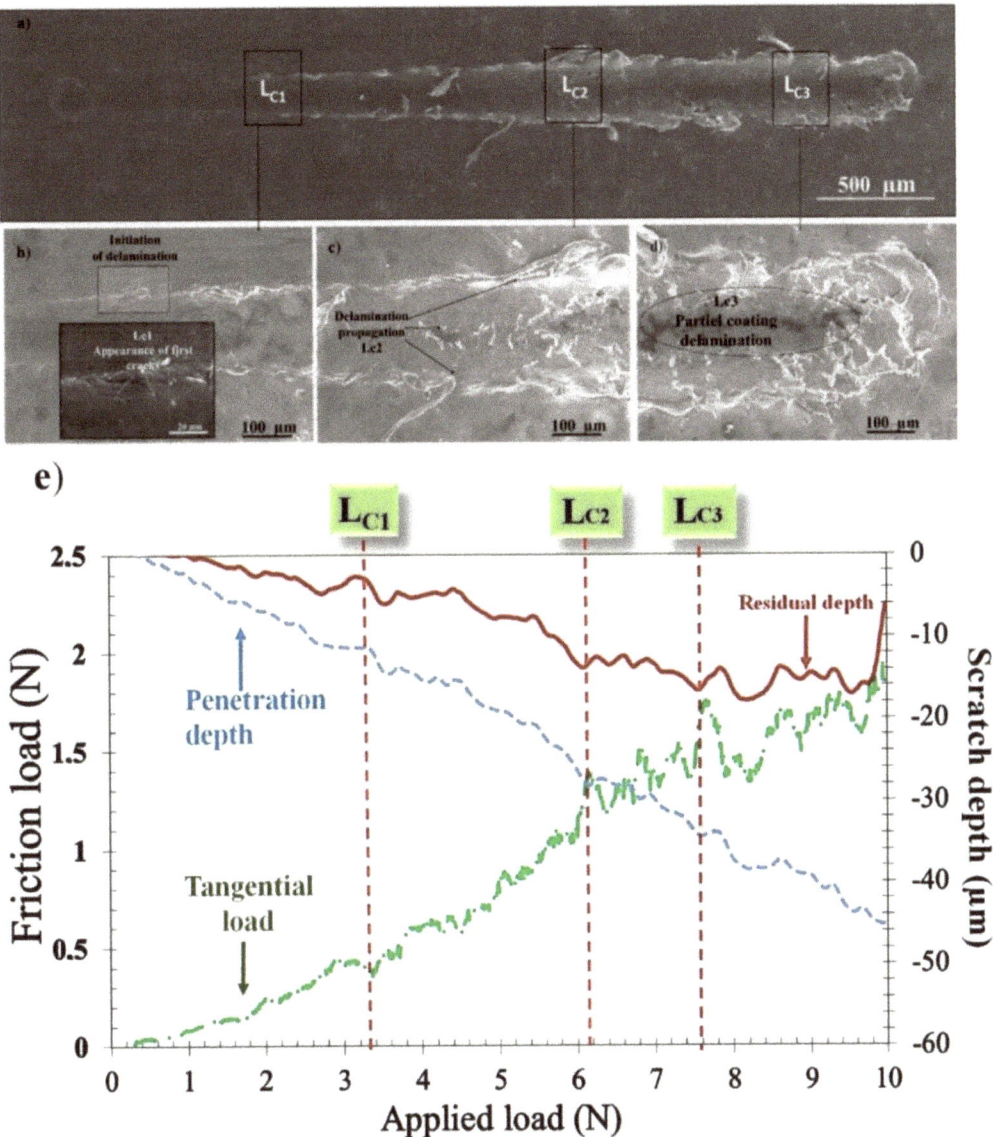

Figure 4. Progressive scratch test on an FEP coating. (**a**) SEM micrographs of scratch tracks. Magnified photo at (**b**) L_{c1}, (**c**) L_{c2}, and (**d**) L_{c3}. (**e**) Tangential load, penetration, and residual depths as functions of the applied load.

3.4. Friction and Wear Resistance

Multi-pass sliding scratch testing was performed on the FEP coatings to evaluate the friction behavior and wear resistance. The SEM micrographs and EDS analyses of the multi-pass scratch test of FEP coatings at different applied loads are presented in Figure 5.

The wear tracks obtained under 100 sliding cycles at constant loads of 0.3, 1, 2, and 3 N were examined using SEM. The applied load was lower than the cohesive failure load (L_{c1}) of the coatings.

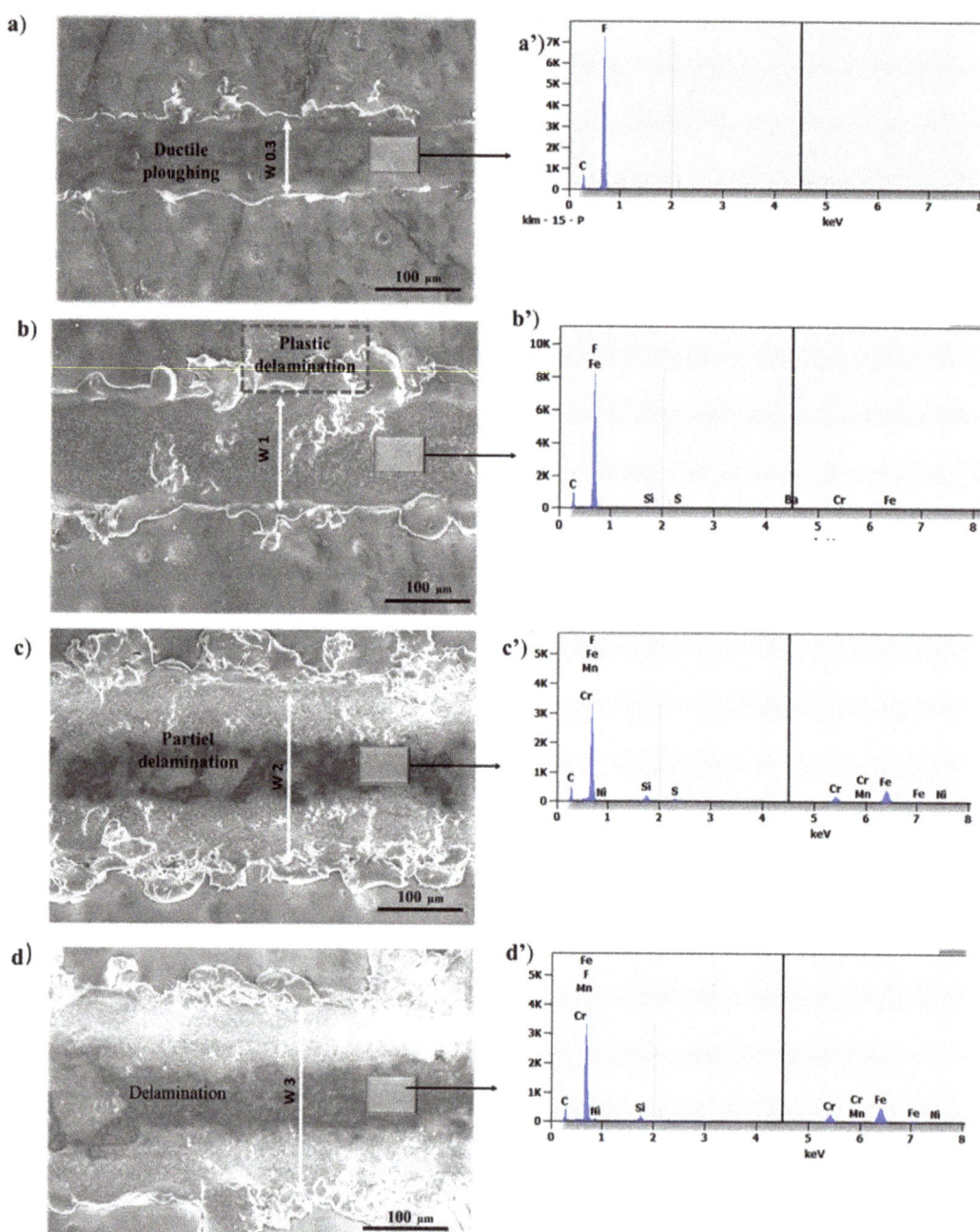

Figure 5. SEM micrographs and corresponding EDS analyses of 100-pass scratch tracks at (**a**) 0.3 N, (**b**) 1 N, (**c**) 2 N, and (**d**) 3 N.

Figure 5 illustrates the SEM images of the scratch tracks at various applied loads (0.3, 1, 2, and 3 N). As demonstrated in Figure 5, after a 100-pass scratching, the width of the scratch tracks increased considerably as the loading was raised. At 0.3 N, the SEM results revealed plastic deformation located at the edge and a ductile ploughing in the center of the track. At 1 N, an increased plastic deformation was noticed at the edge and in the center of the tracks, but the film still adhered to its substrate, and only cohesive failure occurred. However, at higher loads (2 and 3 N), an extensive plastic deformation occurred, leading to a chipping phenomenon. At this stage, the FEP coating demonstrated delamination, which was an indication of adhesive wear.

This result was proved by EDS analysis of the wear tracks, as shown in Figure 5 and Table 2. For the applied load of 0.3 N, the chemical composition present in the wear track (Figure 5a′) was the same as that obtained in the film (see Figure 2b). Then, at 1 N (Figure 5b′), trace amounts of iron were noticed, but nevertheless, after 100 passes and at a load of 1 N, the film behavior remained cohesive.

Table 2. Chemical composition of the film after and before wear tracks.

	Chemical Composition (wt%)							
	C	F	Si	S	Mn	Cr	Fe	Ni
Track of 0.3 N	21.10	78.90						
Track of 1 N	20.09	77.62	0.125	0.015		0.40	1.75	
Track of 2 N	14.45	63.56	0.80	0.020	0.38	4.90	14.3	1.59
Track of 3 N	10.58	54.6	0.86	0.02	0.46	6.86	24.17	2.45

However, at higher loads (2 and 3 N) (cf. Figure 5c,d′), the EDS analysis showed the appearance of the substrate elements C and Fe in the wear tracks during sliding (see Table 2). The elevated loading level and the high mechanical properties led to excessive Hertzian stresses, which promoted coating wear in the middle of the wear track [32].

The results of the wear tests in terms of friction coefficient and wear volume are illustrated in Figure 6. The evolution of the coefficients of friction of the coatings is presented as a function of the number of cycles for various applied loads in Figure 6a and after 100 cycles at various applied loads and velocities in Figure 6b.

The friction coefficient (COF) was recorded for 100 passes at each load. It can be noticed that the increase in the applied load led to a higher average value of the friction coefficient. At the lowest loads (0.3 and 1 N), the film had an exceedingly low friction coefficient that did not exceed 0.13 at 100 passes.

At higher applied loads (2 N and 3 N), the COF was, respectively, 0.18 and 0.24 at 100 passes. The increase in COF value for high applied loads was explained by the appearance of substrate after partial delamination of the FEP coating. The low friction coefficient reflects the high friction resistance of this coating. Such results were like those found by Nemati et al. [33]. Authors found that coating of the fluoropolymer PTFE on stainless steel was effective in lowering the COF from 1.2 to 0.16. The low and stable behavior of the COF was due to the synergistic lubrication effects of PTFE [34].

For each applied load and sliding velocity, the wear volume (V) and dissipated energy (E) were measured. The evolution of the wear volume was plotted versus the applied load for each sliding velocity, as shown in Figure 6c. Raising the applied load led to a significant increase in wear volume and dissipated energy regardless of the sliding velocity. This tendency was confirmed by the analysis of micrographs and the chemical composition of wear tracks, as previously observed. Concerning the velocity effect, the wear volume increased as the sliding velocity increased from 50 to 100 mm min^{-1}.

The evolution of the wear volume (V) was plotted versus the dissipated energy (E) (see Figure 6d). Each point on this plot represents an experiment carried out at a given load (0.3, 1 N, 2 N, and 3 N) and for a given sliding speed (10, 50, and 100 mm min^{-1}). The obtained graph reports a linear relation, and the slope of (E) vs. (V) is defined as the wear coefficient. Such an approach has already been successfully used in other works [26], and herein, the wear coefficient was equal to 3.12×10^{-4} mm^3 N m^{-1}. This was of the same order of magnitude as that reported in other studies using both PFA and PTFE with wear rates corresponding to $\sim 10^{-4}$ mm^3 N m^{-1} [35].

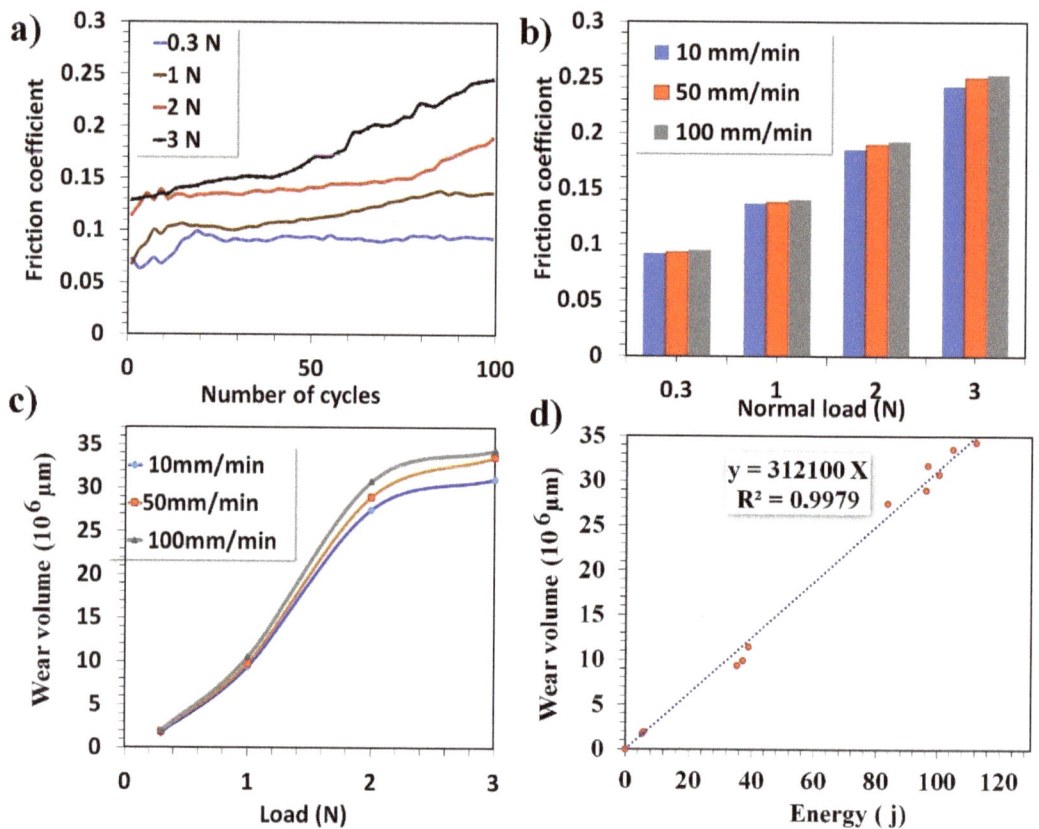

Figure 6. (**a**) COF plot of FEP coatings versus the number of cycles for 0.3, 1, 2, and 3 N at 50 mm min^{-1}, (**b**) COF bar graph and (**c**) wear volume versus load after 100 passes at various velocities, and (**d**) wear volume as a function of the dissipated energy.

3.5. Corrosion Resistance

The cathodic and anodic polarization curves recorded for the SS304 and FEP coating at t = 1 h and after 60 days of immersion in 3.5 wt% NaCl solution are shown in Figure 7, and the specific data (Table 3) were obtained by Tafel fitting of these curves. The corrosion resistance of the coating can be described in detail by the electrodynamic polarization curve, and the lower polarization current indicated a superior corrosion resistance [36].

The extrapolation of the polarization curves was carried out to determine I_{corr} and E_{corr} (Figure 7, Table 3). As seen in Figure 7, the corrosion current density of SS304 decreased from 33×10^{-4} to 1.58×10^{-4} µA/cm^2 with the coating deposition. In addition, the corrosion potential was raised from -450 mV to -240 mV. The enhancement of the SS304 corrosion resistance is associated with the physical-chemical characteristics of the FEP resins [10,37–40].

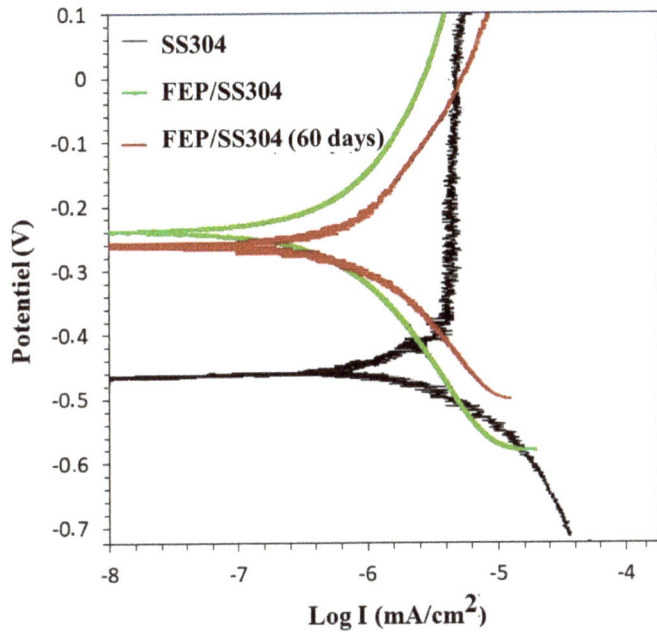

Figure 7. Polarization curve of SS304 and FEP coatings in 3.5% NaCl for 1 h and for 60 days.

Table 3. Potentiodynamic polarization data of SS304 and FEP coating.

Immersion Time	Samples	E_{corr} (mV)	I_{corr} ($\mu A/cm^{-2}$)	Inhibition Efficiency (%)
1 h	SS304	−450	33.1×10^{-4}	-
	FEP/SS304	−240	1.58×10^{-4}	95.22
60 days	FEP/SS304	−264	2.51×10^{-4}	92.4

To further demonstrate the long-term anti-corrosion performance of the FEP coatings, polarization measurements were performed after immersing them in a 3.5 wt% NaCl corrosive environment for 60 days. According to Table 3 and Figure 7, a very small negative shift in E_{corr} and few positive shifts of I_{corr} were observed. This indicated that the anti-corrosion performance of the coatings was consistent after the immersion test. Moreover, the inhibition efficiency of the coatings to SS304 remained remarkable after the 60-day immersion test. The FEP coatings could further extend their protective effect by inhibiting the penetration of the corrosion medium into the stainless-steel substrate when in an aggressive and humid environment. Thus, FEP protected the product from any kind of chemical corrosion.

4. Conclusions

In this work, a protective FEP coating was deposited on a stainless-steel substrate using the air spray process and then cured to obtain a compact and uniform film. The mechanical, adhesion, tribological, and corrosion performance of FEP coatings were investigated by nanoindentation, scratch test, SEM, and cyclic voltammetry. The following conclusions were drawn from the experimental results. The FEP coatings exhibited a dense, poreless, and homogeneous structure with a pseudohexagonal lattice crystalline structure. They showed good mechanicals properties, hardness, and Young's modulus and good scratch resistance to adhesive and cohesive failure, with a high adhesion to SS304. During the multi-pass scratch test of the FEP coating, both the wear volume and the dissipated energy increased when the normal applied load and sliding velocity were raised. It was also deduced that the coating exhibited a ductile behavior. After the wear tests, the wear on the FEP coatings with the increase of the applied load passed from cohesive wear to adhesive wear. The multi-pass scratch provided an easy and quick approach to study FEP wear resistance. The friction coefficient of the FEP coating did not exceed 0.13, and the wear coefficient was around 3.12×10^{-4} mm^3 N m^{-1}. The FEP coating enhanced the corrosion resistance of SS304 and provided a significant protection during 60 days of immersion in a NaCl solution.

Author Contributions: Conceptualization, N.B., K.L. and A.M.; methodology, N.B. and K.K.; validation, K.L. and A.M.; formal analysis, N.B. and K.K.; investigation, N.B. data curation, N.B. and K.K.; writing—original draft preparation, N.B.; writing—review and editing, K.L.; supervision, K.L.; project administration, K.L.; funding acquisition, K.L. and A.M. All authors have read and agreed to the published version of the manuscript.

Funding: This research received no external funding.

Institutional Review Board Statement: Not applicable.

Informed Consent Statement: Not applicable.

Data Availability Statement: Not applicable.

Acknowledgments: This work was carried out under a partnership agreement with the RIET industry Revetement Industriel EURO-Tunisien, Zaghouan, Tunisia. The authors acknowledge their manager Ramzi Hachicha for his assistance.

Conflicts of Interest: The authors declare no conflict of interest.

References

1. Boillot, P.; Peultier, J. Use of Stainless Steels in the Industry: Recent and Future Developments. *Procedia Eng.* **2014**, *83*, 309–321. [CrossRef]
2. Bodur, T.; Cagri-Mehmetoglu, A. Removal of *Listeria monocytogenes*, *Staphylococcus aureus* and *Escherichia coli* O157:H7 biofilms on stainless steel using scallop shell powder. *Food Control.* **2012**, *25*, 1–9. [CrossRef]
3. Ismaïl, R.; Aviat, F.; Michel, V.; Le Bayon, I.; Gay-Perret, P.; Kutnik, M.; Fédérighi, M. Methods for Recovering Microorganisms from Solid Surfaces Used in the Food Industry: A Review of the Literature. *Int. J. Environ. Res. Public Health* **2013**, *10*, 6169–6183. [CrossRef]
4. Cooper, I.; Tice, P. Food contact coatings—European legislation and future predictions. *Surf. Coat. Int. Part B Coat. Trans.* **2001**, *84*, 105–112. [CrossRef]
5. Berman, D.; Erdemir, A.; Sumant, A.V. Reduced wear and friction enabled by graphene layers on sliding steel surfaces in dry nitrogen. *Carbon* **2013**, *59*, 167–175. [CrossRef]
6. Zhang, H.; Zhao, Y.; Jiang, Z. Effects of temperature on the corrosion behavior of 13Cr martensitic stainless steel during exposure to CO_2 and Cl^- environment. *Mater. Lett.* **2005**, *59*, 3370–3374. [CrossRef]
7. Ziemniak, S.; Hanson, M. Corrosion behavior of 304 stainless steel in high temperature, hydrogenated water. *Corros. Sci.* **2002**, *44*, 2209–2230. [CrossRef]
8. Peter, I.; Rosso, M.; Gobber, F.S. Study of protective coatings for aluminum die casting molds. *Appl. Surf. Sci.* **2015**, *358*, 563–571. [CrossRef]
9. Rodríguez-Alabanda, Ó.; Romero, P.E.; Soriano, C.; Sevilla, L.; Guerrero-Vaca, G. Study on the Main Influencing Factors in the Removal Process of Non-Stick Fluoropolymer Coatings Using Nd:YAG Laser. *Polymers* **2019**, *11*, 123. [CrossRef] [PubMed]
10. Barhoumi, N.; Dhiflaoui, H.; Kaouther, K.; Ben Rhouma, A.; Hamdi, F. Study of Wear and Corrosion Performance of Fluoropolymer PFA Electrostatically Deposited on 304 Steel. In *Advances in Mechanical Engineering and Mechanics II*; Lecture Notes in Mechanical Engineering; Bouraoui, T., Ed.; Springer: Cham, Switzerland, 2021; pp. 103–110. [CrossRef]
11. Primc, G. Recent Advances in Surface Activation of Polytetrafluoroethylene (PTFE) by Gaseous Plasma Treatments. *Polymers* **2020**, *12*, 2295. [CrossRef] [PubMed]
12. Ohkubo, Y.; Okazaki, Y.; Shibahara, M.; Nishino, M.; Seto, Y.; Endo, K.; Yamamoto, K. Effects of He and Ar Heat-Assisted Plasma Treatments on the Adhesion Properties of Polytetrafluoroethylene (PTFE). *Polymers* **2021**, *13*, 4266. [CrossRef] [PubMed]
13. Shim, E.; Jang, J.-P.; Moon, J.-J.; Kim, Y. Improvement of Polytetrafluoroethylene Membrane High-Efficiency Particulate Air Filter Performance with Melt-Blown Media. *Polymers* **2021**, *13*, 4067. [CrossRef] [PubMed]

14. Ebnesajjad, S.; Khaladkar, P.R. *Fluoropolymer Applications in the Chemical Processing Industries: The Definitive User's Guide and Handbook*; Elsevier Science: Amsterdam, The Netherlands, 2017.
15. Guerrero-Vaca, G.; Carrizo-Tejero, D.; Rodríguez-Alabanda, Ó.; Romero, P.E.; Molero, E. Experimental Study for the Stripping of PTFE Coatings on Al-Mg Substrates Using Dry Abrasive Materials. *Materials* **2020**, *13*, 799. [CrossRef]
16. Olifirov, L.K.; Stepashkin, A.A.; Sherif, G.; Tcherdyntsev, V.V. Tribological, Mechanical and Thermal Properties of Fluorinated Ethylene Propylene Filled with Al-Cu-Cr Quasicrystals, Polytetrafluoroethylene, Synthetic Graphite and Carbon Black. *Polymers* **2021**, *13*, 781. [CrossRef]
17. Cardoso, V.F.; Correia, D.M.; Ribeiro, C.; Fernandes, M.M.; Lanceros-Méndez, S. Fluorinated Polymers as Smart Materials for Advanced Biomedical Applications. *Polymers* **2018**, *10*, 161. [CrossRef]
18. Teng, H. Overview of the Development of the Fluoropolymer Industry. *Appl. Sci.* **2012**, *2*, 496–512. [CrossRef]
19. Chen, B.; Wang, J.; Yan, F. Friction and Wear Behaviors of Several Polymers Sliding Against GCr15 and 316 Steel Under the Lubrication of Sea Water. *Tribol. Lett.* **2011**, *42*, 17–25. [CrossRef]
20. Akram, W.; Rafique, A.F.; Maqsood, N.; Khan, A.; Badshah, S.; Khan, R.U. Khan Characterization of PTFE Film on 316L Stainless Steel Deposited through Spin Coating and Its Anticorrosion Performance in Multi Acidic Mediums. *Materials* **2020**, *13*, 388. [CrossRef]
21. Chen, X.; Yuan, J.; Huang, J.; Ren, K.; Liu, Y.; Lu, S.; Li, H. Large-scale fabrication of superhydrophobic polyurethane/nano-Al_2O_3 coatings by suspension flame spraying for anti-corrosion applications. *Appl. Surf. Sci.* **2014**, *311*, 864–869. [CrossRef]
22. Ferreira, E.S.; Giacomelli, C.; Spinelli, A. Evaluation of the inhibitor effect of l-ascorbic acid on the corrosion of mild steel. *Mater. Chem. Phys.* **2004**, *83*, 129–134. [CrossRef]
23. Xu, Y.; Qin, J.; Shen, J.; Guo, S.; Lamnawar, K. Scratch behavior and mechanical properties of alternating multi-layered PMMA/PC materials. *Wear* **2021**, *486–487*, 204069. [CrossRef]
24. Panin, S.V.; Luo, J.; Buslovich, D.G.; Alexenko, V.O.; Kornienko, L.A.; Bochkareva, S.A.; Byakov, A.V. Experimental—FEM Study on Effect of Tribological Load Conditions on Wear Resistance of Three-Component High-Strength Solid-Lubricant PI-Based Composites. *Polymers* **2021**, *13*, 2837. [CrossRef]
25. Moon, S.W.; Seo, J.; Seo, J.-H.; Choi, B.-H. Scratch Properties of Clear Coat for Automotive Coating Comprising Molecular Necklace Crosslinkers with Silane Functional Groups for Various Environmental Factors. *Polymers* **2021**, *13*, 3933. [CrossRef] [PubMed]
26. Khlifi, K.; Dhiflaoui, H.; Ben Aissa, C.; Barhoumi, N.; Larbi, A.B.C. Friction and Wear Behavior of a Physical Vapor Deposition Coating Studied Using a Micro-Scratch Technique. *J. Eng. Mater. Technol.* **2022**, *144*, 1–26. [CrossRef]
27. Oliver, W.C.; Pharr, G.M. An improved technique for determining hardness and elastic modulus using load and displacement sensing indentation experiments. *J. Mater. Res.* **1992**, *7*, 1564–1583. [CrossRef]
28. Mendibide, C.; Fontaine, J.; Steyer, P.; Esnouf, C. Dry Sliding Wear Model of Nanometer Scale Multilayered TiN/CrN PVD Hard Coatings. *Tribol. Lett.* **2004**, *17*, 779–789. [CrossRef]
29. Panda, P.K.; Yang, J.-M.; Chang, Y.-H. Water-induced shape memory behavior of poly (vinyl alcohol) and p-coumaric acid-modified water-soluble chitosan blended membrane. *Carbohydr. Polym.* **2021**, *257*, 117633. [CrossRef]
30. Tcherdyntsev, V.V.; Olifirov, L.K.; Kaloshkin, S.D.; Zadorozhnyy, M.Y.; Danilov, V.D. Thermal and mechanical properties of fluorinated ethylene propylene and polyphenylene sulfide-based composites obtained by high-energy ball milling. *J. Mater. Sci.* **2018**, *53*, 13701–13712. [CrossRef]
31. Sulen, W.L.; Ravi, K.; Bernard, C.; Ichikawa, Y.; Ogawa, K. Deposition Mechanism Analysis of Cold-Sprayed Fluoropolymer Coatings and Its Wettability Evaluation. *J. Therm. Spray Technol.* **2020**, *29*, 1643–1659. [CrossRef]
32. Ordoñez, M.F.C.; Paruma, J.S.R.; Osorio, F.S.; Farias, M.C.M. The Effect of Counterpart Material on the Sliding Wear of TiAlN Coatings Deposited by Reactive Cathodic Pulverization. *Sci. Cum Ind.* **2015**, *3*, 59–66. [CrossRef]
33. Nemati, N.; Emamy, M.; Yau, S.; Kim, J.-K.; Kim, D.-E. High temperature friction and wear properties of graphene oxide/polytetrafluoroethylene composite coatings deposited on stainless steel. *RSC Adv.* **2016**, *6*, 5977–5987. [CrossRef]
34. Wang, H.; Sun, A.; Qi, X.; Dong, Y.; Fan, B. Experimental and Analytical Investigations on Tribological Properties of PTFE/AP Composites. *Polymers* **2021**, *13*, 4295. [CrossRef]
35. Sidebottom, M.A.; Pitenis, A.A.; Junk, C.P.; Kasprzak, D.J.; Blackman, G.S.; Burch, H.E.; Harris, K.L.; Sawyer, W.G.; Krick, B. Ultralow wear Perfluoroalkoxy (PFA) and alumina composites. *Wear* **2016**, *362–363*, 179–185. [CrossRef]
36. Kumar, A.; Ghosh, P.K.; Yadav, K.L.; Kumar, K. Thermo-mechanical and anti-corrosive properties of MWCNT/epoxy nanocomposite fabricated by innovative dispersion technique. *Compos. Part B Eng.* **2017**, *113*, 291–299. [CrossRef]
37. Husain, E.; Nazeer, A.A.; Alsarraf, J.; Al-Awadi, K.; Murad, M.; Al-Naqi, A.; Shekeban, A. Corrosion behavior of AISI 316 stainless steel coated with modified fluoropolymer in marine condition. *J. Coat. Technol. Res.* **2018**, *15*, 945–955. [CrossRef]
38. Delimi, A.; Galopin, E.; Coffinier, Y.; Pisarek, M.; Boukherroub, R.; Talhi, B.; Szuneris, S. Investigation of the cor-rosion behavior of carbon steel coated with fluoropolymer thin films. *Surf. Coat. Technol.* **2011**, *205*, 4011–4017. [CrossRef]
39. Liang, J.; Azhar, U.; Men, P.; Chen, J.; Liu, Y.; He, J.; Geng, B. Fluoropolymer/SiO_2 encapsulated aluminum pigments for enhanced corrosion protection. *Appl. Surf. Sci.* **2019**, *487*, 1000–1007. [CrossRef]
40. Leivo, E.; Wilenius, T.; Kinos, T.; Vuoristo, P.; Mäntylä, T. Properties of thermally sprayed fluoropolymer PVDF, ECTFE, PFA and FEP coatings. *Prog. Org. Coat.* **2004**, *49*, 69–73. [CrossRef]

Review

A Journey from Processing to Recycling of Multilayer Waste Films: A Review of Main Challenges and Prospects

Geraldine Cabrera [1], Jixiang Li [1], Abderrahim Maazouz [1,2] and Khalid Lamnawar [1,*]

[1] Univ Lyon, CNRS, UMR 5223, Ingénierie des Matériaux Polymères, INSA Lyon, Université Claude Bernard Lyon 1, Université Jean Monnet, CEDEX, F-69621 Villeurbanne, France; geralcabrera09@gmail.com (G.C.); jixiang.li@insa-lyon.fr (J.L.); abderrahim.maazouz@insa-lyon.fr (A.M.)
[2] Hassan II Academy of Science and Technology, Rabat 10100, Morocco
* Correspondence: khalid.lamnawar@insa-lyon.fr

Abstract: In a circular economy context with the dual problems of depletion of natural resources and the environmental impact of a growing volume of wastes, it is of great importance to focus on the recycling process of multilayered plastic films. This review is dedicated first to the general concepts and summary of plastic waste management in general, making emphasis on the multilayer films recycling process. Then, in the second part, the focus is dealing with multilayer films manufacturing process, including the most common materials used for agricultural applications, their processing, and the challenges of their recycling, recyclability, and reuse. Hitherto, some prospects are discussed from eco-design to mechanical or chemical recycling approaches.

Keywords: circular economy; recycling; eco-design; coextrusion; multilayers

Citation: Cabrera, G.; Li, J.; Maazouz, A.; Lamnawar, K. A Journey from Processing to Recycling of Multilayer Waste Films: A Review of Main Challenges and Prospects. *Polymers* 2022, 14, 2319. https://doi.org/10.3390/polym14122319

Academic Editor: Emin Bayraktar

Received: 16 May 2022
Accepted: 5 June 2022
Published: 8 June 2022

Publisher's Note: MDPI stays neutral with regard to jurisdictional claims in published maps and institutional affiliations.

Copyright: © 2022 by the authors. Licensee MDPI, Basel, Switzerland. This article is an open access article distributed under the terms and conditions of the Creative Commons Attribution (CC BY) license (https://creativecommons.org/licenses/by/4.0/).

1. Introduction

Since the discovery of polyethylene and polypropylene during the 1950s, polymers materials have been become popular and they are widely used for many applications of our daily life. Nowadays, plastic production in Europe fluctuates around 57 million tons [1]. Packaging (39.6%) and the building sector (20.4%) are the biggest end-use markets for plastics. Meanwhile, the agricultural sector represents 3.4% of the total plastic demand [1]. Between all the plastic materials applications, flexible films became very popular mainly due to their versatility, lightness, resistance, and printability. Applications of polymer films are diverse, but they are usually classified into two categories: packaging and non-packaging. The packaging products are divided into consumer (primary packaging) and non-consumer (secondary and tertiary packaging) [2]. The principal function of primary packaging is to protect the product. Then, the secondary and tertiary main purpose is to group different primary packages for easy and safe transportation [3]. Regarding the final structure of the flexible films, they can be divided into monolayer and multilayer films. Monolayer films are commonly used for tertiary and secondary packaging and less often for agricultural and building applications. Meanwhile, the structure of multilayer films is made with different layers that can be polymeric or non-polymeric materials, such as paper or aluminum foils. Using modern technologies such as coextrusion, it is possible to obtain multilayer films from 2 up to +20 layers'-layer films [4], Nowadays, 17% of the world's flexible film production are multilayer films, which the most popular application is food packaging [3].

Meanwhile, the non-packaging sector involves trash bags, labels, films for agriculture, construction, etc. Low-density polyethylene (LDPE) [5–7] and high-density polyethylene (HDPE) [8,9] are the most common polymers used in the consumer packaging sector, followed by polyethylene terephthalate (PET) [10–12] and polypropylene (PP) [13,14]. In the case of agricultural applications and another non-consumer packaging, LDPE and linear low-density polyethylene (LLDPE) [15,16] are the most used materials [17].

The increasing generation and accumulation of non-biodegradable waste are becoming a general popular issue since a huge amount of plastic packaging products are currently designed to have a short service life owing to the low cost and easy production [3]. Most consumers negatively perceive plastic packaging because of the considerable high amount of waste produced in their daily lives.

According to an extrapolation from different EU countries, Plastics Europe et al. explained that in 2018 approximately 24.9% of the plastic packaging waste went to landfill and 42.6% were incinerated. Then, the remaining 32.5% of the plastic waste were recycled or exported. However, as Hestin et al. reported, the amount of waste exported out or within the EU is included in the recycled rate. Then, a recycling rate of 15% is estimated for the EU when the extra-EU exports are excluded. The recent ban on the imports of occidental plastic waste to China provides a solid argument to focus on the development of recycling inside the EU. Thanks to this, from 2016 to 2018 plastic waste exports outside the EU decreased by 39% [18].

The main objective of this review is to serve as a guide for the industrial and academic community through all the stages of plastic waste's valorization, with a focus on the multilayer films, their design, and processing. First, the summary of plastic waste management in general is described, making emphasis on the multilayer film's recycling process. Finally, the multilayer film's manufacturing process and innovative coextrusion technologies are reviewed, including the most common materials used for agricultural applications, their processing, and the challenges of their recycling. As well, the concept of eco-design of multi-micro/nanolayer films is presented as a promising solution to the numerous problems that comes with the valorization of multilayer plastic film's packaging.

1.1. Polymer-Based Flexible Films Waste Management

The start-of-life phase for the polymers (virgin or recycled) used in plastic packaging applications begins with the processing via multiple converting techniques such as extrusion, coextrusion, blown extrusion, etc. Then, the first type of solid plastic waste (SPW) is generated during the manufacturing process, called post-industrial (PI) waste. This type of waste includes waste from production changeovers, fall-out products, cuttings, and trimming. In terms of recycling, the PI waste is the higher-quality grade of polymer waste, since it is clean, and the composition of the polymer is known [19].

Later, at the end-of-life, the product is thrown out and becomes post-consumer (PC) waste. The PC waste can be collected separately or not, depending on the country. This waste is a complex mix of polymers of unknown composition and could be potentially contaminated by organic fractions (food remains) or non-polymer fractions (paper). As expected, this PC plastic waste becomes more difficult to recycle than the PI waste [20].

Given the complexity of the challenges and the different actors that need to be involved in the recycling chain, a multitude of measures has to be implemented such as covering products design, waste collection, sorting, recycling, and end-use [21]. Once it does, the processing options for the end-use are similar for both PI and PC plastic waste. Recycling is the preferred option since it closes the loop back to the now secondary "new raw materials". During recycling, the new raw materials can be obtained via mechanical (leading to granulates) or chemical (leading to monomer building blocks) pathways. Then, energy recovery is the greatest option in the case where the polymer waste cannot be recycled. Finally, landfill is the less-preferred option and should be avoided at all costs [20].

Considering an environmental point of view, the best alternative is to avoid the creation of SPW in general. This involves propositions of smarter packaging with eco-design or alternative materials at the production stage [22–24]. Naturally, all these go together with the efforts related to raising the awareness of the consumer by promoting the re-use of plastics products. However, such effort should be run in parallel to that related developing lop an efficient valorization of the huge amounts of SPW generated more and more every year [20]. Enhanced communication through the whole recycling chain, from packaging designers to the end-users, will support and help to identify possible areas of improvement, too.

1.2. End-of-Life Treatment of Plastic Waste

The use stage and waste treatment are always present in the lifecycle of any material. The total number of cycles that a material can submit will depend on its alteration during all stages of its lifecycle [3]. The European Union (EU) proposes through the Waste Framework Directive (2008/98/EC), the following waste management hierarchy: collection, sorting, and reprocessing [25]. Nevertheless, this waste management procedure will vary depending on the source of waste and the local implemented collection.

Over the last years, plastic waste management has been studied in many research studies [26–28], with rigid and mixed plastics as the main focus [29,30]. This means that the flexible films are usually considered as the non-recyclable fraction of the waste stream and in consequence sent directly to landfills or energy recovery. Most recycling companies consider that the small thickness and low bulk density of these materials can disturb the conventional recycling process. Taking into account the little information available and the not well-documented technological advances, it becomes necessary to develop cost-effective technologies for this plastic flexible waste [3].

Horodytska et al. [3] did an extensive review of the state-of-art films waste management technology, where they identified the shortcomings and established the guidelines for future research. Inspired by them, for this review, the plastic waste will be classified into post-industrial (PI) and post-consumption (PC) plastic waste, since their differences in material characteristics.

In the case of PI plastic waste, the sorting stage between multilayers and monolayers is easy. However, considering the unknown origin of the PC plastic waste, it becomes more difficult to distinguish them between monolayers and multilayers. Hence, in this article, we will classify the post-consumption plastic waste as waste from the agricultural and packaging sector without the distinction of the layer's structure. Furthermore, between the different activities involved in waste management, only collection, sorting, and treatment will be considered, since they show the most differences between rigid and flexible plastics. Then, regarding the end-of-life treatment, the plastics demand for new products gives an idea of the types of polymers that composed the bulk of the collected plastic waste. Figure 1 shows an example of the plastics demand per sector and polymer type in the EU [1]. As observed, the largest share of all PC plastic waste is packaging waste. The "big five" raw materials of high-density polyethylene (HDPE), low-density polyethylene (LDPE), linear low-density polyethylene (LLDPE), polypropylene (PP), and polyethylene terephthalate (PET) are the most common polymers used for packaging applications, which means that these polymers will dominate the composition of plastic waste. This information shows us that product design has an important impact on the recyclability (end-of-life) and the degree to which we can incorporate recycled materials (start-of-life) to new products. In design from recycling, the recycled polymer that originates the secondary raw material will be the starting point of new product development.

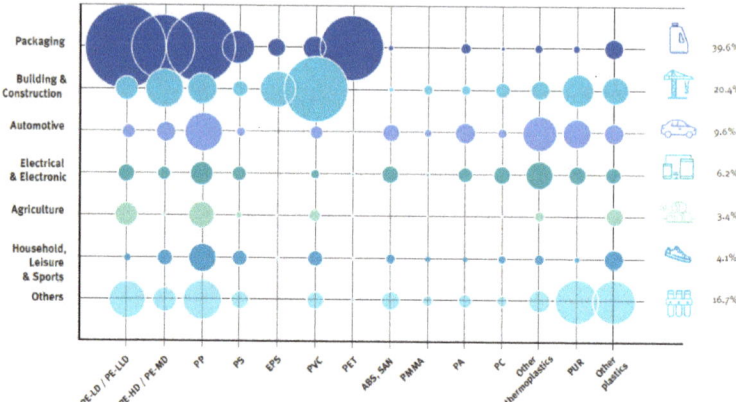

Figure 1. European plastics demand by segment and polymer type 2019. Reprinted from report [31]. Copyright 2020, with the permission from PlasticsEurope. (A color version of this figure can be viewed online).

2. Circular Economy

The plastic's circular economy is a model for a closed system that promotes the reuse of plastics products, generates value from waste, and avoids sending recoverable plastics to landfill. Plastic waste is a valuable resource that can be used to produce new polymer-based raw materials and manufacture plastic parts, or to generate energy when recycling is not viable [32]. In contrast with the linear model, the circular economy is an alternative for the industry to adopt a new plastics economy to enhance both socio-economic performances across the supply chain, while drastically reducing plastic waste. The circular economy model is based on [33]: Reduce plastic waste and pollution through product design, Retain resources and products in use, and Regenerate and preserve natural systems.

Most of the plastic materials are used for packaging and thus designed with an anticipated life expectancy of less than 1 year. Therefore, these choices in combination with the linear economy have been a major source of plastic waste, which is reflected by the accumulation of plastic waste in the ocean. Currently, a total of 150 million tons is the estimation of plastic waste found in the oceans. This proportion is projected to increase to 250 million tons by 2025, if we con-tinue to generate waste at the current rate [33].

Nowadays, the plastic industry is researching alternatives to replace fossil sources with renewable resources and carbon dioxide (CO_2). The new strategy is all along the value chain as displayed in Figure 2: from product design to recycling, then focusing on converting more waste into recyclates, maximizing resource efficiency, and reducing greenhouse gas emissions [32,34].

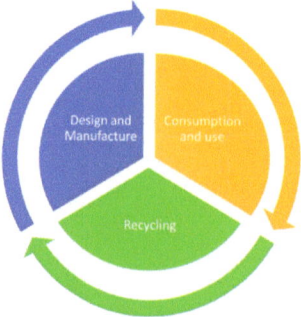

Figure 2. Circular economy flow diagram model.

3. Waste Collection Systems

3.1. Post Industrial and Agricultural Waste Collection

The post-industrial (PI) waste does not require complex collections systems since it is a clean waste of well-known composition and is normally re-used in the same company. To recycle their own scrap without leaving their facilities, many converters use developing technologies. This type of recycling allows the recovery of materials with good properties and are suitable for high-quality products manufacturing. With these types of systems, converters reduce their waste volumes and decrease the required virgin polymer amounts to be used [3].

Regarding agricultural waste, certain companies dedicated to this sector usually collect the waste on-site and then directly transport it to the recycling facilities.

3.2. Household Post-Consumer (PC) Waste Collection

Since the beginning, the recovery of recyclable materials has been divided into two different approaches. The first is based on source separation at individual households, and separate collection systems; the second is the recovery by mechanical processing and sorting of mixed residual waste at central establishments, which receive a larger waste flow. However, despite the efforts of the EU in the 1970s to recover valuable sources from mixed municipal waste, the poor quality of the output product proved that recycling applications were not suitable, and they risked spreading dangerous substances to the environment, such as heavy metals in compost [35].

Nowadays, to achieve the European Commission's goal of increasing the recycling of packaging waste to 80% by 2030 (Packaging Waste Directive 94/62/EC), source separation and separate collection are the best recovery schemes to implement. In general, source separation should prevent the contamination of plastic waste by separating from other wastes at the source, and are considered as the best feasible approaches, technically and environmentally [35]. In Table 1, definitions of the different separate collections are displayed. To give an overview of the separate collection systems in plastic packaging waste, in general, the Kerbside collection (single-stream or co-mingled) and the bring/public systems runs beside the Mixed solid waste collection [35].

Table 1. List of definitions of different collection systems [35].

TERMINOLOGY	DEFINITION
Kerbside Collection	Separate collection system, categorized as property close collection
Mixed Solid Waste (MSW)	Collected in a system in which no other separate collection is present.
Bring Points	Common collection points are shared by a larger number of citizens and involve individual transport to deliver the recyclable waste materials

In the EU, the collection systems differ across the countries. Germany and Austria are considered to have the friendliest collection of flexible packaging in contrast to France and the UK. However, drop-off sites for post-consumer household films collection have become more frequent in the UK. Around 71 local authorities have kerbside collection programs for plastic flexible films. Table 2 shows a summary of different collecting schemes from the EU members. This information displayed represents the general situation in each country. Plastic film waste is generally easier to recuperate in co-mingled systems. However, the recuperation efficacy of each system will depend on the technologies implemented at the sorting facilities after the waste collection [3].

Table 2. Plastic flexible films collection schemes and programs in European countries ([3,35]).

Collection Type		Materials	EU Countries	Films Collection
Kerbside	Single fraction	Plastic	Austria, Netherlands	Comingled flexible and rigid plastic collection
	Co-mingled	Plastic and metal	Germany, Slovenia, Hungary	Collected with mixed plastics
			France	Some collections with mixed plastics
			Italy	Rigid and film plastics are collected separately
		Glass, plastic and metal all in one bin	Ireland	Collected with mixed recyclables
Bring points	Single Fraction	Plastic	Sweden	Collected with mixed plastics
	Co-mingled	Plastic and metal	Spain, Portugal	Collected with mixed plastics
Retail return system (drop-off)	-	Plastic	United Kingdom	Plastics PE films collected separately

4. Sorting of Waste Plastic Films

In order to increase the efficiency of the recycling process, separation of collected waste is a very important step to obtain high-quality materials. Mixed waste coming from co-mingled or single-stream collection system, are generally sorted in Materials Recovery Facilities (MRF) [3].

It is important to mention that the performance of sorting and recycling varies from country to country since this step is particularly affected by the quality and output of the collection schemes and the level of contamination of the collected waste. As well, the recycling performance is related to the quality of the flows collected, especially with the pollutants found at the sorting stage in relation to the quality needed and the final end-use [21].

The technologies and equipment used by the MRFs (Materials Recovery Facilities) depend on the input waste stream received. In this article, the different sorting stages are explained depending on the collection system used to recuperate the plastic waste:

- **Mixed solid waste (MSW) collection system:** In Europe, the plants that receive mixed-waste (MSW) from householders are known as Mechanical-Biological Treatment plants, since mechanical and biological processes are used. The waste stream received consists of organic kitchen waste and recyclable materials. Different sorting stages are frequently carried out. Generally, the trommel is used for size separation and eddy current and magnets systems to remove ferrous and aluminum metals. From these plants, metals, beverage cartons, and plastics as LDPE, HDPE, and PET are the recovered materials obtained for recycling. Unfortunately, in this type of sorting, the flexible films are currently considered as contaminants and sent to landfills with the other rejects from the plants [3,35].
- **Separate collection systems:** The sorting of recyclable materials coming from separate collection programs is performed in single-stream or clean MRFs. For co-mingled recyclables, the input waste consists of plastic, glass, and paper [3]. Generally, plastic bags are removed from the waste stream at the first stage. Nowadays, the development of different mechanical equipment is necessary to facilitate the task of manual sorting by well-trained operators, which continues to be the most common method applied. At the moment, vacuum systems are installed for collecting and covering handpicked material, and bag-splitters are used to open and empty the plastics bags [36]. Then, the next stage of sorting is the separation of the flexible films (2 dimensional flat materials) from rigid heavy products (3-dimensional materials), such as rigid plastic bottles and cans. Ballistic separation is the most common well-known technology

for this purpose. Air separation is an alternative technology to ballistic separation in certain applications, mostly where the feed material is relatively dry. This system relies on the fact that light material can be conveyed in an airstream at relatively low velocity, whereas heavy material will not be conveyed. However, since both ballistic and air separation technologies cannot distinguish between flexible plastic films and fiber-based material such as paper, near-infrared (NIR) optical detections are used to complete these techniques. This optical technique is based on the wavelengths reflected by the material when a NIR light hits its surface. A special sensor can determine the material type from its "fingerprint" of wavelengths [36].

In [3], the authors mentioned that results from recent studies showed that PP, PE, and mixed polyolefin can be identified by NIR. These materials can be extracted using flotation or hydroplaning, but they cannot separate from each other. Then, even if NIR is currently the most efficient technology for flexible film sorting, it has certain limitations. For example, black parts and thin coating layers cannot be detected, and neither surface between reverse printings can be distinguished [3].

Other sorting plants for the classification of recyclable materials from kerbside collection and from bringing point collection exist. For example, in Spain and Portugal, where all types of plastic and metal packaging are collected together, the sorting is performed in specific plants so-called lightweight packaging. The waste pass through different stages: reception and storage, pre-treatment, sorting of materials, and management of rejected waste. In order to remove film sheets and cardboard that can block or damage the sorting line, a pre-treatment is necessary. Then, several types of equipment such as a bag opener and ballistic separator are used, even if manual sorting is still the most common system applied. First, clean film and paper are separated by a pneumatic separator and then a magnetic one is used to remove metal items. Then, products of PET, HDPE, carton, and mixed plastics are separated by optical NIR techniques. Finally, induction separation is applied to remove the non-magnetic metals items [3].

One remaining challenge for sorting facilities is continuously changing the material composition to be sorted, which indeed requires new sorting technology. The initial product's compositions constantly change because of the use of new mate-rials such as bioplastics, changes in the regulatory frameworks, and the constantly evolving patterns of consumption. However, as separate collection programs become more efficient, the material quality of the mixed streams drops, due to the addition of different low-quality and more problematic materials [35].

Recently, Schmidt et al. [37] studied the quantity and composition of multilayer packaging contained in the post-consumer waste stream. They showed that multilayer packaging is not assigned to any specific sorting fraction, since there is a lack of large-scale industrial sorting and recycling process. In consequence, multilayer packaging is dispersed into various recycling paths such as films, mixed plastics, or residual material. Therefore, due to a multitude of different packaging solutions on the market and the sometimes too-small quantities of some multilayer packaging types, there are no economic processes for its recycling.

5. Plastic Films Waste Treatment

After the collection and sorting process, different recycling processes can be applied for the flexible plastic films waste treatment process. Mechanical recycling [19,38,39], chemical recycling [40–42], and energy recovery [43,44] will be discussed in this review.

Methods of recycling are generally divided into four categories: primary, secondary, tertiary, and quaternary (Figure 3). Primary recycling is considered when the materials after recycling present equal or improved properties compared to the initial or virgin materials. On the other hand, when the recycled material obtained presents worsened properties than the virgin material, the method is called secondary recycling or down-cycling method. In the tertiary (also known as chemical or feedstock) recycling method, the waste stream is converted into monomers or chemicals that could be advantageously used in the chemical industries. Finally, the quaternary (also known as thermal recycling, energy recovery, and energy from waste) recycling method correspond to the recovery of plastic as energy and is not considered as recycling in a true circular economy. It is important to note that there is a hierarchy in these four recycling methods, where the mechanical recycling is to be implemented first. Plastics Europe et al. [45] believe that adopting this hierarchy of recycling technolo-gies, from mechanical to chemical recycling, offers the value chain optimal circularity, coupled with lower environmental impact.

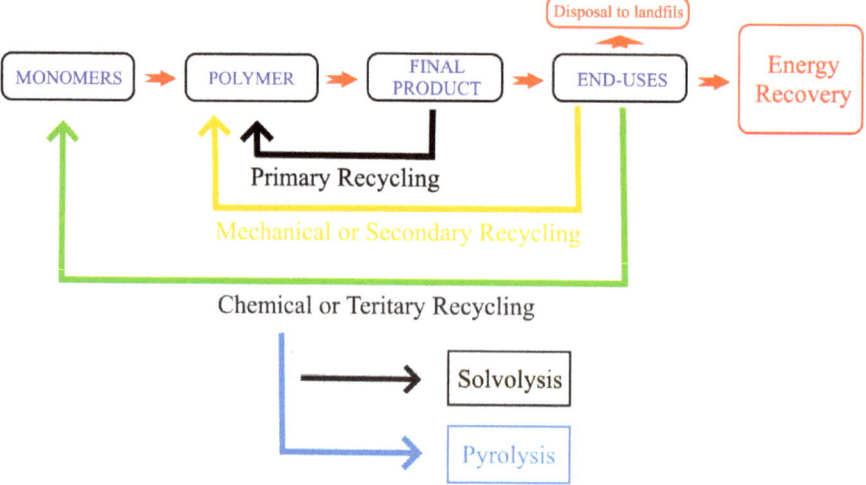

Figure 3. Schematic representation of the recycling methods.

In the context of circular economy thinking, the recycling of materials can be also categorized based on the product, which is manufactured from the secondary raw materials:

- **Closed-loop recycling:** Recycled materials are used to produce the same product from where they were originally recovered. Only recycled plastics or a blend between recycled and virgin plastics can be used to produce a new product. This type of recycling ensures that the product can be recycled continuously, and its recovered material can be added at the same rate [20].
- **Open-loop recycling:** Recycled materials are used for different applications than the product they were originally recovered from. However, this does not imply that the new application is of "lower value" [20].

These two terms will be used in this article to classify the recovery materials obtained from different plastic packaging waste (post-industrial, post-consumption agricultural, and packaging).

5.1. Mechanical Recycling

Mechanical recycling (Figure 4) is the most common recycling technique applied for solid plastic waste (PI and PC), and it is carried out by different mechanical processes where the polymer structure remains unchanged. In terms of quality, the recycled material from closed-loop processes is very close to the original material. Consequently, these materials can be used as secondary raw materials for high added-value products manufacturing. The input waste normally consists of products from a single type of plastic and vaguely contaminated. The recycling process is composed of waste transformation by extrusion, where the plastic is melted and re-granulated. Decontamination methods can be also applied before the re-granulation [3].

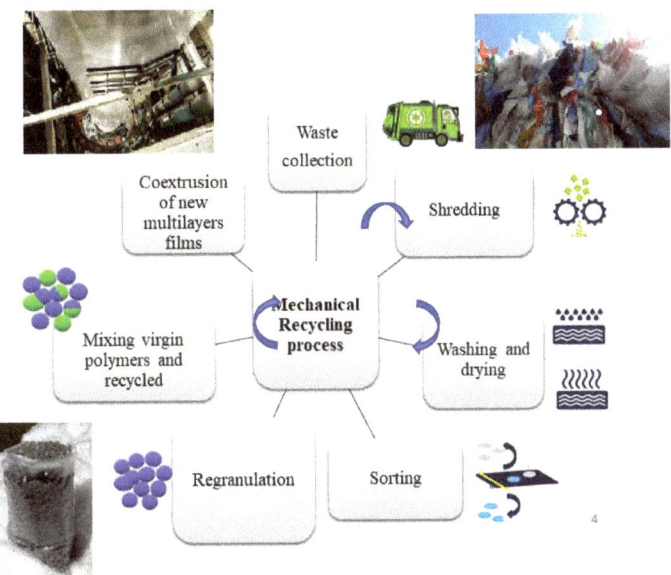

Figure 4. Schema: Example of a multilayer film's mechanical recycling process. Pictures of plastic waste obtained from the website of "Barbier Group" [46]. This figure was generously shared by Mr. Gerard Pichon from "Barbier Group".

In the case of the open-loop mechanical recycling process, the input waste is a single type of polymer material or a blend of compatible polymers. As shown in Figure 4, mechanical recycling consists of different processes such as:

- Separation and sorting: based on shape, density, color, or chemical composition.
- Baling: for ease handling, storage, and transport, the plastic waste is fed into a baler where it is compressed into bales.
- Washing and drying: elimination of contaminants.
- Shredding: size reduction of the product to flakes.
- Compounding and pelletizing: reprocessing of the flakes into granulates.

The previous steps can be applied anywhere between, multiple times, or not at all, depending on the type of plastic waste. For example, sorting is applied more often to post-consumer waste (PC) than to post-industrial waste (PI), since the latter tends to be greater separated in advance in terms of its composition. Then, since the post-consumption (PC) waste stream contains contaminants such as plastic additives, inks, and remnants of incompatible polymers that affect the plastic properties during reprocessing, they are suitable only for less demanding applications: trash bags, pipelines, products for agricultural applications, etc., [3].

5.2. Chemical Recycling

In this recycling methods, plastic waste is considered as raw material for the production of valuable products such as monomers or petrochemical feedstocks. Various processes can be considered for chemical recycling, and they present different levels of maturity. Nowadays, plastic waste is considered a promising raw material for the production of fuels and chemicals. The interest is the production of valuable products such as monomers or petrochemical feedstock. The most known processes are gasification (partial oxidation), pyrolysis, hydrogen technologies, fluid-catalytic cracking, and depolymerization (methanolysis, glycolysis, and hydrolysis). Both types of chemical recycling consist of monomer recycling or feedstock recycling and are considered as ideal methods for the protection of the environment by the reduction of the non-degradable waste volume [20]. For polyolefins, pyrolysis is the most common technique used. The products obtained from pyrolysis are liquid and gas that enclose the substances of interest. However, since the separation costs are very high, the recycled products are used as fuel. A summary of the most used chemical recycling techniques is displayed in Table 3, including their challenges and advantages.

Table 3. Summary of some chemical recycling techniques for plastic waste, incorporating their advantages and challenges [20].

Techniques	Advantages	Challenges
Chemolysis	Generates pure value-added products	To be cost-effective requires high volumes
Pyrolysis	Simple technology and suitable for highly heterogeneous plastics blends	Complexity of reactions
Fluid Catalytic cracking	Economically favorable since the reaction conditions are less strict	Absence of suitable reactor technology
Hydrocracking	Suitable for plastic blends. Good quality of the produced naphta	High operational cost Presence of inorganics High cost of Hydrogen
Gasification	Well-known technology	Generation of noxious NO_x

It is important to mention that most of the chemical recycling techniques of plastic waste are still at early stages and they are not expected to be completely operational before 2025 [21]. However, they have high potential for the valorization of contaminated and heterogeneous plastic waste, where the separation and sorting are not viable economically and technically. Therefore, as reported by Plastic Europe et al., a 7.2 billion euro's investment is expected from great polymer producers by 2030. Subject to European regulatory constraints in favor of waste reduction and the circular economy, producers aim to move from a production of 1.2 million tons of recycled plastics in 2025 to 3.4 million tons in 2030 [45].

5.3. Energy Recovery

Energy recovery consists of the burning of waste to produce energy in the form of electricity, heat, and steam. Due to their high calorific value, plastic products are considered as a promising source of energy, since they are derived from crude oil. The production of water and carbon-dioxide after combustion make them similar to other petroleum based-fuels [47]. The volume of waste can be reduced by 90% after the incineration, which is an advantage when the landfilling is limited, and the lack of space becomes important. There are different incineration methods used for the plastic waste [3]:

- One stage: co-incineration of municipal solid waste with high fractions of plastic waste
- Two-stage: fluidized bed combustion process
- Cement industry: plastic solid waste commonly used as a fuel in cement kilns in order to save energy and reduce costs

Even with notable differences between the member states of the EU, there is a general trend of redirecting plastic waste from landfilling to incineration due to the now-strict measures concerning landfilling imposed by the EU legislation. In consequence, most of the low-quality waste is channeled to energy recovery facilities, whose capacity is in constant increase [21]. However, an important number of environmental concerns are associated with the incineration of plastic waste. The combustion of synthetic polymers such as PET and PE can generate volatile organic compounds, smoke, particulate-bound heavy metals, polycyclic aromatic hydrocarbons, and dioxins, which have been identified in airborne particles from the incineration [47]. Moreover, despite the economic benefits, energy recovery is not in resonance with the circular economy principles. Therefore, the recycling and reuse process are prior for the plastic waste management. Energy recovery should be applied only to the non-recyclable fraction [3].

5.4. Post-Industrial Plastic Film Waste Recycling

In the EU, the recycling rates of the post-industrial (PI) plastic waste are higher compared with the household or post-consumer (PC) waste. The large volumes of industrial waste and the known composition create most cost-effective recycling and facilitate the production of purer recyclable materials [21]. In this context, many companies have been investing into different recycling tech-nologies in order to recycle their own production scrap. This type of recycling system leads to recovery materials with good properties and appropriate for high quality products processing. Then, since their composition is known, the sorting stage becomes easier between monolayers and multilayers systems.

In the last two decades, the formulation and production of biodegradable and/or bio-based polymers, as an alternative to the synthetic counterparts, is a viable strategy towards sustainability in a green future. Despite the interest in the growth of biopolymers, their waste stream is limited. Most of them are polylactic acid (PLA) and PLA-based composites which are usually not collected separately [48]. Biodegradable polymers are susceptible to be broken down into simple compounds because of microbial action, and different bioplastics have been known to undergo this process in a reasonably short time (e.g., six months), and are commonly identified as biodegradable.In some specific cases, mechanical recycling might be chosen as a priority compared with the biodegradability process, since it has been demonstrated that the mechanical recycling process is more environmentally friendly. Cosate de Andrade et al. [49] presented a Life Cycle Analysis of PLA comparing chemical recycling, mechanical recycling, and composting, and they found that mechanical recycling had the least environmental impact, followed by chemical recycling and, lastly, composting, when considering the climate change, human toxicity, and fossil depletion categories. Whereas the recycling of bio-based polymers and their composites must be further investigated and better addressed, considering the target applications for these materials and for a successful recycling process, the bioplastic and their composites waste stream must be collected separately to the other plastic waste stream [23,34]. Furthermore, we have to consider also that those bio-based materials are very sensitive to humidity, thermal, and/or hydrolytic degradation, and the loss of their physical, mechanical properties is noted up to the various steps of mechanical recycling. As reported in the recent review of Morici E. et al. [34], only homogeneous (bio)polymers-based materials could successfully perform the chemolysis through glycolysis, aminolysis, methanolysis, alcoholysis, and hydrolysis. For heterogeneous biobased materials, the cracking and gasification could be considered more appropriate methodologies. The collected (bio)plastic materials, having an extremely heterogeneous nature, could be successfully recycled through energy recovery.

5.4.1. Monolayer Film's Recycling

The post-industrial (PI) waste coming from monolayer films is recovered by the mechanical recycling process. Depending on their characteristics, they can be recycled by closed or opened-loop. The non-printed monolayer film's scraps are recovered by closed-loop recycling since they are clean and homogenous. In the case of printed scrap, only a closed-loop recycling is possible in order to avoid important quantities of waste going to landfilling.

Nowadays, many advanced technologies are developed to obtain recycled materials coming from surface printed plastic waste. Some of these technologies involve filtration, homogenization, and degassing stages, including final extrusion of recycled pellets. *EREMA*, which is a leading plastic recycling company, has patented the latter mentioned technologies. However, because of the poor properties of the final product obtained, the recycled material is only adequate for less demanding applications such as trash bags, plastic lumber, etc.

Horodytska et al. did extensive research concerning the different deinking methods available and their advantages and disadvantages. Focusing on water-based inks, many researchers have investigated how to eliminate the ink from polyethylene films by using different surfactants [3]. At the beginning [50], concluded that cationic surfactants were the most effective (over a pH range of 5–12) at deinking. On the contrary, anionic surfactants had almost no deinking effect even at high pH levels. Then, in the case of nonionic surfactants, the deinking process is possible depending on the pH level of the solution. Eventually [51] investigated the deinking of solvent-based ink from polyethylene films. They found that cationic surfactants were the most effective, which is the same case as water-based inks. The only difference is that a minimum pH level of 11 is required for water-based inks [51].

Later, researchers from the University of Alicante developed an innovative deinking process to remove the ink from plastic's surfaces [50]. Since they proved that this process was economic and technically viable, they settled a semi-industrial deinking plant called Cadel Deinking. Figure 5 describes the deinking process with a water-based solution where no environmentally dangerous chemicals are used. In order to obtain recycled plastic free of ink with a good quality, the printed film goes through different steps such as grinding, deinking, washing, drying, and pelletizing. Finally, the recycled pellets obtained can be used for high added value product processing. Moreover, they designed a water treatment system in order to reduce the consumption and to recover the deinking chemicals [3].

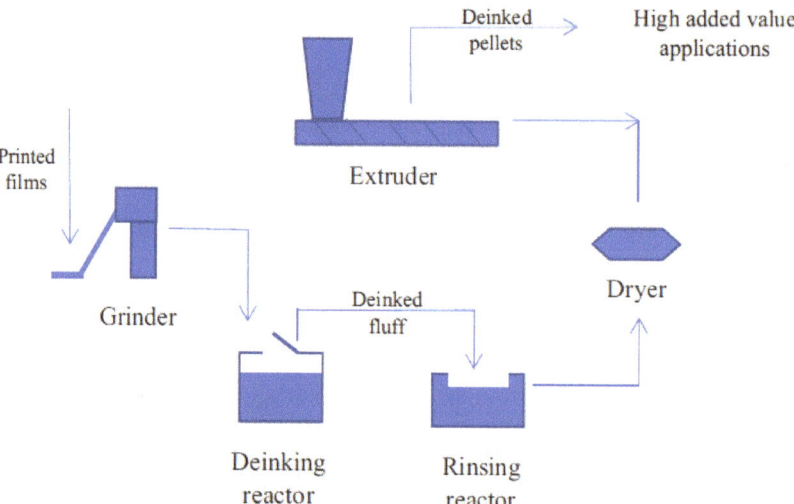

Figure 5. Cadel Deinking process of printed films, developed by the University of Alicante. Reprinted from publication [2]. Copyright 2018, with the permission from Elsevier. (A color version of this figure can be viewed online).

Gamma Meccanica is an Italian company who also developed a deinking technology for plastic flexible films. The process is limited to plastic film rolls, as can be observed in Figure 6, which is not ideal since most of the printed waste films have defective products such as shopping bags, packages, etc. As a result, a grinding stage is necessary to cover a large number of film waste. In Table 4, a summary of the different deinking processes available in the world are described and explained. The *CLIPP+* is a European-funded project where carbon dioxide in supercritical conditions is used for deinking and deodorization of PI polyolefins films. The latest results showed that the recycled films obtained could be used in secondary packaging applications [3].

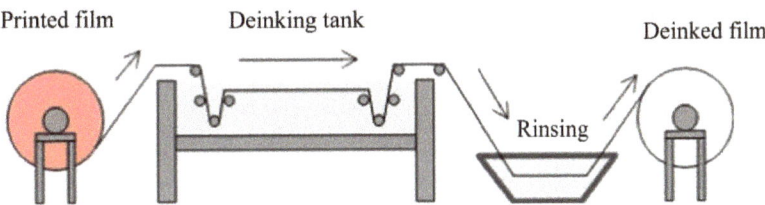

Figure 6. Gamma Meccanica deinking process. Reprinted from publication [3]. Copyright 2018, with the permission from Elsevier. (A color version of this figure can be viewed online).

Table 4. Summary of mechanical recycling and deinking process of post-industrial monolayer waste films [3].

Waste Treatment	Mechanical Recycling System	Type of Extruder System	Description
Closed-loop recycling	Deinking	Cadel Deinking (Spain)	Water-based. High quality recycled pellets obtained for high-added value products
Open-loop recycling	Re-extrusion	Degassing extruders EREMA (Austria)	One stage. Loss of properties. Less demanding applications
	Deinking	Gamma Meccanica (Italy)	Detergents, solvents. Limited to plastic rolls. High costs
		Geo-Tech (US)	Water-based, work with plastic flakes but focused mainly on rigid plastics.
		Metalurgica Rhaaplex (Brazil)	Solvent-based solution and friction system. Work with plastics flakes. High costs and lower quality of recycled material obtained
		CLIPP+(EU)	Carbon dioxide in supercritical conditions. Still in development.

Chemical recycling of post-industrial (PI) monolayer film scrap had also been studied, as pyrolysis in LDPE plastics bags. Liquid fractions with hydrocarbons could be obtained in the domain of commercial gasoline. However even if the technique is environmentally friendly, large amounts of waste are required to reduce the functioning costs [3].

5.4.2. Recycling of Multilayer Films

Multilayer films are made of different types of polymers for synergic properties as well transparency, gas and water barrier properties, stiffness, flexibility, etc. For the clarity purpose, a specific section of this chapter is dedicated to their processing and a wide range of applications. By consequence, the recycling of these films becomes more challenging compared with monolayer films. Through the past years, different recycling methods such as compatibilization, delamination, and dissolution-precipitation have been investigated [2,3,52,53]. Since complex blends are obtained after the recycling process, compatibilizers are added into the blends in order to increase the cohesion between the different polymers. For example, DuPont has developed a variety of compatibilizing resins such as Fusabond® for film applications [54]. In addition, Dow Chemical Company has also developed polymer modifiers such as RETAIN®, in order to facilitate the recycling process of barrier films that contain EVOH or PA [55].

The mechanism of multilayer delamination consists of the dissolution of macromolecules. Physically, chemically, or mechanically, the delamination can be induced by the decomposition of the interlayer or by reactions at the interface [2]. The segregation of the different layers of the films and the recycling of polymer blends are done separately. The polymers used as "tie-layers" to join two layers (usually made with incompatible polymers) are usually removed with a determined solvent. For example [56] investigated the recycling of multilayer films containing PE, aluminum, and PET. In order to delaminate the multilayer film, Acetone was used, and PET was depolymerized with ethanol in supercritical conditions. Then, since the solvents were recuperated by distillation, the process was considered not harmful for the environment [56].

In the case of inked multilayered film, researchers also worked on a combined recycling process of delamination and deinking. They determined that delamination should be applied first, since sometimes the ink deposits between the layers (mostly with food contact applications). However, in order to reach higher recycling rates, a correct sorting of delaminated polymers before the extrusion is necessary [3].

The selective dissolution-precipitation technique is a mechanical recycling method used to separate and recycle the polymers using solvent or non-solvent systems. A separation step is necessary after the dissolution of the polymers, using the differences in material densities [3].

Figure 7 displays the summary of the current advances in research. It describes two paths to recycling multilayered films. The first path is the separation of the different multilayer polymers in separated recycling streams, in order to make them suitable for recycling. The second path is the processing of the used polymers together in one compatibilization stage [2]. As explained before, compatibilization steps consist of the addition of suitable molecules that work as compatibilizers. Since post-industrial waste has a known composition, they are suitable to be recycled by compatibilization.

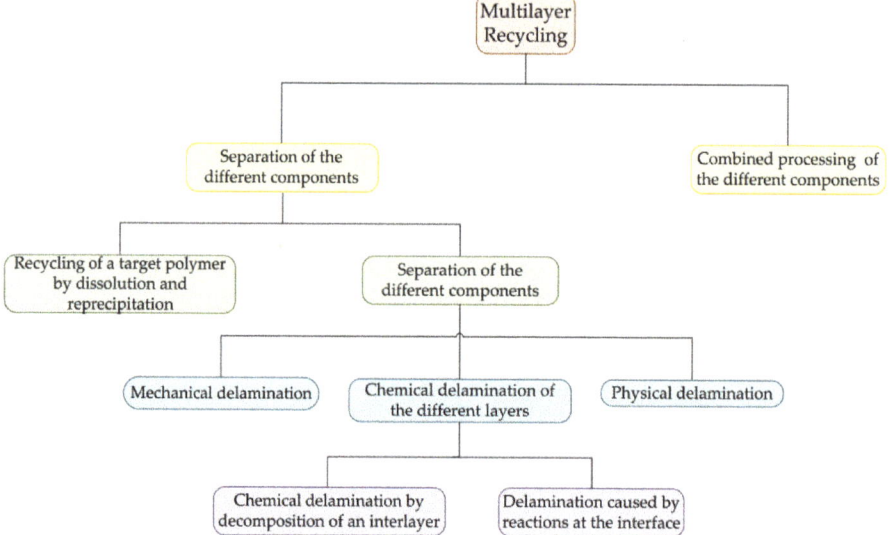

Figure 7. Summary of the introduced recycling methods of multilayers films waste. Adapted from [2]. (Copyright 2022, Elsevier).

The separation methods of the different polymers presented on the multilayer are sub-divided in two strategies: (i) Separation of selective polymers by the dissolution-precipitation method and (ii) physical, mechanical, or chemical delamination method.

In general, physical delamination methods consist of the dissolution of the interlayer (tie-layer) using solvents, water, or aqueous solutions with a specific pH value. The mechanical delamination is a less used method since usually a strong adhesion between the layers is present. In the case of chemical separation methods, there are two types: delamination by the decomposition of an interlayer and by the induction of reaction at the interphase [2].

More recently, Ref. [57] presented a new approach from eco-designed to recycling of the multilayer structures as an alternative way to improve the valorization process, assuming that their mechanical recycling is possible. This recent work describes a future-oriented approach for the recycling of polyethylene-based multilayer films. The method involves going from eco-design to mechanical recycling of multilayer films via forced assembly coextrusion. This study's originality consisted in limiting the number of constituents, reducing/controlling the layer's thickness, and avoiding the use of tie layers as compatibilizers. The ultimate goal was to improve the manufacturing of new products using recycled materials by simplifying their recyclability [57]. Based on the results obtained, a proof of concept was demonstrated with the eco-design approach of multi-micro/nanolayer films as a very promising solution for the industrial issues that arise with the valorization of recycled materials.

5.5. Post-Consumer Plastic Film Waste Recycling

As explained before, considering the diverse and unknown composition of the post-consumer plastic waste, it is difficult to separate them into monolayers and multilayers. Thus, in this article, they will be classified as waste from agricultural and packaging sectors.

5.5.1. Agricultural Plastic Waste (APW)

Plastic products for the agricultural sector represent almost 4% of the total plastic product consumed in Europe and the USA. In the North European countries, agricultural films are mainly used for silage and bale wrapping films; meanwhile, in the South countries, green houses, low and high tunnel and mulching are predominant [58]. However, an increased accumulation of plastic waste in rural areas is increasing thanks to the expanding and extensive use of plastics in agriculture. In the past years, most of this waste was sent to landfill or burnt uncontrollably by the farmers, which released harmful substances to the environment, and affected the human health and the safety of the farming products [59].

The majority of the plastics films for agriculture are made mostly of low-density polyethylene (LDPE), due to its relatively good mechanical and optical properties. Then, high-density polyethylene (HDPE) and ethylene-vinyl acetate (EVA) are common polymers also used for some agricultural applications. As mentioned before, since the plastic production for agriculture applications represents a minor percentage of the total plastic production, this fact facilitates the collection of the plastic waste. The agricultural polymer-based films used in a specific rural region are similar since the same cultivations take place. Then, the APW (agricultural plastic waste) generated at a regional level is homogeneous, since it is concentrated geographically and generated at specific periods each year, with the exception of bale wrapping films, silage films, and other related plastics. Nevertheless, even if the APW waste management is easier than other post-consumer plastic waste, the APW is heavily contaminated with pesticides, soil, stones, vegetation, and other organic waste. Of course, the contamination level depends on the applications, management during their use, removal practices, and storage conditions of the plastic waste in fields [59].

The most justified recycling method for the APW is mechanical recycling from a financial and environmental point of view. In the case of the non-recyclable waste, energy recovery in cement kilns is the most common practice used. In Figure 8, an example of the agricultural plastic film's life cycle in France is displayed, in the context of a circular economy. The disadvantage of the mechanical recycling process is the contamination with dirt, soils, etc., which obliges the recycler to include a washing step and sometimes even a pre-washing before the waste starts to be processed. This washing step increases the costs and the need of water and energy. However, it is possible to reduce the contamination during the baling of the plastic waste. On the other hand, even if the energy recovery process is an efficient alternative to the APW management, it is a controversial subject of public concern due to the potential contributions of gases combustion to the atmospheric pollution. Currently, regulation and technical specifications are being developed among the European countries [58].

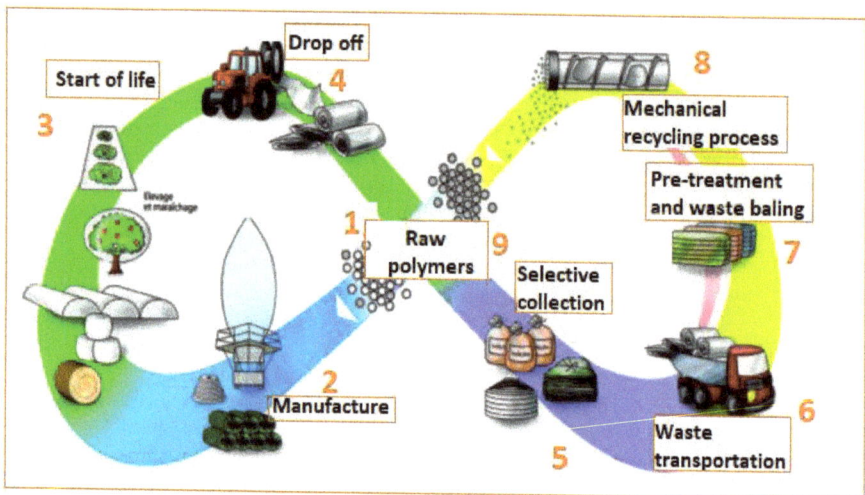

Figure 8. Schematic of the agricultural plastics film's life cycle in the context of a circular economy. Adapted from [60].

Mechanical recycling and the APW quality.

The post-consumer APW mechanical recycling is preferred due to the homogeneity of the films and single polymer waste available. The films go through different steps during the mechanical recycling: washing, shredding, drying, and pelletizing [3]. In order to summarize and evaluate the recycling practices of APW, it is important to understand the factors that limit their recyclability:

- **Inert contaminants:** The contamination by sand and soil of the PE-based APW is the reason why this waste is not commonly recycled with other PE plastic wastes. Besides the high cost of the washing stage, the water is usually trapped in the folds and is not easily removed during drying. Prolongation of the drying cycle degrades the plastic and consumes energy. Additionally, stones and soil can damage the blades for the cutting and processing equipment that increases the maintenance costs [58].
- **Thickness:** Most of the recycling industries in Europe indicated that the thickness of the APW limits the productivity of the mechanical recycling lines. It was shown that the productivity of the recycling process decreases from 1000 kg/h for a 40 μm average thickness film to 500 kg/h for a 20 μm film [58].
- **Co-mingled plastics:** A poor sorting induces the contamination of the PE-based APW with other polymers; for example, PE based films mixed with PP agrochemical bottles. The formulation of the plastic films also affects the sorting. In the case of the multilayer films made with PE and EVA, there is no problem since PE and EVA are compatible and even the latter helps the recycling process by increasing the Melt Flow Index (MFI). However, PE is not compatible with PP, polyamide (PA), polycarbonate (PC), etc., which have different melting points than PE. This reasoning highlights the importance of preventing cross-contamination between the sorted piles throughout storage in the field, transportation, and baling [59].
- **Ageing (ultraviolet radiation):** Photodegradation of the plastics is caused by the exposure to ultraviolet light (UV), which affect the recyclability. Chain scission and crosslinking phenomena affect mostly the rheology of the molten mixture [61]. The chain scission reduces the molecular weight; meanwhile, the crosslinking increases the bonding between polymers chains, which leads to a molecular weight increase. The formation of carbonyl groups and vinyl groups are the major functional groups that accumulate with the photodegradation of PE [61].

Briassoulis et al. [58] did an extensive investigation about the quality of agricultural plastic waste's characteristics after their mechanical recycling. They found that the inadequate sorting of the plastic waste (per thickness, color, and category) is an important factor that affects the quality and commercial value of the APW. They also demonstrated that the exposure of the APW to the sun radiation under normal field conditions (for the typical time periods) is not intense enough to lead the plastic to a severe degradation that make the APW non-recyclable, with an exception in the case of the mulching films. In cases where mechanical recycling is too expensive because of the sorting high costs, drying stages, or the low quality of the waste material, energy recovery is the commonly selected option to avoid landfilling [3].

Then, La Mantia [62] investigated the possibility to use the recycled material obtained from greenhouse covering in closed-loop. They obtained the best results with virgin and recycled material monolayer blends and with coextruded blends where the recycled material was placed between two virgin layers.

Most recently, Cabrera G. et al. [63] studied the valorization of post-consumer agricultural waste films, specifically bale wrapping multilayer films. These waste films go through different processes that involved the following: waste collection, recycling, and reuse of the multilayer waste films as a secondary raw material for new applications. However, after the recycling process, they observed a migration phenomenon of certain additives, which represents a drawback for the industrial process and limits the reuse of the waste films as a raw material. After an exhaustive study, they proved that the use of mineral fillers is an excellent solution to avoid the migration of these additives and therefore increase the recyclability of bale wrapping multilayer films.

5.5.2. Packaging Plastic Waste

According to Horodytska et al. 2015, in Europe approximately 100 billion plastic bags are consumed every year, meanwhile, bonly 7% are recycled [3]. However, enormous efforts have been made to reduce the consumption of retail bags and other wrap films. High quality materials can be obtained from the recycling of retail bags and other wrap films. If the waste is clean and uncontaminated, it can be suitable for the same applications as the original [3].

Unfortunately, flexible film waste coming from Kerbside collection is commonly considered for the recyclers as a contaminant and in consequence is removed from the waste stream. Even if sometimes it is used for energy recovery, most of the time it is sent for landfilling. For example, the household waste includes primary packaging that are usually multilayer films. Nevertheless, since it is not possible to separate monolayer from multilayer, all the flexible films are rejected from the waste stream and considered as a contaminant [3].

The majority of the plastic films are made from high-density polyethylene (HDPE), low-density polyethylene (LDPE), and linear low-density polyethylene (LLDPE). The authors of [64] recovered the data of post-consumer plastic film collected in the United States in 2017. For the 2017 survey, they used the following film categories:

- **PE Clear film:** Clean PE film from commercial sources, including stretch wrap and polybags.
- **PE colored film:** Mixed color PE film from commercial sources, stretch wrap but no post-consumer bags.
- **PE retail bag and film:** Mixed color, clean PE film, stretch wrap and retail collected post-consumer bags, sacks, and wraps.
- **Kerbside collection films:** Post-consumer PE mixed film collected in kerbside and sorted at materials recovery facilities (MRF).
- **Other PE films:** PE film that does not fit in any of the previous categories.
- **Other non-PE films:** Non-PE film that include polypropylene (PP) and polyvinylchloride (PVC).

MOORE Recycling Associates [64] reported that in 2017 the number of plastics bags and film recovered for recycling was 1 billion pounds, which represents an increase of 54% since 2005. They explained that depending on the collection system, recovered film bales can contain a combination of HDPE, LDPE, and LLDPE resins or can contain a single resin. For example, stretch films or pallet wrap can be collected separately and sorted as PE clear film or it may be mixed with other polyethylene films. Then, plastic bags and wrap are commonly mixed with stretch films and other retailer-generated scrap film, in order to obtain an efficient collection at retail collections. Hence, "bags only" bales containing only bags and wrap are not common. Thus, the total amount of recovered post-consumer bags and packaging were defined as the combined total of Kerbside collection films with a specific percentage of the PE retail bags and films bale.

The majority of the US film processing plants are for clean PE films, which can be easily used as raw material for a new product without the washing step, or for a single polymer film (only LDPE). Films, sheet, and composite lumber production are the main uses for recycled post-consumer films, as shown in Figure 9. As observed, composite lumber remains the preferred domestic end use market for the post-consumer films [64].

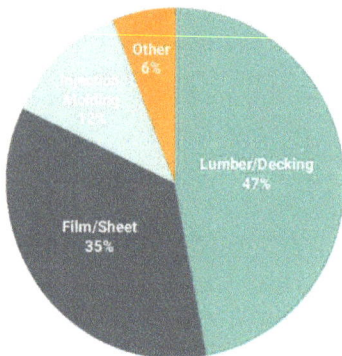

Figure 9. Reclaimed end-uses for U.S post-consumer plastic packaging in 2017. Adapted from [64].

Achilias et al. [65] reported that the dissolution/precipitation technique is an effective process in recycling waste packaging material. The proper experimental conditions as type of solvent/non solvent, polymer concentration, and dissolution temperature were selected based on the model polymers (HDPE, LDPE, PP, PS, etc.). They obtained good polymer recovery in most of the waste samples studied. However, lower recovery values obtained in other samples were attributed to the removal of additives that were present in the original structure [65].

To sum-up, there are two fundamental needs to improve the packaging film's recycling:

- **Higher demand to absorb the material currently collected:** Recyclers are struggling to compete with virgin polymer suppliers. Without the appropriate demand, the improving of the collection and processing in fractures becomes more expensive and difficult [64].
- **More education on recycling:** With the lack of general awareness, the quality of the drop-off film recycling streams and Kerbside recycling struggles. It has been demonstrated that consumers do increase participation if they are conscious that they need to recycle [64].

Concerning chemical recycling, a company named RES Polyflow™ patented a process to obtain petroleum products from mixed plastic waste. As reported by them, the blend between rigid and flexible plastics can feed a reactor. Molten fluids, condensable petroleum gas, and other gases can be produced through the reactor, under certain heat and anaerobic conditions [3].

Nowadays, plastic film producers are committed to a sustainable environmental development. Many efforts have been made to increase the recyclability of the materials. In consequence, the multilayer film design (for different application) is done considering the end of life of the product [35].

6. Processing of Multilayer Films and Applications

This section provides an overview on the research work that has been carried out in the academia in the past years regarding multilayered polymers from coextrusion process. The state-of-the-art methods for fabricating multilayered polymers are firstly introduced and compared. A specific attention is given to their applications in regard to the agriculture sector as well their recyclability.

6.1. Coextrusion: Principle and Technologies

Coextrusion is an industrial process widely used to form multilayered sheets or films that are suitable for various products, ranging from food packaging, the medical area, and recently in microelectronic and nonlinear optics with more than thousands of layers [66,67]. Multilayered polymer systems are produced to satisfy specific requirements for high value-added applications such as gas barrier films [68,69], mechanical robust systems [70–73], and optical applications [74–76]. The coextrusion is a process which combines multiple polymers via two or three extruders using a feedblock system, where the polymers melted (from separate extruders) are brought together [77]. Each component of the multilayer structure provides its own end-use characteristics.

A typical example is oxygen-barrier food packaging. This kind of packaging material has usually a polyethylene (PE) or polypropylene (PP) based structure with an oxygen-impermeable polymer such as ethylene-vinyl-alcohol (EVOH) or polyamide (PA) as a central barrier layer. Since the barrier polymers have normally poor adhesion to the main structure polymers, copolymers are used as tie-layers in order to compatibilize and improve adhesion between the barrier and the external layers [78].

Over the last five decades, continued research and development in the academic and industry domains allowed a continual growth and expansion of the micro and nanolayered film coextrusion technology [79] to commercial relevancy. Meanwhile, many interfacial and rheological phenomenon during the co-extrusion and forced assembly are also deeply studied [80–83]. Nearly 500 issued patents for composition of micro or nanolayered materials applications have been published between 2000 and 2010 [84]. In order to increase the research and commercialization of the advanced microlayer technology, research in polymer processing have been developed as well as the advancement of coextrusion feedblocks and layer multiplying die manufacturing.

6.1.1. Cast Film Coextrusion

A combination of two or more extruders through a multichannel-layered feedblock is used to produce conventional 2 to 17 layered cast films. Polymer materials separated by different streams are combined into parallel layers in the feedblock before exiting to a film, sheet, or annular die. Polymer dies companies such as Cloeren, Nordson, Macro Engineering, etc., produce multilayered polymer feedblocks which are up to 32 layers. Over time, multilayered polymer films with less than 20 layers have been produced in order to tackle complex blends film's extrusion process, due to the performance and cost factors such as those listed by [84]:

- Potential reduction in expensive polymer materials by controlling the polymer domain location, continuity, and thickness.
- Incorporation of recycled materials at the internal layers without degrading the film properties.
- Reduce the film thickness maintaining the mechanical, transport, and/or optical film properties.

The ability to increase the layer number comes from the feedblock design. In 2002, a single feedblock and film die system reached the micro and nanolayer scale. In Figure 10, a Nanolayer™ feedblock designed by Cloeren is displayed. This feedblock was designed to produce directly more than 1000 layers in a single unit. The die connects a selected number of extruders and redistributes the incoming melt streams into hundreds or thousands of layers. These layers are ordered and distributed within the block using a design, which was inspired from vein splitting. Then, the ordered thousand-layer polymer melts flows, and exits the feedblock directly into the die in order to form the product film or sheet [81,84].

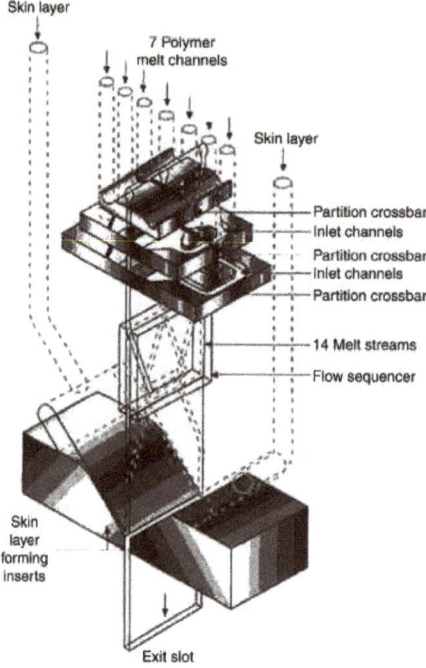

Figure 10. Multi-nano-layers feedblock design by Cloeren Incorporated. Adapted from [84].

Before the fabrication of single shot feedblocks, Dow Chemical Company developed a combination processing technique using a simple two to five layered feedblock with a series of sequential layer multiplication dies. In this approach of sequential layer multiplication, the two to five layered polymer melt flows through a conventional feedblock and then is fed to a series of layer multiplication dies. These layer multiplication dies double the number of layers by a cutting process, spreading and stacking the layered melt stream (Figure 10). As shown in Figure 10, the final number of layers in the polymer film is determined as the function of a number of layer multiplication dies, which are placed in series between the feedblock and final film or sheet exit dies. The number of layers of the film obtained can be calculated as a function of the number of layers in the feedblock and the number of layer multiplying dies [78,81,84].

Figure 11 displays an example of nanolayer film coextrusion. The layer multiplying dies is coupled with a two-layered feedblock that will produce films with a number of layers following a 2^{n+1} model. The "n" represents the number of sequential layer multiplying dies, which are placed in series between the feedblock and film exit die [74]. Layer multiplication enables structures with hundreds or thousands of layers to be produced. A layered melt stream from the feedblock is fed through a series of layer multipliers. In each multiplier, the initial melt stream is divided vertically in two, spread horizontally, and then recombined, while keeping the total thickness of the melt constant, thus doubling the number of layers and reducing the thickness of each layer after each multiplier (Figure 10). Therein, multilayer coextrusion is capable of fabricating films having thousands of layers with individual layer thickness down to the nanoscale at low environmental (solvent-free) and budgetary costs [67,78,81].

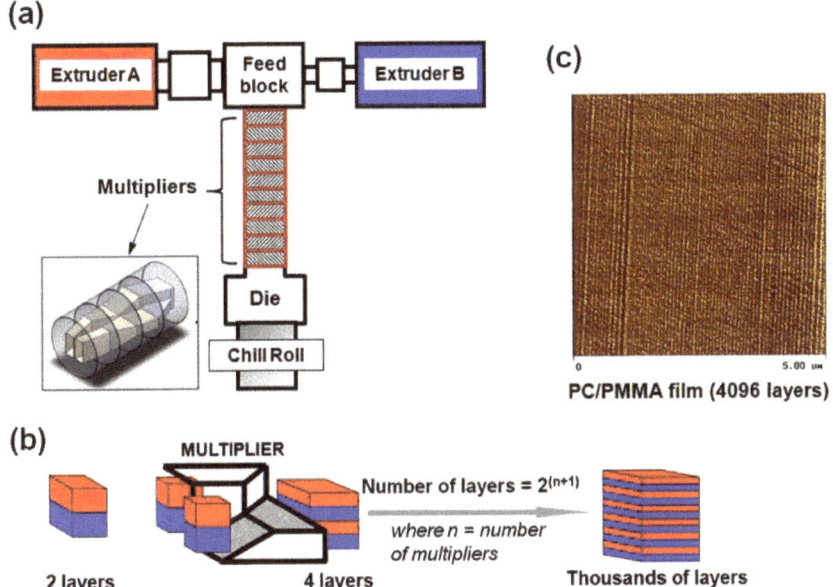

Figure 11. Schematic illustrations of (a) multilayer coextrusion process with two polymeric components with a specific scheme of multiplier element used in our laboratory, and (b) general layer multiplication concept schematic of the layer multiplication by cutting, spreading and recombining. (c) AFM phase image showing the cross-section of a 4096 layers PC/PMMA (50/50) film as an example for coextruded multilayered polymers. Reprinted from publication [67]. Copyright 2020, with the permission from Wiley. (A color version of this figure can be viewed online).

A coextrusion of two or more polymers is possible, creating different layer configurations. A configuration ABC represents coextrusion of three different polymers as alternating layers. Meanwhile, an ATBTA configuration represents a tie-layer polymer (T) alternating between polymers A and B. The latter structure can also be combined with skin layers which are normally added after the layer multiplication dies [74].

Comparing the coextrusion approaches, the layer multiplier die technique is a more flexible and low-cost technique than the single shot feedblock. Nowadays, the single feedblock processing technique is used more often in production of commercial scale products. Thus, the sequential layer multiplying die has been used for research and as a development tool. Most of the time, commercial products formulations and structures are designed and optimized before their commercialization with lower cost equipment and production costs [78,81,84].

6.1.2. Blown Film Coextrusion

Blown film extrusion is widely used to produce packaging films. The majority of these films are multilayered in order to improve its mechanical and thermal properties as required by the medical or food industry. In Figure 12, a schematic diagram of the blown film processes is displayed. As observed in this process, an extruder is used to melt and forward molten polymer into an annular film die. Then, air is injected into the center of the annular die to inflate the polymer bubble. This bubble is cooled down by an air ring, which blows air on the bubble surface to decrease its temperature until the polymer becomes solid. A stabilizing cage is commonly used to minimize the bubble movement as it collapses in the collapsing frame to make a flat film. The film is then pulled over and fed into a film winder to obtain a finished film roll [85].

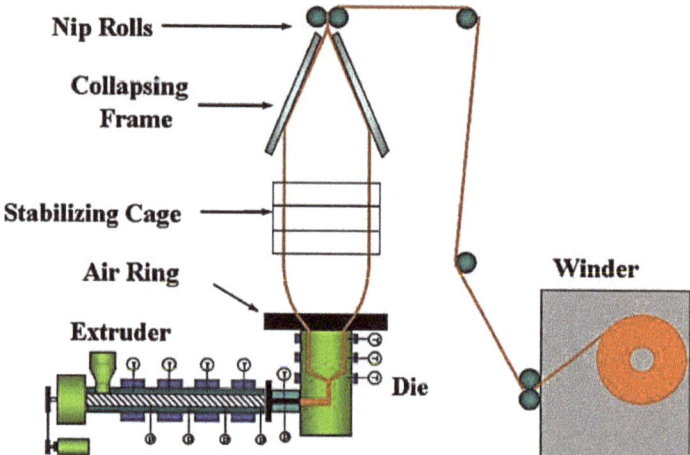

Figure 12. Blown film process schema. Adapted from [85].

In recent years, research efforts have been made by Dow Chemical Company and Cryovac/Sealed Air Corporation to develop new technologies in the context of micro and nanolayer coextrusion for blown film processing. The early version of blown film technology used spiral mandrel dies. The layers are made by separate spiral manifolds, which are present at different radial distances [84]. Then, the melt from different manifolds is joined together near the die exit to form a layered structure. In this type of die, to increase the number of layers in the structure, the diameter of the die has to be increased in order to make room for more spiral mandrel manifolds for each new layer. Thus, this tends to limit the number of layers, since larger diameters dies have longer residence times which can lead to the degradation of the polymers being processed [85].

Another style of spiral mandrel die has been developed, where the spiral channels are cut on the surface of a flat plate rather than on the cylinder surface. This design allows multiple overlapping spirals cut into the same plate. The use of stacked plates with spiral channels on surface plates allow stacking multiple plates to create layered structures. Because of the dimensions of these large flat plates, these dies are commonly referred to as "pancake" style dies [85].

The advantage of using a flat die for the coextrusion of multilayer films is the ability to stack plates on top of each other. In Figure 13, a schematic diagram of the multiple stacked plates bolted together is shown, where each set of plates produces one layer. The layers are added sequentially to the previous layer as the structure flows up the die towards the exit. An example of a commercial stacked plate die is also displayed in Figure 11, in which it is used to produce coextruded films. Structures containing up to 11 layers have been demonstrated using this type of design. However, increasing the number of layers after 11 layers is challenging, due to the pressure drop and lack of space for more extruders [85]. Multilayered blown film lines are currently commercially available from different equipment manufacturers such as Davis Standard, Macro Engineering, Alpha Marathon, Bandera, Windsor, etc., [84].

The challenge of adapting the feedblock and layer multiplier dies technologies from flat film to annular structures involves ensuring the layers continuity around the circumference of the bubble. No uniform layer thicknesses in the films can be obtained if there are breaks and weld lines during the layer wrapping around the circular dies [84].

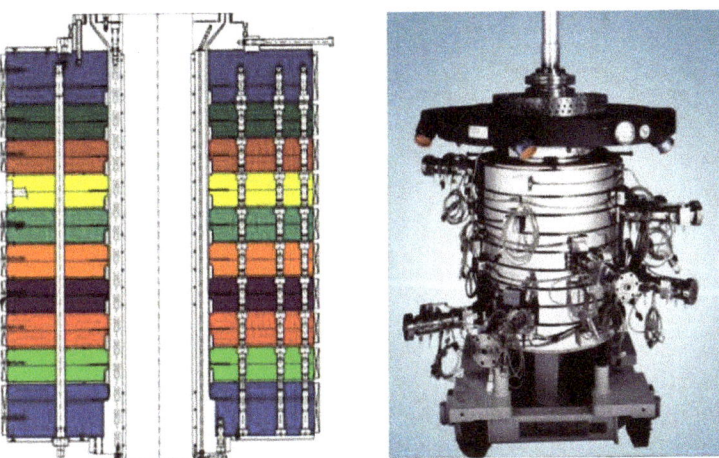

Figure 13. Schematic and commercial example of a multilayer stacked plate or "pancake" die. Adapted from [85].

Using a feedblock and layer multipliers in combination with a specific film die is the new concept applied to produce microlayers by blown extrusion. There are two important characteristics of the film die to consider:

- Protection of the thinner microlayers as they flow from the feedblock to and around the die.
- Geometry design that allows the layers to flow slowly through the die while maintaining the microlayered structure

In order to protect the microlayered structure as it flows through the process, another layer is added to encapsulate the microlayers. In Figure 14, a schema of the encapsulation die is displayed (on the left), which produce a circular encapsulated structure (on the right). This example shows a single core material being encapsulated by another layer. However, in theory, the single core material could be replaced by a microlayer structure [85].

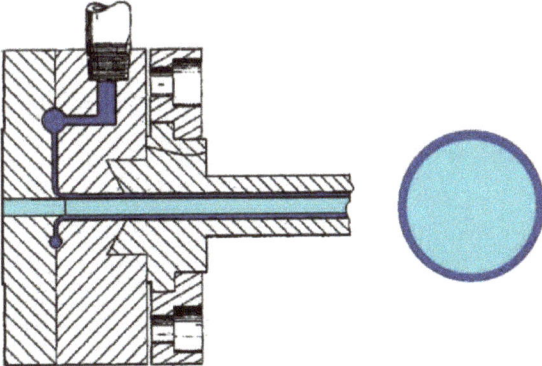

Figure 14. Schema of an encapsulation die (on the left) producing an encapsulated structure (on the right). Adapted from [85].

Further research innovations confirmed the processing of 100+ microlayered films in a blown film structure as [85] demonstrated. Developing uniform wrap of sequentially layered films will continue to challenge the layer multiplier die approach for blown film coextrusion [84].

6.2. Multilayer Films in Agricultural Applications

In the last decades, the agricultural evolution has been parallel to the technological development associated to the use of multilayer plastic films. For the agricultural market, the use of multilayer plastics films has been essential, allowing a remarkable increase of agricultural productions, earlier harvesting, and reduction of plagues presence [86]. The growing use of plastics in agriculture has been helping the farmers to increase their crop production. Nowadays, the use of plastics allows to increase yields, earlier harvests, less dependence on herbicides and pesticides, better protection of food products, and more efficient water conservation [59]. An extensive expanding of plastic films in agriculture is reported worldwide since the middle of the last century. The most common applications of plastics in agriculture are displayed in Table 5.

Table 5. Most common applications of plastic films in agriculture. Reprinted with the permission from [87]. (Copyright 2012, Pagepress).

Protective Cultivation Films	Livestock Farming Films
Low tunnel Greenhouse and tunnel Mulching Direct covering Nursery film Covering vineyards	Silage films Bale wrapping films Protection films

Plastic films can improve product quality by mitigating extreme weather changes, optimizing growth conditions, extending the growing season, and reducing plant diseases. An estimated 2–3 million tons of plastics are used each year in agricultural applications. Almost half of the total plastics produced each year for agricultural applications is used in protective cultivation as mulching, greenhouses, small tunnels, temporary covering for fruits trees, etc., [88]. For this application, the most common polymers used are LLDPE followed by LDPE, and EVA [3].

Plastic films used in agriculture are made by blowing coextrusion process. Coextrusion offers many possibilities by combining the properties of different polymers in order to satisfy each agricultural application [86]:

- **Greenhouse covering:** Greenhouse are films used for crop protection (Figure 15). Films with a range of 100 μm to 1 μm that answer the basic requirements of thermal protection, light transmission, and direct or artificial flood lighting.
- **Mulching:** This technique consists in covering the soil (where the cultivation has been planted) with a layer that protects seedlings and young plants (Figure 16). The films used are transparent or opaque, white, colored, or black with range thickness of 20–50 μm, which are mechanically laid on the soil.
- **Low tunnels:** They consist of small arch-shaped support structures covered by plastic films with the objective to create a microclimate suitable for cultivation (Figure 17). Films of about 60–100 μm thickness are generally installed.
- **Silage films:** Silage is a technique used for conservation of wet forage by acidification of the environment protected from the ambient air (Figure 18). Depending on their specific application, films with a range between 35 and 100 μm are installed.
- **Bale wrap films:** Wrapping films are used for individually or continuously wrapping cylindrical or square bales of fodder in order to obtain an airtight envelop, which allow the anaerobic fermentation process necessary for the production of silage (Figure 19).

Films can be black, white, and other colors (green and brown) with thickness range of 20–30 μm.

Figure 15. Greenhouse covering. Reprinted from website [89].

Figure 16. Black Mulch films. Reprinted from website [90].

Figure 17. Low tunnel transparent film. Reprinted from website [91].

Figure 18. Protection Silage films. Reprinted from website [46].

Figure 19. Wrapping stretch films. Reprinted from website [46].

As mentioned before, the most common polymers used in agriculture plastic films are linear low-density polyethylene (LLDPE), linear density polyethylene (LDPE), and Ethylene vinyl acetate copolymer (EVA). The most important properties for the agricultural application of these polymers are shown in Table 6.

Table 6. Main properties of the polymers used in agricultural plastic films applications. Adapted from [86].

Properties	LDPE	LLDPE	EVA
Photostability	+/−	+/−	+/−
Transparency	−	−	+
Mechanical properties	+/−	+	+
Creep	+	+	−
Welding	+	+	+
Extrusion	+	−	+

(+) good; (+/−) intermediate; (−) bad

Additionally, agriculture films contain additives such as antioxidant, UV-rays, absorbers, fillers, etc., to make them suitable to the application they are aimed for. The main functions of additives are [86]:

- **Plasticizers, lubricants, sliding agents, etc.:** Facilitate the extrusion process.
- **Compatibilizers and coupling agents:** Improve the polymer blends adhesion.
- **Thermal stabilizers and light stabilizers:** Increase the resistance to degradation during the extrusion process or the product life cycle.
- **Cling additives:** Provide adhesive properties to the surface of the films, used mostly for stretch wrap films.
- **Mineral fillers, reinforcements, impact absorbers:** Improve mechanical properties.
- **Pigments, colorants, nucleant agent:** Modify product appearance.
- **Smoke inhibitors, biocides, anti-fog agents, etc.:** Improve the performance of the product.

6.3. Polymers Commonly Used in Agricultural Multilayer Films

6.3.1. Linear Low-Density Polyethylene (LLDPE)

Linear low-density polyethylene resin are molecules with linear polyethylene backbone to which short alkyl groups are attached at random intervals. They are produced by the copolymerization of ethylene with 1-alkenes. Ethyl, butyl, or hexyl groups are the most common branches found in these copolymers. However, a variety of other alkyl groups can be also found, both linear and branched. The typical range of average branches separation along the main chain is around 25–100 carbon atoms. Additionally, LLDPE can present low levels of long chain branching (LCB), but not as complex as the ones found in low-density polyethylene [92]. The comonomers used are typically 1-butene, 1-hexene, and 1-octene. The carbon atoms participating in the comonomer vinyl group are incorporated into the polymer backbone. The remaining carbon atoms form pendant groups, which then are referred to as short-branches (SCB).

One of the attractive attributes of LLDPE is its high film tear strength. This property coupled to its relatively high clarity compared to HDPE (high density polyethylene), generate high acceptance as a polymer grade for film packaging and other applications such as agriculture [92].

6.3.2. Low-Density Polyethylene (LDPE)

The LDPEs are polymers, which contain high branching concentration that hinder the crystallization process, resulting in relatively low densities. Ethyl and butyl groups are the branches found in the LDPE backbone together with some long branches. Since these polymers are produced by high pressure polymerization process, the ethyl and butyl branches are frequently clustered together, separated by long unbranched backbones. Long chain branches (LCB) appear at random intervals along the main chain length. The LCB can also turn into branches themselves. Low-density polyethylene polymers have densities typically in the range of 0.90–0.94 g/cm^3 [92].

Furthermore, the high content of short branches (SCB) found in LDPE reduce its degree of crystallinity, which result in a flexible product with a low melting point. On the other hand, the LCB proportionate desirable processing characteristics, high melt strengths, and relatively low viscosities. These characteristics are attractive for film-blowing processing. The most popular applications of LDPE are low load commercial and retail packaging, and trash bags. Other uses include shrink-wrap, vapor barriers, agricultural ground cover, and greenhouse covers.

6.3.3. Ethylene Vinyl Acetate (EVA)

Ethylene Vinyl Acetate (EVA) is the largest volume of polar ethylene copolymers that are most commercialized. Similar free radical polymerization processes used for LDPE homopolymers are used to obtain EVA polymers. The comonomer vinyl acetate (VA) is distributed randomly along the ethylene backbone and long-chain branches. VA comonomer low crystallinity and high polarity gives the polymer adhesive characteristics and low seal initiation temperature. EVA resins are processed below 230 °C since they are thermally unstable [93].

Since the VA has higher density than the ethylene monomer, it disrupts the polymer crystallinity. This means that the VA increase the polymer density and decrease the crystallinity [93]. The resulting materials have low modulus and good clarity. Additionally, their high branch content results in low lamellar thickness, which translates into low melting temperatures. By contrast, the LCB provide these copolymers with melt characteristics similar to the ones of LDPE. EVA is used generally in packaging films, because of their flexibility, toughness, elasticity, and clarity desirable characteristics. The other main use of ethylene vinyl acetate is as a component for adhesive's formulation.

6.3.4. Polyisobutylene (PIB)

Polyisobutylene (PIB) is a vinyl polymer produce from a monomer isobutylene (IB) by cationic polymerization (Figure 20). PIB is usually classified as a synthetic rubber or elastomer, despite its linear structure. PIB is a colorless to light yellow, elastic, semisolid, or viscous substance. It has unique properties such as low moisture and gas permeability, good thermal and oxidative stability, chemical resistance, and high tack in adhesive formulations. Additionally, it is odorless, tasteless, and nontoxic. Because of their nonpolar structure, PIBs are soluble in aliphatic and aromatic hydrocarbon solvents. The PIB amorphous characteristics and low glass transition temperature ($T_g \approx -62$ °C) give its high flexibility and permanent tack [94].

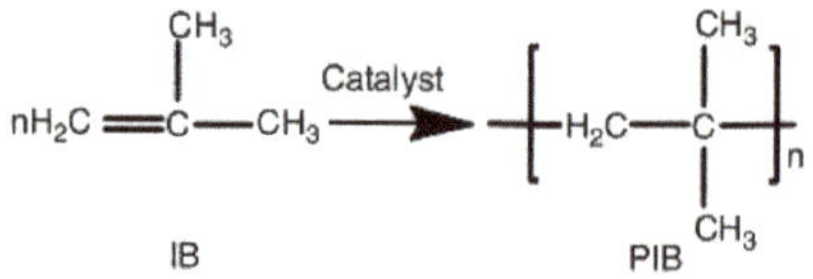

Figure 20. Scheme of the polymerization of isobutylene (IB) to form polyisobutylene (PIB). Adapted from [94].

PIB applications include adhesives, agricultural chemicals, sealants, cling film, personal care products and pigments concentrates, etc.

PIB is widely used as a basic substance in the compounding of pressure-sensitive adhesives (PSAs). Low-molecular weight PIBs are soft and liquid-like, which makes them suitable tacktifiers. The two most important parameters of PSAs are tack and holding power. The cohesive strength of PIBs is relatively low; it can be increased with the addition of high-molecular weight PIB or fillers. PSAs formulated with PIBs are aging resistant and are employed to give adhesion to a variety of substrates such as polyethylenes films for cling films applications [94]. In order to obtain the desired PSA properties, different approaches can be obtained. A combination of low molecular weight and high molecular weight PIBs is used to reach a balance between the cohesive and tack strength. Typically, 80 wt% of low molecular weight PIB is combined with high molecular weight PIB, in order to obtained PSAs with fairly mild adhesive characteristics.

PIB based PSAs are widely used in well-establish application areas such as building, packaging, and tapes. Thanks to their nontoxicity, it is approved for food/food contact, and thus the special interest for food packaging and agricultural applications [94].

7. Main Conclusions and Prospects

The increasing generation and accumulation of non-biodegradable waste is becoming a high-profile public issue, since huge amounts of plastic packaging products are currently designed to have a short service life owing to their low cost and easy production. Unfortunately, most consumers have a negative perception of plastic packaging because of the considerable amount of waste produced in their daily lives. All things considered, it is of great importance to find new solutions for the valorization of multilayered films.

It is a daunting task to summarize the recent and intensive papers published in the last 3 years regarding the subject. However, to the best of our knowledge, there are few works dedicated to (i) eco-design in order to improve recyclability by mechanical recycling and (ii) the recycling of multilayer films in general. This is a reason why a special attention is given to those approaches to complete the pool of the recent papers in this field. Hence, this review focused on the summary of plastic waste management in general, making emphasis on the multilayer film's recycling process. Then, the multilayer film's manufacturing process was described, including the most common materials used for agricultural applications, their processing, and the challenges of their recycling.

Recycling is a key factor to close the loop in the circular economy. However, the fact that multilayer films are made of different types of polymers causes their recycling to be even more challenging. Considering that complex blends are obtained after the mechanical recycling process, compatibilizers need to be added into the blends, but this process can create new issues that need to be further solved. Hence, a novel approach from eco-design to recycling of multilayer structures should be considered as an alternative if mechanical recycling can be assumed.

In addition to chemical recycling, which is one of the promising solutions in the future, the use of strategies such as structure simplification and eco-design are key factors in increasing the recyclability of multilayer plastic films.

Author Contributions: Conceptualization, K.L.; methodology, K.L.; validation, K.L.; formal analysis, G.C. and J.L.; investigation, G.C.; data curation, G.C.; writing—original draft preparation, G.C. and J.L.; writing—review and editing, K.L.; supervision, K.L.; project administration, K.L.; funding acquisition, K.L. and A.M. All authors have read and agreed to the published version of the manuscript.

Funding: This research was funded through financial support from the Région Auvergne-Rhône-Alpes Council (ARC, AURA 2017–2020), the MESRI (Ministère de l'Enseignement Supérieur, de la Recherche et de l'Innovation) and the French National Research Agency (ANR, grant no. ANR-11-RMNP-0002 and ANR, grant no. ANR-20-CE06-0003). L.J. thanks CSC program for support.

Institutional Review Board Statement: Not applicable.

Informed Consent Statement: Not applicable.

Data Availability Statement: Not applicable.

Acknowledgments: The authors thank the funding contributors to support this work.

Conflicts of Interest: The authors declare no conflict of interest.

References

1. Plastics Europe—Association of Plastic Manufacturers. *Plastics—The Facts 2020*; PlasticEurope: Brussels, Belgium, 2020; pp. 1–64.
2. Kaiser, K.; Schmid, M.; Schlummer, M. Recycling of Polymer-Based Multilayer Packaging: A Review. *Recycling* **2017**, *3*, 1. [CrossRef]
3. Horodytska, O.; Valdés, F.J.; Fullana, A. Plastic Flexible Films Waste Management—A State of Art Review. *Waste Manag.* **2015**, *77*, 413–425. [CrossRef] [PubMed]
4. Zhang, X.; Xu, Y.; Zhang, X.; Wu, H.; Shen, J.; Chen, R.; Xiong, Y.; Li, J.; Guo, S. Progress on the Layer-by-Layer Assembly of Multilayered Polymer Composites: Strategy, Structural Control and Applications. *Prog. Polym. Sci.* **2019**, *89*, 76–107. [CrossRef]
5. Mangaraj, S.; Goswami, T.K.; Mahajan, P.V. Applications of Plastic Films for Modified Atmosphere Packaging of Fruits and Vegetables: A Review. *Food Eng. Rev.* **2009**, *1*, 133. [CrossRef]
6. Shemesh, R.; Krepker, M.; Goldman, D.; Danin-Poleg, Y.; Kashi, Y.; Nitzan, N.; Vaxman, A.; Segal, E. Antibacterial and Antifungal LDPE Films for Active Packaging. *Polym. Adv. Technol.* **2015**, *26*, 110–116. [CrossRef]
7. Beigmohammadi, F.; Peighambardoust, S.H.; Hesari, J.; Azadmard-Damirchi, S.; Peighambardoust, S.J.; Khosrowshahi, N.K. Antibacterial Properties of LDPE Nanocomposite Films in Packaging of UF Cheese. *LWT-Food Sci. Technol.* **2016**, *65*, 106–111. [CrossRef]
8. Gumiero, M.; Peressini, D.; Pizzariello, A.; Sensidoni, A.; Iacumin, L.; Comi, G.; Toniolo, R. Effect of TiO_2 Photocatalytic Activity in a HDPE-Based Food Packaging on the Structural and Microbiological Stability of a Short-Ripened Cheese. *Food Chem.* **2013**, *138*, 1633–1640. [CrossRef]
9. López-Rubio, A.; Almenar, E.; Hernandez-Muñoz, P.; Lagarón, J.M.; Catalá, R.; Gavara, R. Overview of Active Polymer-Based Packaging Technologies for Food Applications. *Food Rev. Int.* **2004**, *20*, 357–387. [CrossRef]
10. Nisticò, R. Polyethylene Terephthalate (PET) in the Packaging Industry. *Polym. Test.* **2020**, *90*, 106707. [CrossRef]
11. Gomes, T.S.; Visconte, L.L.Y.; Pacheco, E.B.A.V. Life Cycle Assessment of Polyethylene Terephthalate Packaging: An Overview. *J. Polym. Environ.* **2019**, *27*, 533–548. [CrossRef]
12. Nayak, S.; Khuntia, S.K. Development and Study of Properties of Moringa Oleifera Fruit Fibers/ Polyethylene Terephthalate Composites for Packaging Applications. *Compos. Commun.* **2019**, *15*, 113–119. [CrossRef]
13. Siracusa, V.; Blanco, I. Bio-Polyethylene (Bio-PE), Bio-Polypropylene (Bio-PP) and Bio-Poly(Ethylene Terephthalate) (Bio-PET): Recent Developments in Bio-Based Polymers Analogous to Petroleum-Derived Ones for Packaging and Engineering Applications. *Polymers* **2020**, *12*, 1641. [CrossRef] [PubMed]
14. Mooninta, S.; Poompradub, S.; Prasassarakich, P. Packaging Film of PP/LDPE/PLA/Clay Composite: Physical, Barrier and Degradable Properties. *J. Polym. Environ.* **2020**, *28*, 3116–3128. [CrossRef]
15. Tornuk, F.; Hancer, M.; Sagdic, O.; Yetim, H. LLDPE Based Food Packaging Incorporated with Nanoclays Grafted with Bioactive Compounds to Extend Shelf Life of Some Meat Products. *LWT-Food Sci. Technol.* **2015**, *64*, 540–546. [CrossRef]
16. Mulla, M.; Ahmed, J.; Al-Attar, H.; Castro-Aguirre, E.; Arfat, Y.A.; Auras, R. Antimicrobial Efficacy of Clove Essential Oil Infused into Chemically Modified LLDPE Film for Chicken Meat Packaging. *Food Control* **2017**, *73*, 663–671. [CrossRef]
17. WRAP (Waste and Resource Action Programme). *Plastics Market Situation Report*; WRAP Publishing: Banbury, UK, 2016; pp. 1–29.
18. Plastics Europe. *Plastics—The Facts 2019*; PlasticEurope: Brussels, Belgium, 2019.
19. Schyns, Z.O.G.; Shaver, M.P. Mechanical Recycling of Packaging Plastics: A Review. *Macromol. Rapid Commun.* **2021**, *42*, 2000415. [CrossRef]
20. Ragaert, K.; Delva, L.; Van Geem, K. Mechanical and Chemical Recycling of Solid Plastic Waste. *Waste Manag.* **2017**, *69*, 24–58. [CrossRef]
21. Hestin, M.; Mitsios, A.; Said, S.A.; Fouret, F.; Berwald, A.; Senlis, V. *Deloitte Sustainability Blueprint for Plastics Packaging Waste: Quality Sorting & Recycling*; Deloitte: London, UK, 2017.
22. Biji, K.B.; Ravishankar, C.N.; Mohan, C.O.; Srinivasa Gopal, T.K. Smart Packaging Systems for Food Applications: A Review. *J. Food Sci. Technol.* **2015**, *52*, 6125–6135. [CrossRef]
23. Zhong, Y.; Godwin, P.; Jin, Y.; Xiao, H. Biodegradable Polymers and Green-Based Antimicrobial Packaging Materials: A Mini-Review. *Adv. Ind. Eng. Polym. Res.* **2020**, *3*, 27–35. [CrossRef]
24. Eilert, S.J. New Packaging Technologies for the 21st Century. *Meat Sci.* **2005**, *71*, 122–127. [CrossRef]
25. European Parliament and Council. European Parliament and Council Directive 2008/98/EC of the European Parliament and of the Council of 19 November 2008 on Waste and Repealing Certain Directives (Waste Framework). Available online: https://eur-lex.europa.eu/legal-content/EN/TXT/PDF/?uri=CELEX:32008L0098&rid=9 (accessed on 15 May 2022).
26. Rigamonti, L.; Grosso, M.; Møller, J.; Martinez Sanchez, V.; Magnani, S.; Christensen, T.H. Environmental Evaluation of Plastic Waste Management Scenarios. *Resour. Conserv. Recycl.* **2014**, *85*, 42–53. [CrossRef]
27. Singh, P.; Sharma, V.P. Integrated Plastic Waste Management: Environmental and Improved Health Approaches. *Waste Manag. Resour. Util.* **2016**, *35*, 692–700. [CrossRef]
28. Idumah, C.I.; Nwuzor, I.C. Novel Trends in Plastic Waste Management. *SN Appl. Sci.* **2019**, *1*, 1402. [CrossRef]
29. Pan, D.; Su, F.; Liu, C.; Guo, Z. Research Progress for Plastic Waste Management and Manufacture of Value-Added Products. *Adv. Compos. Hybrid Mater.* **2020**, *3*, 443–461. [CrossRef]
30. Turku, I.; Kärki, T.; Rinne, K.; Puurtinen, A. Characterization of Plastic Blends Made from Mixed Plastics Waste of Different Sources. *Waste Manag. Res.* **2017**, *35*, 200–206. [CrossRef]

31. Plastics Europe. An Analysis of European Plastics Production, Demand and Waste Data. In *Plasts—Facts 2015*; PlasticEurope: Brussels, Belgium, 2015; p. 33. [CrossRef]
32. Plastics Europe. *The Circular Economy for Plastics—A European Overview*; PlasticEurope: Brussels, Belgium, 2019; pp. 1–9.
33. Payne, J.; McKeown, P.; Jones, M.D. A Circular Economy Approach to Plastic Waste. *Polym. Degrad. Stab.* **2019**, *165*, 170–181. [CrossRef]
34. Morici, E.; Carroccio, S.C.; Bruno, E.; Scarfato, P.; Filippone, G.; Dintcheva, N.T. Recycled (Bio)Plastics and (Bio)Plastic Composites: A Trade Opportunity in a Green Future. *Polymers* **2022**, *14*, 2038. [CrossRef]
35. Cimpan, C.; Maul, A.; Jansen, M.; Pretz, T.; Wenzel, H. Central Sorting and Recovery of MSW Recyclable Materials: A Review of Technological State-of-the-Art, Cases, Practice and Implications for Materials Recycling. *J. Environ. Manag.* **2015**, *156*, 181–199. [CrossRef]
36. Haig, S.; Morrish, L.; Morton, R.; Wilkinson, S. *Film Reprocessing Technologies and Collection Schemes*; WRAP (Waste & Resources Action Programme): Banbury, UK, 2012.
37. Schmidt, J.; Grau, L.; Auer, M.; Maletz, R. Multilayer Packaging in a Circular Economy. *Polymers* **2022**, *14*, 1825. [CrossRef]
38. Maris, J.; Bourdon, S.; Brossard, J.-M.; Cauret, L.; Fontaine, L.; Montembault, V. Mechanical Recycling: Compatibilization of Mixed Thermoplastic Wastes. *Polym. Degrad. Stab.* **2018**, *147*, 245–266. [CrossRef]
39. Suzuki, G.; Uchida, N.; Tuyen, L.H.; Tanaka, K.; Matsukami, H.; Kunisue, T.; Takahashi, S.; Viet, P.H.; Kuramochi, H.; Osako, M. Mechanical Recycling of Plastic Waste as a Point Source of Microplastic Pollution. *Environ. Pollut.* **2022**, *303*, 119114. [CrossRef] [PubMed]
40. Thiounn, T.; Smith, R.C. Advances and Approaches for Chemical Recycling of Plastic Waste. *J. Polym. Sci.* **2020**, *58*, 1347–1364. [CrossRef]
41. Rahimi, A.; García, J.M. Chemical Recycling of Waste Plastics for New Materials Production. *Nat. Rev. Chem.* **2017**, *1*, 0046. [CrossRef]
42. Davidson, M.G.; Furlong, R.A.; McManus, M.C. Developments in the Life Cycle Assessment of Chemical Recycling of Plastic Waste—A Review. *J. Clean. Prod.* **2021**, *293*, 126163. [CrossRef]
43. Lombardi, L.; Carnevale, E.; Corti, A. A Review of Technologies and Performances of Thermal Treatment Systems for Energy Recovery from Waste. *Waste Manag.* **2015**, *37*, 26–44. [CrossRef]
44. Punčochář, M.; Ruj, B.; Chatterj, P.K. Development of Process for Disposal of Plastic Waste Using Plasma Pyrolysis Technology and Option for Energy Recovery. *Procedia Eng.* **2012**, *42*, 420–430. [CrossRef]
45. Plastics Europe. *Association of Plastic Manufacturers Communiqué de Presse*; PlasticEurope: Brussels, Belgium, 2021.
46. Barbier Group. Available online: https://www.barbiergroup.com/en/secteur/agriculture-en/mulch-film/#range=very-high-resistance-wrapping-films (accessed on 8 March 2020).
47. Al-Salem, S.M.; Lettieri, P.; Baeyens, J. Recycling and Recovery Routes of Plastic Solid Waste (PSW): A Review. *Waste Manag.* **2009**, *29*, 2625–2643. [CrossRef]
48. Abrha, H.; Cabrera, J.; Dai, Y.; Irfan, M.; Toma, A.; Jiao, S.; Liu, X. Bio-Based Plastics Production, Impact and End of Life: A Literature Review and Content Analysis. *Sustainability* **2022**, *14*, 4855. [CrossRef]
49. Cosate de Andrade, M.F.; Souza, P.M.S.; Cavalett, O.; Morales, A.R. Life Cycle Assessment of Poly(Lactic Acid) (PLA): Comparison Between Chemical Recycling, Mechanical Recycling and Composting. *J. Polym. Environ.* **2016**, *24*, 372–384. [CrossRef]
50. Fullana, A.; Lozano, A. Method for Removing Ink Printed on Plastic Films. EP2832459A1, 26 April 2017.
51. Gecol, H.; Scamehorn, J.F.; Christian, S.D.; Riddell, F.E. Use of Surfactants to Remove Solvent-Based Inks from Plastic Films. *Colloid Polym. Sci.* **2003**, *281*, 1172–1177. [CrossRef]
52. Tartakowski, Z. Recycling of Packaging Multilayer Films: New Materials for Technical Products. *Resour. Conserv. Recycl.* **2010**, *55*, 167–170. [CrossRef]
53. Walker, T.W.; Frelka, N.; Shen, Z.; Chew, A.K.; Banick, J.; Grey, S.; Kim, M.S.; Dumesic, J.A.; Van Lehn, R.C.; Huber, G.W. Recycling of Multilayer Plastic Packaging Materials by Solvent-Targeted Recovery and Precipitation. *Sci. Adv.* **2020**, *6*, eaba7599. [CrossRef] [PubMed]
54. SPI. *Compatibilizers: Creating New Opportunity for Mixed Plastics*; SPI: Washington, DC, USA, 2015.
55. Packaging, S. Enhancing the Value of Barrier Film Recycle Streams with Dow's Compatibilizer Technology Introducing Effective Recycle Compatibilizer Technology. *Food Spec. Packag.* **2014**, *18*. Available online: https://silo.tips/download/food-specialty-packaging-october-2014-volume-18-issue-2 (accessed on 15 May 2022).
56. Fávaro, S.L.; Freitas, A.R.; Ganzerli, T.A.; Pereira, A.G.B.; Cardozo, A.L.; Baron, O.; Muniz, E.C.; Girotto, E.M.; Radovanovic, E. PET and Aluminum Recycling from Multilayer Food Packaging Using Supercritical Ethanol. *J. Supercrit. Fluids* **2013**, *75*, 138–143. [CrossRef]
57. Cabrera, G.; Touil, I.; Masghouni, E.; Maazouz, A.; Lamnawar, K. Multi-Micro/Nanolayer Films Based on Polyolefins: New Approaches from Eco-Design to Recycling. *Polymers* **2021**, *13*, 413. [CrossRef]
58. Briassoulis, D.; Hiskakis, M.; Babou, E.; Antiohos, S.K.; Papadi, C. Experimental Investigation of the Quality Characteristics of Agricultural Plastic Wastes Regarding Their Recycling and Energy Recovery Potential. *Waste Manag.* **2012**, *32*, 1075–1090. [CrossRef]
59. Briassoulis, D.; Hiskakis, M.; Scarascia, G.; Picuno, P.; Delgado, C.; Dejean, C. Labeling Scheme for Agricultural Plastic Wastes in Europe. *Qual. Assur. Saf. Crops Foods* **2010**, *2*, 93–104. [CrossRef]

60. APE France Agricultural Plastic Waste and the Circular Economy. Available online: http://www.plastiques-agricoles.com/ (accessed on 5 February 2020).
61. Gulmine, J.V.; Janissek, P.R.; Heise, H.M.; Akcelrud, L. Degradation Profile of Polyethylene after Artificial Accelerated Weathering. *Polym. Degrad. Stab.* **2003**, *79*, 385–397. [CrossRef]
62. La Mantia, F.P. Closed-Loop Recycling. A Case Study of Films for Greenhouses. *Polym. Degrad. Stab.* **2010**, *95*, 285–288. [CrossRef]
63. Cabrera, G.; Charbonnier, J.; Pichon, G.; Maazouz, A.; Lamnawar, K. Bulk Rheology and Surface Tribo-Rheometry toward the Investigation of Polyisobutylene Migration in Model and Recycled Multilayer Agricultural Films. *Rheol. Acta* **2020**, *59*, 821–847. [CrossRef]
64. MOORE Recycling Associates. *2017 National Postconsumer Plastic Bag & Film Recycling Report*; MOORE Recycling Associates: Sonoma, CA, USA, 2019.
65. Achilias, D.S.; Giannoulis, A.; Papageorgiou, G.Z. Recycling of Polymers from Plastic Packaging Materials Using the Dissolution-Reprecipitation Technique. *Polym. Bull.* **2009**, *63*, 449–465. [CrossRef]
66. Bondon, A.; Lamnawar, K.; Maazouz, A. Experimental Investigation of a New Type of Interfacial Instability in a Reactive Coextrusion Process. *Polym. Eng. Sci.* **2015**, *55*, 2542–2552. [CrossRef]
67. Lu, B.; Alcouffe, P.; Sudre, G.; Pruvost, S.; Serghei, A.; Liu, C.; Maazouz, A.; Lamnawar, K. Unveiling the Effects of In Situ Layer–Layer Interfacial Reaction in Multilayer Polymer Films via Multilayered Assembly: From Microlayers to Nanolayers. *Macromol. Mater. Eng.* **2020**, *305*, 2000076. [CrossRef]
68. Messin, T.; Follain, N.; Guinault, A.; Sollogoub, C.; Gaucher, V.; Delpouve, N.; Marais, S. Structure and Barrier Properties of Multinanolayered Biodegradable PLA/PBSA Films: Confinement Effect via Forced Assembly Coextrusion. *ACS Appl. Mater. Interfaces* **2017**, *9*, 29101–29112. [CrossRef]
69. Messin, T.; Follain, N.; Guinault, A.; Miquelard-Garnier, G.; Sollogoub, C.; Delpouve, N.; Gaucher, V.; Marais, S. Confinement Effect in PC/MXD6 Multilayer Films: Impact of the Microlayered Structure on Water and Gas Barrier Properties. *J. Membr. Sci.* **2017**, *525*, 135–145. [CrossRef]
70. Xu, Y.; Qin, J.; Shen, J.; Guo, S.; Lamnawar, K. Scratch Behavior and Mechanical Properties of Alternating Multi-Layered PMMA/PC Materials. *Wear* **2021**, *486–487*, 204069. [CrossRef]
71. Xu, S.; Wen, M.; Li, J.; Guo, S.; Wang, M.; Du, Q.; Shen, J.; Zhang, Y.; Jiang, S. Structure and Properties of Electrically Conducting Composites Consisting of Alternating Layers of Pure Polypropylene and Polypropylene with a Carbon Black Filler. *Polymer* **2008**, *49*, 4861–4870. [CrossRef]
72. Liu, S.; Li, C.; Wu, H.; Guo, S. Novel Structure to Improve Mechanical Properties of Polymer Blends: Multilayered Ribbons. *Ind. Eng. Chem. Res.* **2020**, *59*, 20221–20231. [CrossRef]
73. Mackey, M.; Hiltner, A.; Baer, E.; Flandin, L.; Wolak, M.A.; Shirk, J.S. Enhanced Breakdown Strength of Multilayered Films Fabricated by Forced Assembly Microlayer Coextrusion. *J. Phys. Appl. Phys.* **2009**, *42*, 175304. [CrossRef]
74. Ponting, M.; Burt, T.M.; Korley, L.T.J.; Andrews, J.; Hiltner, A.; Baer, E. Gradient Multilayer Films by Forced Assembly Coextrusion. *Ind. Eng. Chem. Res.* **2010**, *49*, 12111–12118. [CrossRef]
75. Li, Z.; Olah, A.; Baer, E. Micro- and Nano-Layered Processing of New Polymeric Systems. *Prog. Polym. Sci.* **2020**, *102*, 101210. [CrossRef]
76. Carr, J.M.; Langhe, D.S.; Ponting, M.T.; Hiltner, A.; Baer, E. Confined Crystallization in Polymer Nanolayered Films: A review. *J. Mater. Res.* **2012**, *27*, 1326–1350. [CrossRef]
77. Ponting, M.; Hiltner, A.; Baer, E. Polymer Nanostructures by Forced Assembly: Process, Structure, and Properties. *Macromol. Symp.* **2010**, *294*, 19–32. [CrossRef]
78. Lamnawar, K.; Zhang, H.; Maazouz, A. Coextrusion of Multilayer Structures, Interfacial Phenomena. *Encycl. Polym. Sci. Technol.* (Ed.) **2013**. [CrossRef]
79. Baer, E.; Zhu, L. 50th Anniversary Perspective: Dielectric Phenomena in Polymers and Multilayered Dielectric Films. *Macromolecules* **2017**, *50*, 2239–2256. [CrossRef]
80. Lu, B.; Lamnawar, K.; Maazouz, A.; Sudre, G. Critical Role of Interfacial Diffusion and Diffuse Interphases Formed in Multi-Micro-/Nanolayered Polymer Films Based on Poly(Vinylidene Fluoride) and Poly(Methyl Methacrylate). *ACS Appl. Mater. Interfaces* **2018**, *10*, 29019–29037. [CrossRef]
81. Lu, B.; Zhang, H.; Maazouz, A.; Lamnawar, K. Interfacial Phenomena in Multi-Micro-/Nanolayered Polymer Coextrusion: A Review of Fundamental and Engineering Aspects. *Polymers* **2021**, *13*, 417. [CrossRef]
82. Lu, B.; Bondon, A.; Touil, I.; Zhang, H.; Alcouffe, P.; Pruvost, S.; Liu, C.; Maazouz, A.; Lamnawar, K. Role of the Macromolecular Architecture of Copolymers at Layer–Layer Interfaces of Multilayered Polymer Films: A Combined Morphological and Rheological Investigation. *Ind. Eng. Chem. Res.* **2020**, *59*, 22144–22154. [CrossRef]
83. Lu, B.; Lamnawar, K.; Maazouz, A. Influence of in Situ Reactive Interphase with Graft Copolymer on Shear and Extensional Rheology in a Model Bilayered Polymer System. *Polym. Test.* **2017**, *61*, 289–299. [CrossRef]
84. Langhe, D.; Ponting, M. Coextrusion Processing of Multilayered Films. In *Manufacturing and Novel Applications of Multilayer Polymer Films*, 1st ed.; William Andrew: New York, NY, USA, 2016; pp. 16–45, ISBN 9780323374668.
85. Crabtree, S.; Dooley, J.; Robacki, J.; Lee, P.C.; Wrisley, R.; Pavlicek, C. Producing Microlayer Blow Molded Structures Using Layer Multiplication and Unique Die Head Technology. In Proceedings of the Society of Plastics Engineers—27th Annual Blow Molding Conference, ABC 2011, Chicago, IL, USA, 12–13 October 2011.

86. López, J.C.; Pérez-Parra, J.; Morales, M.A. *Plastics in Agriculture-Applications and Usages Handbook*; CEPLA–Plastics Europe: Almería, Spain, 2009; ISBN 978-84-95531-47-6.
87. Scarascia-Mugnozza, G.; Sica, C.; Russo, G. Plastic Materials in European Agriculture: Actual Use and Perspectives. *J. Agric. Eng.* **2012**, *42*, 15. [CrossRef]
88. Kyrikou, I.; Briassoulis, D. Biodegradation of Agricultural Plastic Films: A Critical Review. *J. Polym. Environ.* **2007**, *15*, 125–150. [CrossRef]
89. Shouman. Available online: http://www.shouman.com/greenhouse.html (accessed on 8 March 2020).
90. Bioplastics, E. New EU Standard for Biodegradable Mulch Films in Agriculture Published. Available online: https://www.european-bioplastics.org/new-eu-standard-for-biodegradable-mulch-films-in-agriculture-published/ (accessed on 8 March 2020).
91. AgriExpo. Available online: https://www.agriexpo.online/prod/daiosplastics/product-184967-87932.html (accessed on 8 March 2020).
92. Peacock, A. *Handbook of Polyethylene*; Dekker, M., Ed.; CRC Press: Boca Raton, FL, USA, 2000; Volume 53, ISBN 9781482295467.
93. Patel, R.M. Polyethylene. In *Multilayer Flexible Packaging*; Elsevier: Amsterdam, The Netherlands, 2016; pp. 17–34, ISBN 9780323371001.
94. Willenbacher, N.; Lebedeva, O.V. Polyisobutene-Based Pressure-Sensitive Adhesives. In *Technology of Pressure-Sensitive Adhesives and Products*; CRC Press: Boca Raton, FL, USA, 2008; pp. 1–18.

Article

Multi-Micro/Nanolayer Films Based on Polyolefins: New Approaches from Eco-Design to Recycling

Geraldine Cabrera [1], Ibtissam Touil [1], Emna Masghouni [1], Abderrahim Maazouz [1,2] and Khalid Lamnawar [1,3,*]

- [1] Ingénierie des Matériaux Polymères, UMR 5223 INSA Lyon, Université de Lyon, CNRS, F-69621 Villeurbanne, France; geralcabrera09@gmail.com (G.C.); Ibtissam.touil@insa-lyon.fr (I.T.); emna.masghouni@insa-lyon.fr (E.M.); abderrahim.maazouz@insa-lyon.fr (A.M.)
- [2] Hassan II Academy of Science and Technology, Rabat 10100, Morocco
- [3] Fujian Key Laboratory of Polymer Science, Fujian Normal University, Fuzhou 350007, China
- * Correspondence: khalid.lamnawar@insa-lyon.fr

Citation: Cabrera, G.; Touil, I.; Masghouni, E.; Maazouz, A.; Lamnawar, K. Multi-Micro/Nanolayer Films Based on Polyolefins: New Approaches from Eco-Design to Recycling. *Polymers* **2021**, *13*, 413. https://doi.org/10.3390/polym13030413

Academic Editor: Tamás Bárány
Received: 31 December 2020
Accepted: 26 January 2021
Published: 28 January 2021

Publisher's Note: MDPI stays neutral with regard to jurisdictional claims in published maps and institutional affiliations.

Copyright: © 2021 by the authors. Licensee MDPI, Basel, Switzerland. This article is an open access article distributed under the terms and conditions of the Creative Commons Attribution (CC BY) license (https://creativecommons.org/licenses/by/4.0/).

Abstract: This paper describes a future-oriented approach for the valorization of polyethylene-based multilayer films. The method involves going from eco-design to mechanical recycling of multilayer films via forced assembly coextrusion. The originality of this study consists in limiting the number of constituents, reducing/controlling the thickness of the layers and avoiding the use of tie layers. The ultimate goal is to improve the manufacturing of new products from recycled multilayer materials by simplifying their recyclability. Within this framework, new structures were developed with two polymer systems: polyethylene/polypropylene and polyethylene/polystyrene, with nominal micro- and nanometric thicknesses. Hitherto, the effect of the multi-micro/nanolayer architecture as well as initial morphological and mechanical properties was evaluated. Several recycling processes were investigated, including steps such as: (i) grinding; (ii) monolayer cast film extrusion; or (iii) injection molding with or without an intermediate blending step by twin-screw extrusion. Subsequently, the induced morphological and mechanical properties were investigated depending on the recycling systems and the relationships between the chosen recycling processes or strategies, and structure and property control of the recycled systems was established accordingly. Based on the results obtained, a proof of concept was demonstrated with the eco-design of multi-micro/nanolayer films as a very promising solution for the industrial issues that arise with the valorization of recycled materials.

Keywords: recycling; eco-design; coextrusion; multilayers

1. Introduction

Since the discovery of polyethylene and polypropylene during the 1950s, polymers have become integrated into all areas of our daily life. Nowadays, annual plastics production in Europe fluctuates around 60 million tonnes [1]. The packaging (39%) and building (19.7%) sectors are the biggest end-use markets for plastics, whereas the agricultural sector represents 3.3% of total plastic demand [1]. Among all plastic materials applications, flexible films have become very popular, mainly due to their versatility, lightness, resistance and printability. Currently, 17% of the world's flexible film production consists of multilayer films [2]. These structures are obtained by coextrusion, which is an industrial process widely used to form films that are suitable for various products including food packaging, medical applications and, recently, microelectronics and nonlinear optics with several thousands of layers [3,4].

Multilayered polymer systems are produced to satisfy specific requirements for high value-added applications such as gas barrier films, robust mechanical systems and optical applications [5], and coextrusion is a process that combines multiple polymers. This is done via two or three extruders using a feedblock system in which melted polymers (from separate extruders) are brought together. Each component of the multilayer structure provides its own end-use characteristics. For example, low-density polyethylene (LDPE) and high-density polyethylene (HDPE) are the most common polymers used in the consumer packaging sector, followed by polyethylene terephthalate (PET), polypropylene (PP) and polystyrene (PS). For agricultural applications and other non-consumer packaging, LDPE and linear low-density polyethylene (LLDPE) are the most employed materials [6].

Nevertheless, the increasing generation and accumulation of non-biodegradable waste is becoming a high-profile public issue, since huge amounts of plastic packaging products are currently designed to have a short service life owing to their low cost and easy production [2]. In the European Union, around 25 million tonnes of post-consumer plastic waste are generated every year. An increase of the total volume of flexible consumer packaging is predicted from 27.4 (2017) to 33.5 million tonnes (2020) is predicted [2,6]. Most consumers have a negative perception of plastic packaging because of the considerable amount of waste produced in their daily lives. All things considered, it is of great importance to find new solutions for the valorization of multilayered films. Hence, a novel approach from eco-design to recycling of multilayer structures should be considered as an alternative if mechanical recycling can be assumed.

A design for recycling has been intensively promoted by the European Union in the last few years, in the context of the Circular Economy [7]. The strategy consists in developing new products so that they can be recycled at the end of their service life. The design of flexible polymer films has a large impact on both end-of-life (recyclability) and the degree to which they can incorporate recycled materials. Design for recycling is encouraged via the implementation of extended producer responsibility (EPR) schemes, in which the end-of-life cost is an economic motivation for the producers [8].

Recycling is a key factor to close the loop in the circular economy. However, the fact that multilayer films are made of different types of polymers causes their recycling to be even more challenging. Considering that complex blends are obtained after the recycling process, compatibilizers need to be added into the blends, but this process can create new issues that need to be further solved. The use of strategies such as structure simplification and eco-design are key factors in increasing the recyclability of the materials [2].

The starting point for the eco-design dynamic is the secondary raw material obtained from the recycled polymer waste that comes from a product that has reached the end of its service life. Nonetheless, as Ragaert et al. [8] have explained, design for recycling must consider multiple factors. This starts with an extensive characterization of the recycled polymers in order to identify their strengths and weaknesses, followed by finding potential new or existing products in which the recycled materials can be incorporated. Then, the design of the new products should be adapted for manufacturing using recycled materials. In addition, for some product requirements, cost-effective strategies can be applied in order to upgrade the quality of the recycled material. This can involve the addition of small amounts of additives such as compatibilizers or stabilizers.

Most European recycling companies work with two types of waste streams: polyethylene- and polyethylene terephthalate (PET)-based streams. However, the waste stream of PE-based films can be contaminated by the presence of other types of films containing polypropylene and polystyrene (PS). Considering the latter, for the present investigation, two polymer blend systems were studied and discussed: polyethylene (PE)/polypropylene (PP) and polyethylene (PE)/polystyrene (PS), since the combinations of these polymers are frequently found in the waste streams of recycling companies. Often, they are used together in the manufacturing of products and cannot be easily separated from each other. Upon waste reception, optical, ballistic densitometer and infrared sorting is applied. That said, most of the time, none of these automatic sorting procedures are efficient enough to completely separate polyethylene from polypropylene and polystyrene. Typical waste streams contain between 5 and 10 wt% polypropylene or polystyrene. Generally, a deterioration of the properties caused by the incompatibility between PE/PP and PE/PS is one of the major problems in processing mixed plastic wastes.

For PE/PP blends, numerous studies have been performed on the mechanical properties of mixed plastics, usually with a focus on neat materials. It has been found that the presence of PP in a PE matrix with a ratio between 5 and 30 wt% promotes an increase of the Young's modulus and tensile strength at yield. However, a significant decrease of the elongation at break and the impact strength has also been observed, since polyethylene becomes rigid and more fragile upon breaking [9]. The PE/PP incompatibility has been proved mostly by microscopic and calorimetric analyses [10,11]. Since PE and PP are incompatible in the melt, the blend behaves as a two-phase mixture. The weak interfacial bond between both phases explains the poor mechanical properties, which are directly linked to the blend morphology. Therefore, since PE and PP are immiscible, a theoretical model for the properties of mixtures cannot be effectively applied [9].

Numerous efforts have been made to improve the compatibility between polyethylene and polypropylene. As found in the literature, the incorporation of a compatibilizer is the most successful technique for improving the morphology and phase dispersion by reducing the interfacial tension and enhancing the adhesion between polymer domains, which leads to an improvement of the mechanical properties. Kazemi et al. [12] summarized the most common graft or block copolymers used as compatibilizers, which include segments similar to the blend components. Styrenic block copolymers (SBC), ethylene-propylene elastomers (EPR), including ethylene-propylene-diene copolymer (EPDM), ethylene-vinyl acetate (EVA) and ethylene-octene copolymers (EOC) are the most commonly used in industry [12].

From a thermodynamic point of view, PE/PS systems are immiscible when blended. Indeed, blending such polymer pairs usually leads to materials with poor mechanical properties and a heterogeneous morphology. To improve the final properties of immiscible blends such as PE/PS without the presence of compatibilizers, multilayer coextrusion has been used to combine several systems with semi-crystalline polymers as confined materials against amorphous polymers with layer thicknesses ranging from 100 to 10 nm [13–16]. With the decrease of compatibilizers which seems to be complex in the classical multiphase polymer blends [17], forced assembly layer coextrusion seems to be a promised route to obtain micro-/nanostructured immiscible polymer systems. Owing the geometrical and macromolecular confinement and with the decrease of a layer's thickness, the crystalline morphology of the system gradually evolves from a three-dimensional (3D) spherulite morphology into a one-dimensional (1D) lamellar morphology [15,16,18–20].

A systematic study of HDPE nanolayers sandwiched between thicker polystyrene (PS) in multilayer coextruded systems has been performed in previous investigations [16]. Interestingly, as the layer thickness of HDPE decreased from the micro- to the nanoscale, a significant reduction in crystallinity was observed from 60% to 33%. This result was associated with discoidal morphologies in the microscale (>100 nm) that transformed into long bundles of edge-on lamellae in HDPE nanolayers (<100 nm). Furthermore, the confined crystallization of polypropylene (PP) against a hard polymer (PS) with variable layer thicknesses and compositions in multilayer systems has been investigated [15]. The changes in layer thickness and composition of the component polymers strongly affected the mechanical properties and morphology of the multilayered film. When the film thickness decreased, however, the morphology passed from continuous and homogeneous to irregular, followed by a significant change in the mechanical properties, specifically the elongation at break [15].

The final objective of this investigation was to study the complex PE/PP and PE/PS systems with a new approach from design to recycling. For this purpose, a design of multi-micro/nanolayer films with a different number of layers is proposed, in order to improve the manufacturing of novel products using recycled materials and to simplify their own recyclability.

In the case of PE/PP compatibilization, multi-micro/nanolayer films were prepared by cast forced assembly coextrusion, in a process intended to vary the number of layers. A selected compatibilizer was incorporated into the central layer, in order to improve the adhesion between polyethylene and polypropylene. Next, the multilayer films were recycled with three different mechanical recycling processes. Finally, the influence of the number of layers and the recycling systems on the morphology and mechanical properties of the PE/PP films was evaluated.

Moreover, multi-micro/nanolayer films with polyethylene (PE) and polystyrene (PS) were also obtained by coextrusion, varying the number of layers. Following this, the structures were recycled with two different recycling processes, and the effect of the number of layers as well as the PE/PS compositions on the mechanical properties and morphology was studied.

2. Experimental Section

2.1. Materials and Sample Preparation Methodology

The properties of the materials used in this work are described in Table 1. To study the compatibilization between PE and PP, multilayer films with linear low-density polyethylene (LLDPE) and 10 wt% PP were first prepared. Then, in order to improve the adhesion between PE and PP, an ethylene-octene copolymer (EOC) was selected as a compatibilizer (Table 1) with a fixed rate of 7 wt%. As a second step, multilayer films composed of 50 wt% linear polyethylene (LDPE) and 50 wt% polystyrene (PS) were prepared in order to study the PS/PE compatibilization and the effect of confinement on crystallization.

Table 1. Summary of the polymers used for the PE/PP and PE/PS systems.

Polymer	LLDPE	PP	EOC	LDPE	PS
Manufacturer	Chevron Phillips	LyondellBasell	Dow Chemical	ExxonMobil 165	Crystal 637
Monomer type	Ethylene/1-Hexene Copolymer	Propylene/Ethylene Copolymer	Ethylene/Octene Copolymer		
Density (g/cm^3)	0.918	0.900	0.875	0.922	1.05
MFI (g/10 min)	1.0 [a]	0.85 [b]	1.0 [a]	0.33 [a]	4 [c]

[a] 190 °C/2.16 kg; [b] 230 °C/2.16 kg; [c] 200 °C/5 kg.

2.2. Cast Forced Assembly Multilayer Coextrusion

In this project, multilayer films were fabricated using a homemade multilayer coextrusion setup as displayed in Figure 1. From the feedblock, the three initial layers flowed through a sequence of layer multiplication dies. These dies divided the melt vertically, then spread it horizontally to the original width, and finally stacked the layered melt stream, while maintaining a constant melt thickness [4]. The final number of layers was determined as a function of the number of multiplication dies, which were placed in series between the feedblock and the final film or sheet exit dies (Figure 1). A set of "n" multipliers led to a film of 3×2^n, since a three-layered feedblock was used for the PE/PP systems. For the PE/PS systems, however, the use of a two-layered feedblock and a set of "n" multipliers led to a film with 2×2^n layers. The temperature used during the coextrusion process was 230 °C for the extruders, multipliers and die. The chill roll was set to 40 °C for the PE/PP materials and to 60 °C for the PE/PS materials with a drawing speed of 1.27 m/min for both systems, in order to obtain a film thickness of 200 ± 50 μm. The feedblock configuration used for the PE/PP systems was B/A/B, whereas for PE/PS it was B/A, with "A" and "B" corresponding to the extruders displayed in Figure 1. It is important to mention that only symmetrical configurations were used in this work.

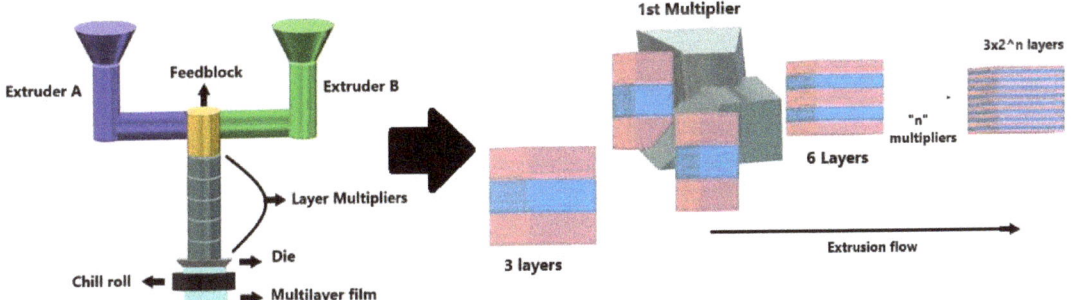

Figure 1. Schematic illustration of layer multiplication in the homemade multilayer coextrusion setup.

For the B/A/B feedblock configuration, two PE/PP systems were extruded: (a) PP/PE/PP, and (b) PP/PE + 7 wt% EOC/PP, with a volumetric flow of 90/10. Meanwhile, for the PE/PS systems, with the B/A feedblock configuration, films with volumetric flows of 50/50 and 10/90 were obtained, keeping the total thickness of the film constant around 200 μm.

All investigated multilayer films are listed in Table 2, where "n" is the number of multipliers and "N" the corresponding number of layers. For the PE/PP systems, the estimated nominal thickness for each layer with a B/A/B film configuration was calculated using Equations (1) and (2), in which "φ_A" and "φ_B" represent the volume fraction of A and B, respectively, "h_{total}" is the total film thickness and "n" is the number of multipliers:

$$h_{nomA} = \varphi_A \frac{h_{total}}{2^n} \tag{1}$$

$$h_{nomB} = \varphi_B \frac{h_{total}}{2 \times 2^n} \tag{2}$$

Table 2. Characteristics of the PE/PP and PE/PS multilayer films obtained by coextrusion.

Films	Volumetric Flow (v/v)	Number of Multipliers (n)	Number of Layers (N)	Total Thickness (μm)	Nominal Layer Thickness (nm) B/A/B Configuration	
					A	B
			PE/PP systems			
PP/LLDPE/PP	90/10	1	6	200	50,000	25000
		3	24	210	23,625	1313
		7	384	250	1758	98
PP/LLDPE + EOC/PP		3	24	210	23,625	1313
		7	384	250	1758	98
Films	Volumetric Flow (v/v)	Number of Multipliers (n)	Number of Layers (N)	Total Thickness (μm)	Nominal Layer Thickness (nm) B/A Configuration	
					A	B
			PE/PS systems			
LDPE/PS (10/90)	10/90	4	32	200	1250	11,250
		7	256		156	1406
		10	2048		20	176
		13	16,384		3	22
LDPE/PS (50/50)	50/50	4	32		6250	6250
		7	256		781	781
		10	2048		97	97
		13	16,384		12	12

273

However, for the PE/PS systems with a B/A configuration, the nominal layer thickness was calculated using only Equation (1), for both A and B components with different volume fractions (50/50 and 10/90 for A/B, respectively), with "h_{total}" as the total film thickness and "n" as the number of multipliers used.

2.3. Mechanical Recycling Processes

Three laboratory-scale mechanical recycling processes were performed for the multilayer films obtained by coextrusion. These recycling processes included different steps such as:

- Grinding: The multilayer films were ground with a blade grinder, leading to the production of small flakes.
- Cast-film extrusion: The flakes obtained from the grinding of multilayer films were used as raw material to feed the extruder. Monolayer films with a thickness of 200 µm were obtained as a result, with an extrusion temperature of 210 °C. The same extruder as the one shown in Figure 1 was used without the feedblock and multiplier elements, in order to obtain monolayer films.
- Extrusion (mini twin-screw extruder): Flakes obtained from the grinding of multilayer films were used as raw material to feed the extruder. The extrusion process was performed at 210 °C. Then, the obtained compounds were pelletized in a cutting machine. The extrusion process was performed using a HAAKE mini co-rotating twin-screw (mini CTW) machine (Thermo-Scientific, Waltham, MA, USA).
- Injection molding: Depending on the recycling process used, flakes (obtained by grinding) or pellets (prepared with the mini twin-screw extruder) were used as raw material to feed the injection-molding machine. A piston injection-molding HAAKE™-Mini-jet was used in order to obtain flat rectangular specimens for tensile testing. Sample dimensions of 2 × 4 mm² (thickness × width) were obtained. The temperature of injection was 220 °C and the injection pressure was 550 bars for 8 s. Then a packing pressure of 450 bars was applied for 8 s.

The multi-micro/nanolayer films studied herein were recycled according to the following:

- R1→ Steps applied: Grinding + Monolayer cast film extrusion
- R2→ Steps applied: Grinding + Injection molding
- R3→ Steps applied: Grinding + Mini twin-screw extrusion+ Injection molding

2.4. Mechanical and Morphological Characterization
Tensile Testing

The tensile tests were performed with an electromechanical testing machine (INSTRON, Norwood, MA, USA). The specimens used had a rectangular shape with dimensions of 4 (±0.5) mm in width, and 50 (±0.1) mm in length. Each sample was clamped into the machine with one end held at a fixed position and the other end displaced at a constant rate of 50 mm/min, using a load cell of 100 N. Data was collected with a chart that monitored the force as a function of the displacement. However, the use of calibrated instrumentation to measure the changes in cross-sectional area during deformation was not accurate enough for thin films such as the multilayer materials in question as Li and al. from the Baer group described in their previous studies [18]. Hence, only engineering stress and strain data were analyzed in this study. The engineering stress of the samples was calculated using Equation (3) (in which "F" is the force causing a given deformation, and "A" is the area). Then, using Equation (4) we calculated the engineering strain, with "l_o" as the initial length and "Δl" the change in length. The elongation and tensile strength at break were determined from the stress-strain plot. Tests were performed only in the machine direction:

$$\sigma_{eng} = \frac{F}{A} \ (\frac{N}{m^2}) \tag{3}$$

$$\varepsilon_{eng} = \frac{\Delta l}{l_o} \times 100 \tag{4}$$

2.5. Scanning Electron Microscopy

The influence of the recycling system on the morphology of the PE/PP and PE/PS multilayers was studied by scanning electron microscopy. For the PE/PP systems, the observations were performed with the specimens obtained after recycling processes R2 and R3. The samples were fractured in liquid nitrogen at a temperature below the glass transition temperature (Tg), and the observations were performed directly without further treatment on a QUANTA 250 FEG microscope (FEI, Hillsboro, OR, USA) in high-vacuum mode.

In the case of the PE/PS multilayer systems, the specimens were first stained by ruthenium tetroxide vapor (RuO_4) for two days and then placed between two epoxy resin plates until consolidation was achieved. Then, the samples were cut and cross-sectioned perpendicularly to their surfaces but parallel to the extrusion direction using a cryo-ultramicrotome (EM UC7, LEICA, Wetzlar, Germany) at room temperature.

3. Results and Discussion
3.1. PE/PP Multilayer Systems
3.1.1. Characterization of the Mechanical Properties
Effect of the Number of Layers

This section is devoted to the effect of the layer thicknesses on the tensile properties of PP/LLDPE/PP multilayer films with 6, 24 and 384 layers. It is important to remember that the PP/LLDPE/PP multilayer films were prepared with a 90/10 composition of LLDPE/PP. The nominal layer thicknesses of the PP/LLDPE/PP films for a B/A/B coextrusion configuration are displayed in Table 2. As can be observed in Figure 2 and Table 3, the number of layers greatly influenced the tensile properties: there was a direct relation to the thicknesses of the nominal layers. The PP/LLDPE/PP-6L and PP/LLDPE/PP-24L were microlayered films. The thicknesses of the nominal layers for PP/LLDPE/PP-6L film were 90 µm and 5 µm for the internal (LLDPE) and external layers (PP), respectively. The PP/LLDPE/PP-24L had nominal thicknesses of 24 µm and 1.3 µm for the internal and external layers, respectively. When the number of layers was increased, however, their nominal thicknesses were decreased, which led us to obtain a nanolayered film. This was the case for the PP/LLDPE/PP-384L films for which the external layer thickness was reduced down to 97 nm.

In the case of the microlayered films, PP/LLDPE/PP-6L and PP/LLDPEPE/PP-24L showed relatively comparable tensile strengths at yield and break. However, the elongation at break exhibited a slight increase with the number of layers due to the presence of thinner PP layers, which according to the literature exhibit superior behavior as compared with LLDPE [9]. It was thus observed that the nanolayered PP/LLDPE/PP-384L presented the highest tensile strength at yield and break (i.e., 8.3 MPa and 25.6 MPa) compared with the microlayered films, as well as the highest elongation at break. These results are quite interesting, since the nanolayered films exhibited a brittle and ductile behavior at the same time. The significant number of layers (384 L) and the thinner PP confined layers (97 nm) were the reasons behind the surprising mechanical response of PP/LLDPE/PP-384L.

Moreover, the tensile properties of the PP/LLDPE/PP multilayer films were compared with the monolayer multicomponent (PE/PP) films as described in Appendix A (Table A1). We observed that the microlayered films (PP/LLDPE/PP-6L and PP/LLDPE/PP-24L) presented tensile strength and elongation at break values closer to the monolayer PE/PP blend (Table A1). This reflects the poor interaction between the polyethylene and polypropylene, despite the microlayered configuration used. The nanolayered PP/LLDPE/PP-384L film, on the other hand, displayed a greater tensile strength and elongation at break than the compatibilized PE/PP monolayer film (with 7 wt% EOC), much closer to the tensile properties of the pure PE-based film (Table A1).

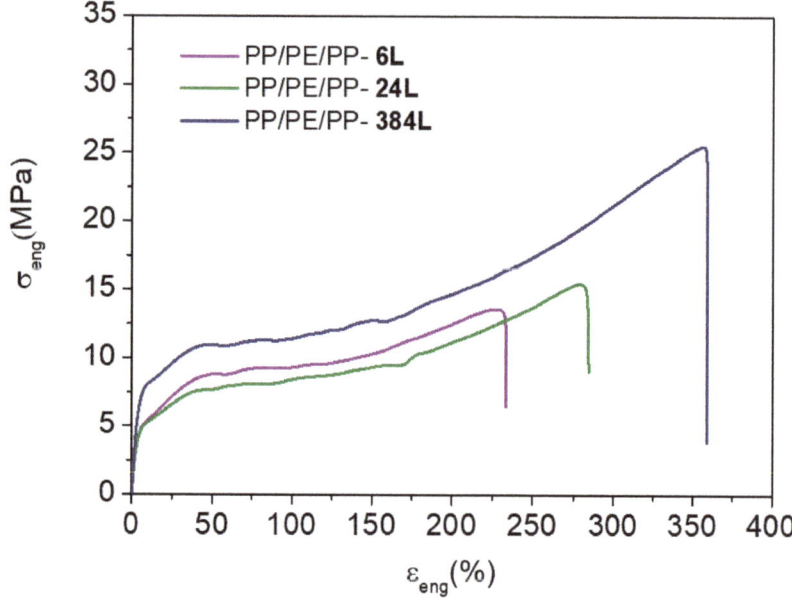

Figure 2. Engineering stress-strain plot of the PP/LLDPE/PP multilayer films with different numbers of layers in the machine direction (MD).

Table 3. Tensile strength properties in the machine direction (MD) of the PP/LLDPE/PP and PP/LLDPE + EOC/PP multilayer films with different numbers of layers.

Multilayer Films	Tensile Strength at Yield (MPa)	Tensile Strength at Break (MPa)	Elongation at Break (%)
PP/LLDPE/PP-6L	5.5 ± 1.7	13.5 ± 3.3	234.7 ± 20.0
PP/LLDPE/PP-24L	4.5 ± 0.5	15.5 ± 1.3	284.5 ± 27.4
PP/LLDPE/PP-384L	8.3 ± 1.2	25.6 ± 4.0	355.5 ± 28.0
PP/LLDPE + EOC/PP-24L	4.3 ± 1.0	20.6 ± 2.5	331.5 ± 18.1
PP/LLDPE + EOC/PP-384L	5.9 ± 1.6	22.4 ± 2.5	351.1 ± 23.0

These results are very promising, since they confirm the importance of the design for recycling concept, as well as the number of confined layers, for improving physical compatibilization. Choosing a suitable configuration of layer thickness and composition makes it possible to achieve the desired mechanical properties without using compatibilizers.

Effect of a Compatibilizer

The influence of an EOC compatibilizer on the mechanical properties of the PP/LLDPE/PP multilayer films is discussed in this section. The tensile properties of PP/LLDPE + EOC/PP-24L and PP/LLDPE + EOC/PP-384L are displayed in Figure 3 and Table 3. Considering the case of the microlayered films, the PP/LLDPE + EOC/PP-24L presented higher elongation at break (331.5 ± 18%) than the uncompatibilized PP/LLDPE/PP-24L film (285 ± 27%). This result was predictable considering the elastomeric nature of the EOC and the influence of this compatibilizer in enhancing the adhesion between PE and PP (Figure A2, Appendix A). However, the slight increase in tensile strength with the presence of EOC was interesting and required further investigation.

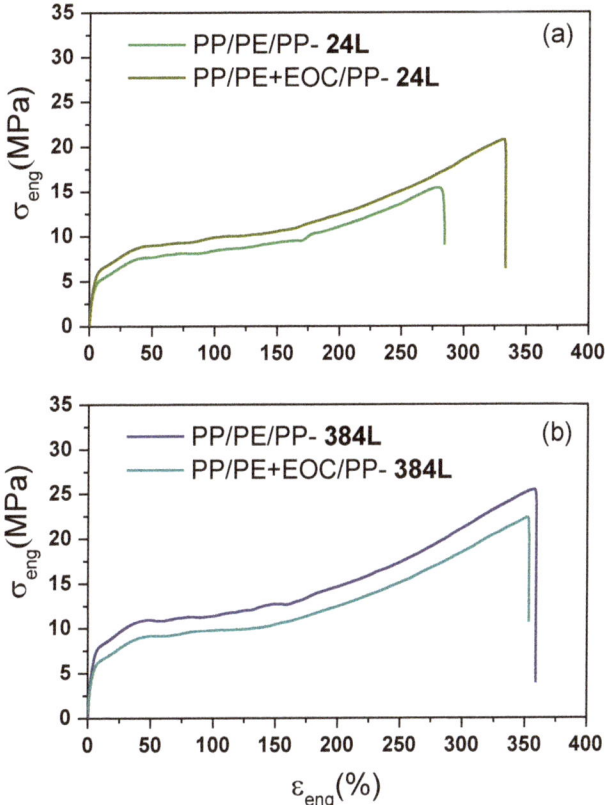

Figure 3. Engineering stress-strain plot of the PP/LLDPE/PP and PP/LLDPE + EOC/PP multilayer films (MD). Influence of the EOC compatibilizer on the films with: (**a**) 24 layers and (**b**) 384 layers.

In the case of the nanolayered films with 384 layers, the effect of the EOC on the elongation at break was almost unnoticeable, compared with the microlayered PP/LLDPE/PP-24L and PP/LLDPE + EOC/PP-24L films. However, a decrease of the tensile strength at yield and break was observed, due to the elastic nature of the EOC. Additionally, we observed that the compatibilized PP/LLDPE + EOC/PP-24L showed similar tensile properties to the PP/LLDPE/PP-384L system without compatibilizer (Table 3).

These results confirmed our previous statement regarding the design and appropriate selection of the multilayer configuration. With multi-nanolayer films, it was possible to enhance the mechanical properties between incompatible polymers such as PE/PP without using compatibilizers. This would give rise to savings in the industrial manufacturing of recycled blends since compatibilizers represent an additional cost.

Influence of the Recycling Process System

The tensile properties of the PP/LLDPE/PP-24L neat multilayer film were compared with those of its recycled monolayer version obtained from the recycling process R1 described in the Experimental Section 2.3. This mechanical recycling process involved grinding and extrusion steps, after which the PP/LLDPE/PP neat film layers were fractured. As can be seen in Figure 4a and Table 4, the elongation at break of the PP/LLDPE/PP-24L neat film (285 ± 27%) decreased by 25% after the recycling process (214 ± 20%), which was expected considering the degradation of the film during the recycling.

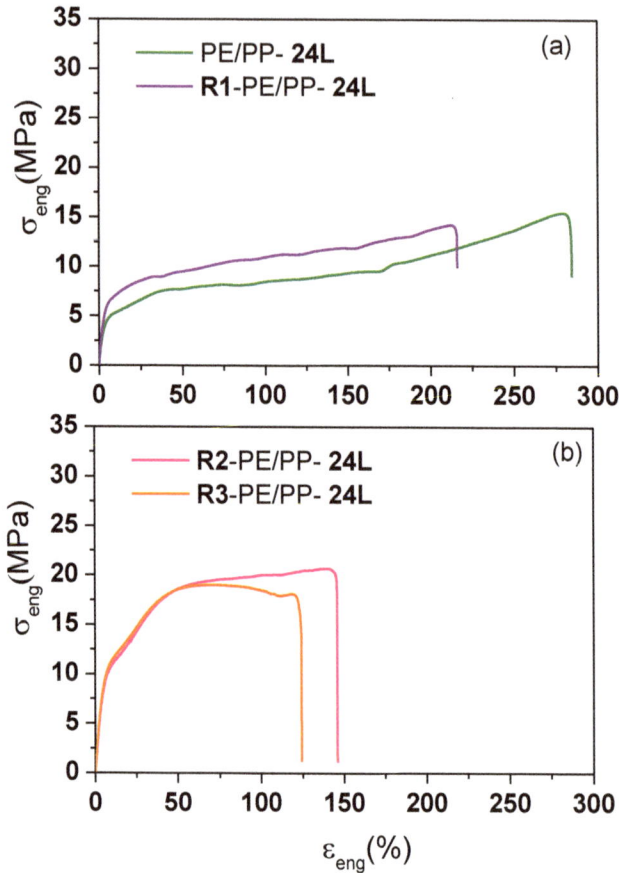

Figure 4. Engineering stress-strain plot of the recycled PP/LLDPE/PP-24L multilayer films (MD) obtained from different recycling processes: (**a**) R1 and (**b**) R2 and R3.

Table 4. Ensile strength properties in the machine direction of the recycled PP/LLDPE/PP-24L multilayer films following different recycling processes.

Recycled Samples	Tensile Strength at Yield (MPa)	Tensile Strength at Break (MPa)	Elongation at Break (%)
	Monolayer film		
R1-PP/LLDPE/PP-24L	6.3 ± 1.3	15.5 ± 3.5	213.7 ± 20.0
	Tensile test rectangular specimen		
R2-PP/LLDPE/PP-24L	10.0 ± 0.5	20.3 ± 0.2	140.5 ± 14.2
R3-PP/LLDPE/PP-24L	10.8 ± 0.3	18.1 ± 0.1	120.3 ± 11.0

For this reason, studies were performed on the specimens obtained from the recycling processes R2 and R3 of PP/LLDPE/PP-24L. The difference between recycling processes R2 and R3 is the additional extrusion step between the grinding and the injection molding. As displayed in Figure 4b, the shape of the stress vs. strain plot was very different from the others since these latter (R2 and R3) specimens had been obtained from injection molding. Consequently, these latter specimens could not be directly compared with monolayer films. Considering the effect of the type of recycling system, we found that the tensile strength at yield and at break was practically the same. However, the elongation at break of R3-

PP/LLDPE/PP-24L decreased by 15% (120 ± 11%) compared with R2-PP/LLDPE/PP-24L (141 ± 14%). This was probably due to the degradation that theR3-PP/LLDPE/PP-24L film suffered during the additional extrusion step before the injection molding. These results highlight the impact of the recycling process on the mechanical properties of the multilayer films.

3.1.2. Morphological Characterization of the Recycled Multilayer Films

The PP/LLDPE/PP-24L and PP/LLDPE/PP-384L films were selected to evaluate the effect of the number of layers and the nominal thickness on the PP/LLDPE/PP film morphology after recycling process R2. As can be observed in Figure 5a, the microlayered recycled film PP/LLDPE/PP-24L maintained its homogeneous multilayered morphology even after recycling. The different PE/PP layers were easily observable, as indicated in Figure 5a, where the smoother surface was believed to correspond to the PE phase.

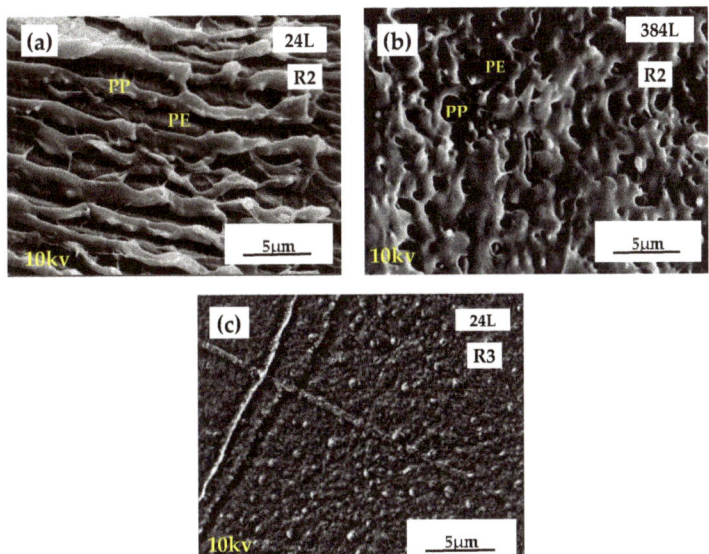

Figure 5. SEM micrographs of (**a**) the PP/LLDPE/PP-24L and (**b**) the PP/LLDPE/PP-384L films recycled with the R2 recycling process, and (**c**) the PP/LLDPE/PP-24L film recycled with R3.

As for the morphology of the nanolayered PP/LLDPE/PP-384L recycled film, the layers were no longer visible after the recycling process (Figure 5b). Since PP/LLDPE/PP-384L had a nominal layer thickness at the nanometric scale, the grinding stage was sufficient to completely destroy the layered morphology. These observations were very interesting, because as seen in Figure 5b, R2-PP/LLDPE/PP-24L exhibited a dispersed morphology, as presented by PE/PP monolayer blends in Appendix A (Figure A1).

Next, the influence of the mechanical recycling system was evaluated for the PP/LLDPE/PP-24L recycled films. As observed in Figure 5c, the effect of the extrusion step carried out after the grinding and before the injection molding greatly affected the morphology, cf. Figure 5a. The extrusion step of recycling process R3 significantly fractured the microlayered structure of PP/LLDPE/PP-24L. A dispersed morphology was thus obtained, similar to the dispersed morphology observed with PE/PP monolayer films (Figure A1). These observations highlight the interesting microstructure of the blends created by the twin-screw extrusion. All these results confirm our previous statement regarding the impact of the recycling process on the mechanical properties and the morphology of the multilayered films.

3.2. PE/PS Multilayer Systems

3.2.1. Study of the Mechanical Properties

Effect of the Number of Layers and Composition

Films of LDPE/PS with 50/50 and 10/90 compositions and different numbers of layers (from 32 to 16,380 layers), as shown in Table 2, were studied with regard to the effect of the layer thickness and the composition on the mechanical properties. For these PE/PS multilayer systems, the thickness of each component and the number of layers was varied from the micro- to the nanometric scale, with a nominal thickness of around 100 nm for the 50/50 composition. This is very close to the thickness of lamellae. or the clarity purpose of the present paper, the effect of layer numbers and LDPE geo-metrical confinement on the crystalline properties are detailed in Appendix B and C. Both 2D WAXS and 1D-WAXS profiles are detailed in Appendix B in Figures A3 and A4. The obtained results corroborate those obtained by DSC as described in Table A2, Appendix C.

The stress-strain curves of multilayer LDPE/PS 50/50 films are displayed in Figure 6. We can see that the tensile strength at yield of the multi-micro/nanolayer structures was similar to that of PS and remained constant over 22MPa. Nevertheless, the number of layers had an influence on the elongation at break of the multilayered films, as shown in Table 5 and Figure 6. For example, the LDPE/PS-32L structure, with a nominal thickness of 6250 nm, exhibited an intermediate behavior between those of LDPE and PS. LDPE/PS-32L was found to be brittle and comparable to neat PS. However, for LDPE/PS-16380L, with a nominal thickness of 97 nm, the mechanical response of the multilayer film showed a higher ductility, which indicates that the brittle behavior observed with LDPE/PS-32L was due to the poor adhesion between the LDPE/PS interfaces at the microscale. Hence, as we increased the number of layers, the adhesion at the LDPE/PS interface improved due to the geometric confinement of the layers, which explains the increased ductility of the film. These results agree with previous studies of multilayered immiscible PP/PS films with varying compositions (from 90/10 to 10/90) reported by Scholtyssek et al., 2010 [15], who found that the elongation at break increased significantly with the decrease of nominal layer thickness at constant composition.

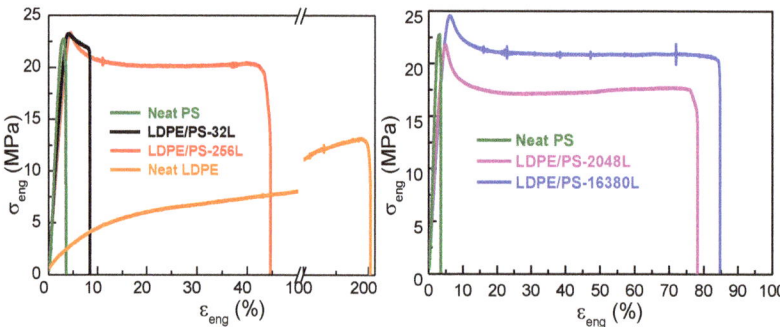

Figure 6. Engineering stress-strain plot of LDPE/PS multilayer 50/50 (wt/wt) films with different numbers of layers in the machine direction.

Table 5. Mechanical properties of the LDPE/PS 50/50 (wt/wt) films with different numbers of layers.

Films	Tensile Strength at Yield (MPa)	Tensile Strength at Break (MPa)	Elongation at Break (%)
LDPE/PS-32L	22 ± 2.8	22 ± 2.2	8 ± 2.4
LDPE/PS-256L	22 ± 6	20 ± 2	45 ± 18
LDPE/PS-2048L	22 ± 2.6	18 ± 2.6	78 ± 27
LDPE/PS-16380L	24 ± 2.2	20 ± 2	84 ± 20

For the multilayered structures with a 90/10 composition, the samples showed an intermediate behavior between the neat LDPE and PS, as seen in Figure 7 and Table 6. However, for the LDPE/PS-16380L film with a nominal layer thickness of ~95 nm, we observed an increase of the tensile strength at yield as well as at break. This can be explained by the effect of geometric confinement of LDPE crystals against the amorphous PS and also by the effect of on-edge orientation during the coextrusion process, as explained in detail in Appendices B and C.

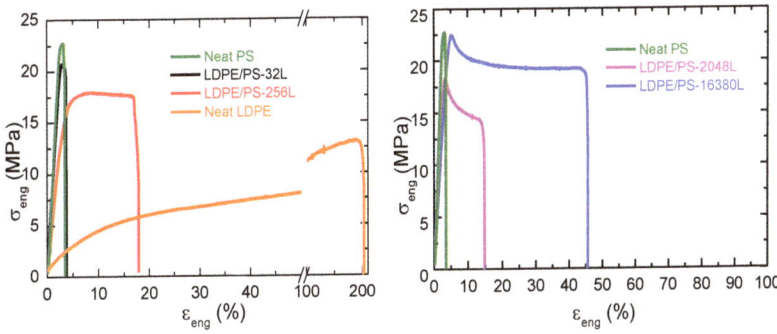

Figure 7. Engineering stress-strain plot of multilayered LDPE/PS 10/90 films (in the machine direction) with different numbers of layers.

Table 6. Mechanical properties of the LDPE/PS 10/90 (wt/wt) films with different numbers of layers.

Films	Tensile Strength at Yield (MPa)	Tensile Strength at Break (MPa)	Elongation at Break (%)
LDPE/PS-32L	20 ± 3.2	20 ± 3	4 ± 0.5
LDPE/PS-256L	18 ± 0.5	17.5 ± 1	16 ± 14
LDPE/PS-2048L	18 ± 1.1	14 ± 1.5	14 ± 3
LDPE/PS-16380L	22 ± 2.6	19 ± 2	45 ± 14

Regarding the effect of the LDPE/PS composition, clear differences were found between the LDPE/PS films with 50/50 and 10/90 compositions, as shown in Figures 6 and 7. The LDPE/PS 50/50 sample showed a higher elongation at break regardless of the number of layers, compared with LDPE/PS 10/90. These results demonstrate that the film's ductility was not only influenced by the nominal layer thickness but also by the composition. A similar conclusion was put forward by Scholtyssek et al. [15], who stated that the composition has a significant effect on the ductility of multilayered films.

Influence of the Recycling Process on the Mechanical Properties

Next, we analyzed the effect of the different recycling processes (described in the experimental section) on the mechanical properties of the multilayer films. Figure 8 displays the effect of the number of layers on the recycled films denoted R2-LDPE/PS (10/90). A significant reduction of the elongation at break was seen for the recycled multilayer film with 2048 layers compared with its 256-layer counterpart. This was expected, since the nominal thickness of the LDPE/PS-2048L (10/90) film was thinner (~100 nm) than the multilayer film with 256 layers (~1 μm), thus reflecting the fragility of the layers during the grinding and injection steps of recycling process R2. Nevertheless, in terms of tensile strength at yield and at break, we observed the opposite effect with the number of layers. This can be explained by a geometric confinement effect of LDPE crystals against the amorphous PS, even after the R2 recycling process.

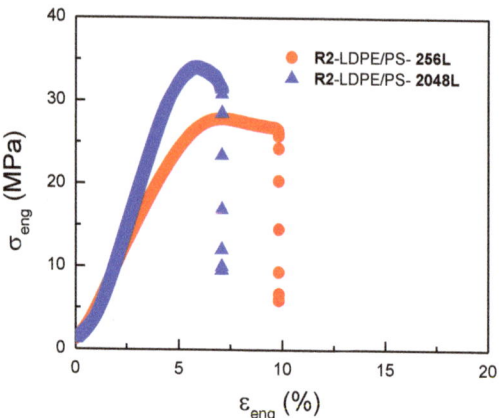

Figure 8. Engineering stress-strain plot of the recycled R2-LDPE/PS (10/90) multilayer films (MD) with different numbers of layers.

Next, Figure 9 compares the effect of recycling processes R2 and R3 on the mechanical properties of the LPDE/PS (90/10) with 32 layers. As a reminder, the difference between recycling processes R2 and R3 is the additional step of twin-screw extrusion between the grinding and injection molding steps. As displayed in Figure 9, both recycled films showed equivalent strengths at yield and at break, and the same was true for the elongation at break. These results indicate that the additional extrusion step for the R3 recycling process did not degrade the multilayer films as much as for the PE/PP systems (Figure 5). This can be due to the brittle behavior of the PS as compared with PE and PP.

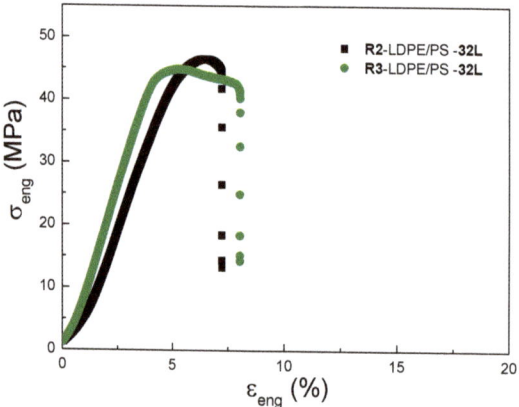

Figure 9. Engineering stress-strain plot of the recycled LDPE/PS (10/90)-32L multilayer films (MD) with the R2 and R3 recycling systems.

3.2.2. Morphological Characterization of the Recycled and Neat LDPE/PS Films

SEM micrographs of neat and recycled LDPE/PS-256L with a 50/50 composition are displayed in Figure 10. These structures were analyzed over a cross-section in the machine direction in the aim of studying the effect of the recycling system on the morphology of the multilayered structure.

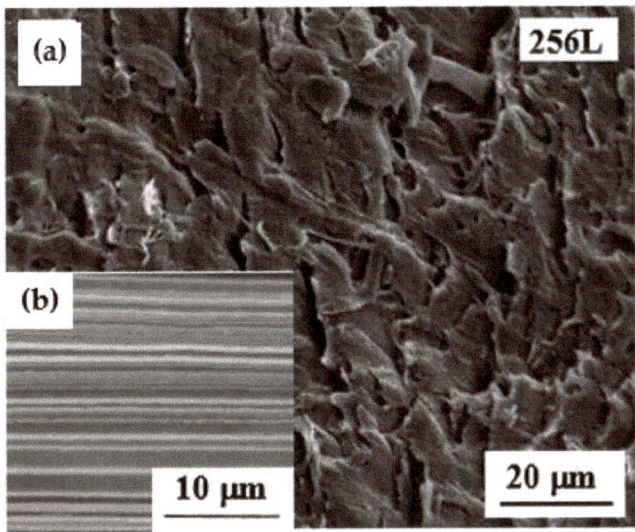

Figure 10. SEM micrographs of (**a**) the recycled R2-LDPE/PS-256L multilayer film and (**b**) the neat LDPE/PS-256L multilayer film with 50/50 composition.

As shown in Figure 10b, the neat LDPE/PS-256L presented a continuous and regularly structured morphology. It is important to mention that the continuity as well as the uniformity of the multilayered films was greatly affected by the film thickness and composition, which is in agreement with the results obtained by Scholtyssek et al. [15]. However, with the recycled R1-LDPE/PS-256L film (Figure 10b) we could no longer differentiate between the layers. This proved that the grinding and extrusion steps were sufficient for destroying the layered morphology presented by the neat LDPE/PS-256L.

4. Conclusions

This paper presents a new approach to design for recycling. In order to improve the manufacturing of new products using recycled materials and to simplify their recyclability, we proposed a design of multi-micro/nanolayer films with different numbers of layers.

Considering that combinations of polyethylene/polypropylene and polyethylene/polystyrene are frequently found in recycling companies' waste streams, these two polymer blend systems were analyzed and discussed in the present investigation. Generally, the deterioration of mechanical properties caused by an incompatibility between polyethylene/polypropylene and polyethylene/polystyrene is one of the major problems in processing mixed plastic wastes.

In order to study the compatibilization of polyethylene/polypropylene, PE/PP multi-micro/nanolayer films were prepared by coextrusion on a laboratory scale, by varying the number of layers. The EOC compatibilizer was incorporated in the central layer, so as to improve the adhesion between the PE/PP layers. Then, the multilayer films were recycled with three different mechanical recycling systems. With the results obtained from the mechanical characterization, we demonstrated that choosing the most suitable configuration for the multilayer films is a valuable way to boost the mechanical properties of the incompatible PE/PP systems. Indeed, with a multi-nanolayer film configuration, it was possible to enhance the mechanical properties between PE/PP without using compatibilizers. Moreover, a comparison between the various recycling systems demonstrated the impact of the recycling process on the mechanical properties of the multilayer films.

Next, a characterization of the morphology carried out with SEM showed that the nanolayered recycled films (from recycling process R2) presented a dispersed morphology with polyethylene as the continuous matrix. However, the microlayered recycled films (from recycling process R2) kept their layered morphology even after recycling. Meanwhile, the additional extrusion stage carried out prior to injection molding, in recycling process R3, greatly fractured the layered structure of the microlayered film. A dispersed morphology was observed when passing from micro- to nanolayer systems. All these results confirm our statement regarding the impact of the recycling process on the mechanical properties and the morphology of the multilayer films.

Moreover, structures with alternating layers of low-density polyethylene (LDPE) and polystyrene (PS) were also prepared by coextrusion of a varying number of layers and compositions. For the symmetrical composition (50/50), the multilayered films presented a brittle behavior with a nominal thickness at the microscale. However, as we decreased the nominal layer thickness down to the nanoscale, the multilayer films became more ductile, as the elongation at break was increased. This can be explained by the geometrical confinement and the orientation of LDPE chains during the coextrusion process. Additionally, the same trend was found for the asymmetrical composition (10/90), suggesting that the ductility of the films was not only influenced by the nominal thickness but also by the composition of the blends.

A continuous and regularly structured morphology of the neat multilayer films was observed with SEM analysis. However, after the R2 recycling process, the layered structure could no longer be observed, indicating that only the grinding and injection molding steps were sufficient to destroy the layered structure presented by the neat multilayer film.

Regardless of the polymer system studied, we have demonstrated that the design of multi-micro/nanolayer films is a very promising solution to the industrial issues that accompany the valorization of recycled materials, without the use of compatibilizers. It is, however, important to mention that the present study was performed exclusively at the laboratory scale as a proof of concept for further scale-up investigations. Moreover, a deeper understanding of the nanostructure mechanisms and morphological involved changes during recycling of the present systems will require further investigation in the future.

Author Contributions: Conceptualization, K.L. and A.M.; methodology, K.L.; validation, K.L.; formal analysis, G.C., E.M. and I.T.; investigation, G.C., E.M. and I.T.; data curation, G.C. and I.T.; writing—original draft preparation, G.C.; writing—review and editing, K.L.; supervision, K.L.; project administration, K.L.; funding acquisition, K.L. and A.M. All authors have read and agreed to the published version of the manuscript.

Funding: This research was funded through financial support from the Région Auvergne-Rhône-Alpes Council (ARC, AURA 2017–2020), the MESRI (Ministère de l'Enseignement Supérieur, de la Recherche et de l'Innovation) and the French National Research Agency (ANR, grant no. ANR-11-RMNP-0002 and ANR, grant no. ANR-20-CE06-0003).

Institutional Review Board Statement: Not applicable.

Informed Consent Statement: Not applicable.

Data Availability Statement: Not applicable.

Acknowledgments: This manuscript is written in honor of the 50th anniversary of the French Polymer Group (Groupe Français des Polymères—GFP). The authors acknowledge the MSc students for their help.

Conflicts of Interest: The authors declare no conflict of interest.

Appendix A. Mechanical Properties and Morphological Characterization of PE/PP Monolayer Films

Model PE/PP monolayer films were prepared by extrusion in order to study the compatibilization between PE and PP with simpler structures. Model blends were prepared with neat LLDPE and PP selected for this study (Table 1) as well as the compatibilizer EOC, in order to improve the adhesion between PE and PP.

Monolayer PE/PP films with and without compatibilizer EOC were obtained by blown film extrusion using a co-rotating twin-screw extruder with an annular blow die with a 50-mm diameter coupled to a melt pump. The temperature profile during extrusion was gradually increased from 190 °C to 220 °C from the feeding, tothe compression, to the pumping zone. The molten polymer in the form of a tube exited from the die at 220 °C and was then drawn upward by the take-up device. When the process started up, air was introduced at the bottom of the die to inflate the tube in order to form a bubble. The tube was then flattened in the nip rolls and taken up by the winder. An air ring was also used to rapidly cool the hot bubble and solidify it above the die exit. The film dimensions were determined by the blow-up ratio (BUR) and the take-up ratio (TUR). BUR is defined as the ratio between the final bubble diameter (Df) and the die diameter (D0), which can be controlled by varying the air pressure. TUR, on the other hand, is defined as the ratio between the speed of the take-up device (Vf) and the extruded material velocity (V0) at the die exit. For the present study, a drawing speed of 3.7 m/min with a fixed BUR and TUR was used in order to obtain films with similar thicknesses of around 30 ± 5 μm. All the films were prepared with the same extrusion parameters.

Appendix A.1. Tensile Properties

Tensile testing of the monolayer films was performed under the same conditions and with the same parameters as explained in the experimental section. The samples had a rectangular shape, with dimensions of 4 ($\pm$0.5) mm in width, and 50 ($\pm$0.1) mm in length (ISO 527-1).

As observed in Table A1, the presence of PP in the PE/PP (90/10) film increased the tensile stress at yield, compared with the properties of the 100% PE film. However, it was also found that inclusion of PP in the PE continuous phase of the PE/PP (90/10) film led to lower deformability or elongation at break and tensile stress at break. The elongation at break of the 100%PE film (562.6 $\pm$ 30%) decreased by 59% compared with the PE/PP (90/10) film (228.1 $\pm$ 35%). As previously reported in numerous studies [17,18], blends of polyethylene and polypropylene are incompatible, resulting in poor mechanical properties. Next, we observed that the inclusion of the EOC compatibilizer reduced the tensile stress at yield and at break of the PE/PP (90/10) film. This was expected, due to the elastomeric nature of the EOC (Table 1). Conversely, the PE/PP (90/10) elongation at break increased by 5% with the addition of 7 wt% EOC. These observations confirm that EOC was a suitable candidate for enhancing the compatibility of our PE/PP blends.

Table A1. Tensile strength properties of the PE/PP monolayer films in the machine direction (MD).

Films (wt/wt)	Tensile Strength at Yield (MPa)	Tensile Strength at Break (MPa)	Elongation at Break (%)
100% PE	6.7 $\pm$ 0.6	44.3 $\pm$ 6	562.6 $\pm$ 30
90% PE 10% PP	8.2 $\pm$ 0.5	19.5 $\pm$ 2	228.1 $\pm$ 35
83%PE 10%PP 7% EOC	2.19 $\pm$ 0.2	16.4 $\pm$ 4	240.0 $\pm$ 48

Appendix A.2. Morphological Characterization with SEM

The morphology of the PE/PP monolayer films was studied using scanning electron microscopy (SEM). The samples shown in Figure A1 were prepared with the aid of a cryoultramicrotome and were chemically stained with ruthenium tetroxide (RuO$_4$) vapor for 2 h, in order to enhance the contrast of the samples' different phases.

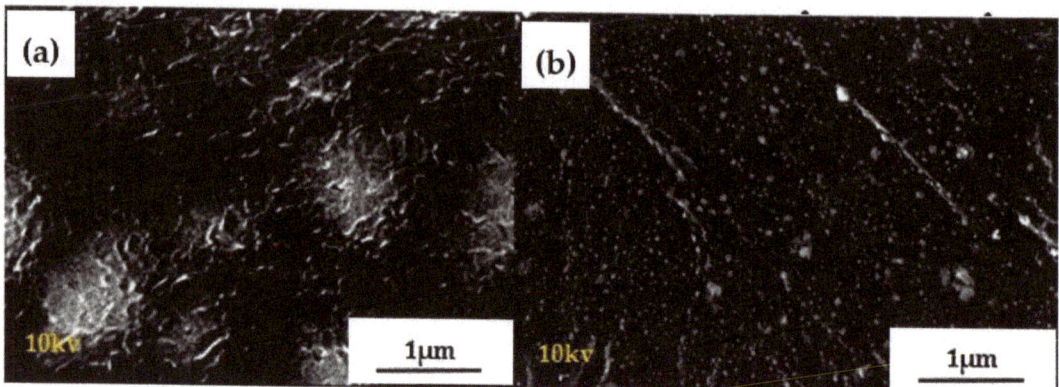

Figure A1. High-vacuum SEM images of PE/PP monolayer films: (**a**) PE/PP (90/10) (without compatibilizer) and (**b**) PE/PP/EOC (83/10/7).

As can be seen in Figure A1a, the PE and PP phases were clearly visible. The continuous phase (black) corresponded to the polyethylene, while the dispersed phase (gray) represented the PP domains, which exhibited more brittle behavior compared with the polyethylene [9]. The SEM micrograph in (Figure A1a shows the phase morphology of the PE/PP monolayer film as a dispersion of PP droplets in the PE matrix. Figure A1b displays the morphology of the PE/PP with 7 wt% EOC as a compatibilizer. As can be observed, the addition of EOC promoted finer morphologies. Figure A2 reveals the PP droplet size distributions of the PE/PP films with and without compatibilizer. The addition of 7 wt% EOC into the PE/PP blend was sufficient to reduce the domain sizes, resulting in a finer morphology. The uncompatibilized PE/PP blend presented an average PP droplet size of 0.1 µm, whereas the EOC-compatibilized blend had an average PP droplet size of 0.01 µm. The compatibilizer resided at the interface between PE and PP and reduced their interfacial tension [12]. These observations confirm the improvement of some of the tensile properties of the PE/PP monolayer films by the presence of 7 wt% EOC as a compatibilizer.

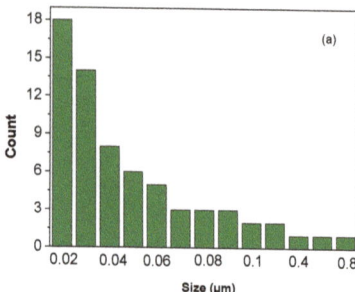 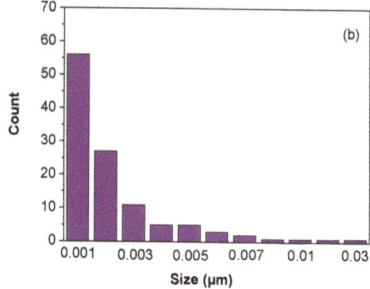

Figure A2. Size distribution of the PP droplets in the PE matrix: (**a**) PE/PP (without compatibilizer) and (**b**) PE/PP/EOC (83/10/7).

Appendix B. LDPE/PS Systems: Effect of the Number of Layers on the Crystalline Orientation of LDPE

The orientation and crystalline structure of LDPE in multilayered LDPE/PS films were characterized by 2D-WAXS patterns with an X-ray incident beam in the extruded and normal direction to the film plane, as seen in Figure A3. The LDPE/PS films exhibited two clear diffraction rings at 2θ = 21.33° and 23.66°, corresponding to (110) and (200) from the orthorhombic crystal structure of polyethylene superposed with the wide amorphous halo of PS (2θ = 18.9°), as seen from the 1-D profile for all multilayered structures in Figure A4. WAXS images of both normal and extruded directions of these multilayered structures revealed a slight equatorial concentration of the (200) planes and became more intense as we increased the number of layers, especially in the case of 16380L. This was an indication that the chains had aligned parallel to the film plane resulting in an on-edge orientation, in which the confined spherulite changed from three dimensions (3-D) to a compressed two-dimensional spherulite. This is mentioned in the literature as discoids nucleating at the LDPE/PS interface. The confinement effect of LDPE chains against amorphous PS in the normal direction(ND) gave rise to an on-edge orientation, in which the LDPE crystals were set parallel to the film plane, whereas in the extruded direction the on-edge oriented chains were stretched and extended along the equator during the coextrusion process.

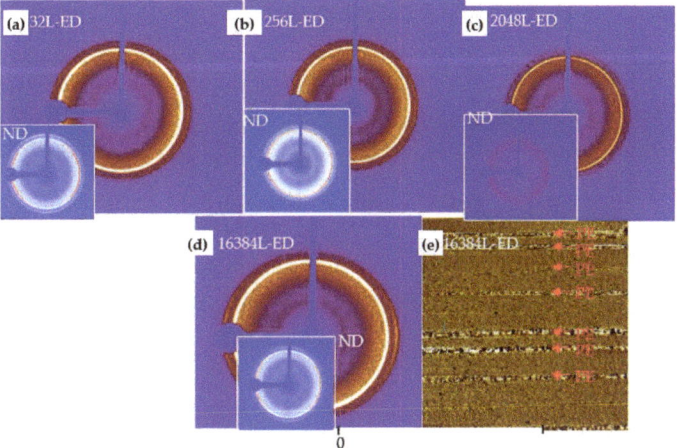

Figure A3. 2D-WAXS profiles recorded with an X-ray beam in normal and parallel directions to the film plane for LDPE/PS multilayers: (**a**) 32L, (**b**) 256L, (**c**) 2048L, (**d**) 16380L and (**e**) example of an AFM image for a sample with 16,380 layers.

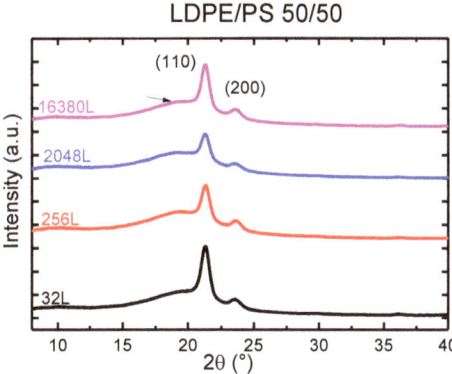

Figure A4. 1D-WAXS profiles for LDPE/PS multilayers with different numbers of layers: 32L, 256L, 2048L and 16380L. The intensity was normalized with the thickness of the studied films. The scattering angle was denoted 2θ.

Appendix C. Thermal Characterization of the LDPE/PS Systems with Dynamic Scanning Calorimetry (DSC)

The DSC parameters for neat LDPE, multilayered films, and their blends with weight fractions of 50/50 are summarized in Table A2. The melting temperature remained relatively constant for all multilayered samples and lay between the melting temperature of neat LDPE and that of the blend. For the melting enthalpy of LDPE, a slight increase was observed, ranging from 24.5 J/g for 32L to 31 for 2048L, corresponding to the nominal thickness of 6250 nm and 195 nm, respectively. However, with a nominal layer thickness of 95 nm for 16380L, the melting enthalpy decreased down to 26 J/g, which was similar to the enthalpy obtained for the blend.

Indeed, a similar trend was observed for the crystallinity of the multilayered films, which increased from 8.5% for 32L to 10.5% for 2048L, but then decreased to 9% for the film with 16380 layers. This was very similar to the crystallinity of the blend, which meant that films with more than 2048 layers presented a crystallinity similar to that of a blend.

Table A2. DSC parameters for the LDPE/PS multilayers, the blend (50/50) and the neat polymer.

Sample	LDPE/PS—50/50				
	T_m (°C)	ΔH_m (J/g)	X_c, DSC (%)	h_N	L (nm)
LDPE	112	117	40	-	9.47
32	110.5	24.5	8.5	6250 nm	9.01
256	111.7	29.7	10	781 nm	9.38
2048	110.6	31	10.5	195 nm	9.04
16380	110	26.5	9	97 nm	8.81
Blend	113	25	8.5	-	9.81

An example of the morphology obtained by AFM for a sample with 16380 layers is presented in Figure A3. As shown, the LDPE nanolayers had a bright color, while the PS appeared as dark layers. These observations showed a morphology with continuous layers of PE and PS, with a large size distribution in the LDPE thickness, where the amorphous PS layers confined the 3-D spherulite of LDPE and were forced to bend and create two-dimensional discoids. The thickness of LDPE measured from the AFM micrographs was higher than the estimated value calculated from Equation (1).

References

1. Plastics Europe and EPRO. "Plastics—The Facts 2016". Available online: www.plasticseurope.de/informations (accessed on 31 March 2020).
2. Horodytska, O.; Valdés, F.; Fullana, A. Plastic flexible films waste management—A state of art review. *Waste Manag.* **2018**, *77*, 413–425. [CrossRef] [PubMed]
3. Bondon, A.; Lamnawar, K.; Maazouz, A. Experimental investigation of a new type of interfacial instability in a reactive coextrusion process. *Polym. Eng. Sci.* **2015**, *55*, 2542–2552. [CrossRef]
4. Lu, B.; Alcouffe, P.; Sudre, G.; Purvost, S.; Serghei, A.; Chuntai, L.; Maazouz, A.; Lamnawar, K. Unveiling the Effects of In Situ Layer–Layer Interfacial Reaction in Multilayer Polymer Films via Multilayered Assembly: From Microlayers to Nanolay-ers. *Macromol. Mater. Eng.* **2020**, *305*, 2000076.
5. Ponting, M.; Burt, T.M.; Korley, L.T.J.; Andrews, J.; Hiltner, A.; Baer, E. Gradient Multilayer Films by Forced Assembly Coextrusion. *Ind. Eng. Chem. Res.* **2010**, *49*, 12111–12118. [CrossRef]
6. Cabrera, G.; Charbonnier, J.; Pichon, G.; Maazouz, A.; Lamnawar, K. Bulk rheology and surface tribo-rheometry toward the investigation of polyisobutylene migration in model and recycled multilayer agricultural films. *Rheol. Acta* **2020**, *59*, 1–27. [CrossRef]
7. Plastics Europe and EPRO. "Plastics—The Facts 2019". Available online: www.plasticseurope.de/informations (accessed on 31 March 2020).
8. Ragaert, K.; Delva, L.; Van Geem, K.M. Mechanical and chemical recycling of solid plastic waste. *Waste Manag.* **2017**, *69*, 24–58. [CrossRef] [PubMed]
9. Bertin, S.; Robin, J.J. Study and characterization of neat and recycled LDPE/PP blends. *Eur. Polym. J.* **2002**, *38*, 2255–2264. [CrossRef]
10. Teh, J.; Rudin, A.; Yuen, S.; Keung, J.C.; Pauk, D.M. LLDPE/PP Blends in Tubular Film Extrusion: Recycling of Mixed Films. *J. Plast. Film Sheeting* **1994**, *10*, 288–301. [CrossRef]

11. Fang, L.C.; Nie, S.; Liu, R.; Yu, N.; Li, S. Characterization of polypropylene-polyethylene blends made of waste ma-terials with compatibilizer and nano-filler. *Compos. Part B Eng.* **2013**, *55*, 498–505. [CrossRef]
12. Kazemi, Y.; Kakroodi, A.R.; Rodrigue, D. Compatibilization efficiency in post-consumer recycled polyeth-ylene/polypropylene blends: Effect of contamination. *Polym. Eng. Sci.* **2015**, *55*, 2368–2376. [CrossRef]
13. Offord, G.T.; Armstrong, S.R.; Freeman, B.D.; Baer, E.; Hiltner, A.; Paul, D.R. Gas transport in coextruded multilayered membranes with alternating dense and porous polymeric layers. *Polymer* **2014**, *55*, 1259–1266. [CrossRef]
14. Wang, H.; Keum, J.K.; Hiltner, A.; Baer, E.; Freeman, B.; Rozanski, A.; Galeski, A. Confined Crystallization of Polyeth-ylene Oxide in Nanolayer Assemblies. *Science* **2009**, *323*, 757–760. [CrossRef] [PubMed]
15. Bernal-Lara, T.; Liu, R.; Hiltner, A.; Baer, E. Structure and thermal stability of polyethylene nanolayers. *Polymer* **2005**, *46*, 3043–3055. [CrossRef]
16. Scholtyssek, S.; Adhikari, R.; Seydewitz, V.; Michler, G.H.; Baer, E.; Hiltner, A. Evaluation of morphology and defor-mation micromechanisms in multilayered PP/PS films: An electron microscopy study. *Macromol. Symp.* **2010**, *294*, 33–44. [CrossRef]
17. Krache, R.; Benachour, D.; Pötschke, P. Binary and ternary blends of polyethylene, polypropylene, and polyamide 6,6: The effect of compatibilization on the morphology and rheology. *J. Appl. Polym. Sci.* **2004**, *94*, 1976–1985. [CrossRef]
18. Li, Z.; Olah, A.; Baer, E. Micro- and nano-layered processing of new polymeric systems. *Prog. Polym. Sci.* **2020**, *102*, 101210. [CrossRef]
19. Vervoort, S.; den Doelder, J.; Tocha, E.; Genoyer, J.; Walton, K.L.; Hu, Y.; Munro, J.; Jeltsch, K. Compatibilization of poly-propylene–polyethylene blends. *Polym. Eng. Sci.* **2018**, *58*, 460–465. [CrossRef]
20. Lu, B.; Lamnawar, K.; Maazouz, A.; Sudre, G. Critical Role of Interfacial Diffusion and Diffuse Interphases Formed in Multi-Micro-/Nanolayered Polymer Films Based on Poly(vinylidene fluoride) and Poly(methyl methacrylate). *ACS Appl. Mater. Interfaces* **2018**, *10*, 29019–29037. [CrossRef] [PubMed]

MDPI
St. Alban-Anlage 66
4052 Basel
Switzerland
Tel. +41 61 683 77 34
Fax +41 61 302 89 18
www.mdpi.com

Polymers Editorial Office
E-mail: polymers@mdpi.com
www.mdpi.com/journal/polymers